Mikrochimica Acta

Supplementum 12

A. Boekestein and M. K. Pavićević (eds.)

Electron Microbeam Analysis

Springer-Verlag Wien GmbH

Dr. Abraham Boekestein
Head Department Instrumental Analysis
Agricultural Research Department (DLO-NL)
State Institute for Quality Control of Agricultural Products (RIKILT-DLO)
Wageningen, The Netherlands

Prof. Miodrag K. Pavićević
Institute of Mineralogy, University of Salzburg, Austria
Permanent address: Faculty of Mining and Geology, University of Belgrade, Yugoslavia

Typesetting: Asco Trade Typesetting Limited, Hong Kong

With 157 Figures

ISSN 0026-3672
ISBN 978-3-211-82359-0 ISBN 978-3-7091-6679-6 (eBook)
DOI 10.1007/978-3-7091-6679-6

Preface

This supplement of *Mikrochimica Acta* contains selected papers from the Second Workshop of the European Microbeam Analysis Society (EMAS) on "Modern Developments and Applications in Microbeam Analysis", which took place in May 1991 in Dubrovnik (Yugoslavia).

EMAS was founded in 1987 by members from almost all European countries, in order to stimulate research, applications and development of all forms of microbeam methods. One of the most important activities EMAS is the organisation of biannual workshops for demonstrating the current status and developing trends of microbeam methods. For this meeting, EMAS chose to highlight the following topics: electron-beam microanalysis (EPMA) of thin films and quantitative analysis of ultra-light elements, Auger electron spectroscopy (AES), electron energy loss spectrometry (EELS), high-resolution transmission electron microscopy (HRTEM), quantitative analysis of biological samples and standard-less electron-beam microanalysis.

Seven introductory lectures and almost seventy poster presentations were given by speakers from twelve European and two non-European (U.S.A. and Argentina) countries were made. One cannot assume that all fields of research in Europe were duly represented, but a definite trend is discernible. EPMA with wavelength-dispersive spectrometry (WDS) or energy-dispersive spectrometry (EDS) is the method with by far the widest range of applications, followed by TEM with EELS and then AES. There are also interesting suggestions for the further development of new apparatus with new fields of application. Applications are heavily biased towards materials science (thin films in microelectronics and semiconductors), ceramics and metallurgy, followed by analysis of biological and mineral samples.

This issue contains the full texts of five introductory lectures and 25 brief articles. Sixteen contributions relate to the refinement of methods and procedures and nine of these are concerned with important applications in various fields. All submissions have been refereed according to the usual procedures.

At the end of this issue is an overview of all the published work. We hope that these contributions to the field of electron microbeam analysis will be found to be useful.

February 1992 *A. Boekestein* and *M. K. Pavićević*

Participants in the EMAS '91 Meeting in Dubrovnik

Contents

Listed in Current Contents

Mikrochim. Acta (1992) [Suppl.] 12: 1–17

EPMA—A Versatile Technique for the Characterization of Thin Films and Layered Structures

Peter Willich

Fraunhofer-Institut für Schicht- und Oberflächentechnik, P.O. Box 540645,
D-W-2000 Hamburg 54, Federal Republic of Germany

Abstract. Electron probe microanalysis (EPMA) is presented as a quantitative technique of near-surface chemical characterization. Three principle operation modes are discussed: (1) "Thick" films (>100 $\mu g/cm^2$, 0.2 μm for a density of 5 g/cm^3) are studied by use of a sufficiently low electron energy. (2) The combined determination of film thickness and composition is applied to "thin" films and multilayers (<250 $\mu g/cm^2$, 0.5 μm for a density of 5 g/cm^3). Relatively fast analysis at a single electron energy is possible under certain restrictions. (3) The universal approach of non-destructive in-depth analysis is based on combining experiments performed at different electron energies. The operation modes are described with respect to experimental procedures, data reduction models, precision, accuracy and the range of practical applications. EPMA is also related to other techniques of thin film and surface analysis.

Key words: electron probe microanalysis (EPMA), thin films, multilayers, film thickness, non-destructive in-depth analysis, ultralight elements.

For almost 40 years electron probe microanalysis has been developed as a powerful tool for the local chemical characterization of solids. The majority of applications, in combination with scanning electron microscopy, is directed to analysis with a lateral resolution of about 1 μm, whereas the aspect of near-surface analysis is frequently neglected. Although interesting attempts in this direction were started as early as the beginning of the 1960s, as listed by Heinrich [1], these procedures and models did not succeed as routine tools for the characterization of thin film samples. This was due to the fact that none of them proved to be sufficiently general and quantitative. A new era of EPMA applied to thin film materials started in 1984, when Pouchou and Pichoir [2] presented a thin film correction model based on a realistic description of $\Phi(\rho z)$ depth distribution functions. Moreover, Pouchou and Pichoir proposed a general procedure for analysis of samples of which the composition varied in depth. In the following years, similar thin film applications have been discussed using completely different $\Phi(\rho z)$ data reduction models [3–5], originally developed to improve the accuracy of conventional "bulk" EPMA.

In addition to the development of general and sufficiently accurate correction models, EPMA of thin films and layered structures has been stimulated by recent instrumental and experimental improvements. This concerns the reliable operation of EPMA instruments at low electron energies, typically below 10 keV, the handling of contamination phenomena [6], and, with respect to wavelength dispersive X-ray spectrometry (WDS), the significantly improved sensitivity for the determination of soft X-rays by use of synthetic multilayer monochromators. Quantitative EPMA of ultralight elements (boron—oxygen), which play an important role in thin film technology, has now reached a remarkable degree of precision and accuracy [7].

Thin film analysis has to be regarded as an essential part of thin film technology, the importance of which is rapidly increasing in many fields of application. An ideal technique of analysis should be applicable to a wide range of materials and layered configurations, easy to operate, quantitative, and sufficiently local with respect to lateral resolution and depth of analysis. From this point of view the recent developments of EPMA have to be considered. This paper reviews, by examples of technological interest, the various operation modes of EPMA applied to thin films and stratified materials. The characteristics of EPMA are discussed in relation to the features of other techniques frequently applied to thin film and surface analysis.

EPMA in Comparison with Other Techniques of Thin Film Analysis

Table 1 gives an overview of a number of techniques to carry out analysis of surfaces and near-surface regions. The data of Table 1 are mainly drawn from a recent review of Werner and Torrisi [8], supplemented by experience of our own laboratory. In

Table 1. Comparison of some techniques applied to local analysis of thin films and surfaces

Technique	Depth resolution	Lateral resolution	Elements	LLD[2] [$\mu g/g$]	Depth profiling	Quantitative analysis[1]
XRF	$>5\ \mu m$	>5 mm	$Z>9$	>5	no	$<3\%$
EPMA-WDS	0.1–2 μm	$>0.5\ \mu m$	$Z>3$	>20	(limited)	$<5\%$
AES	1 nm	$>0.1\ \mu m$	$Z>3$	>1000	sputtering	10–20%
XPS	2 nm	$>100\ \mu m$	$Z>2$	>1000	sputtering	10–20%
RBS	2–20 nm	$>100\ \mu m$	$Z>5$	>1000	non-destr.	$<5\%$
ERD	2–20 nm	$>100\ \mu m$	hydrogen	>1000	non-destr.	$<10\%$
SIMS	2 nm	$>0.1\ \mu m$	all	<1	sputtering	difficult
SNMS/SALI	2 nm	$>1\ \mu m$	all	>1	sputtering	5–20%
LAMMA	$>0.5\ \mu m$	$>3\ \mu m$	all	<1	no	difficult

[1] Accuracy without matched standards

[2] LLD = Lower limit of detection

Acronyms:

XRF X-ray fluorescence, EPMA-WDS electron probe microanalysis-wavelength dispersive spectrometry,

AES Auger electron spectroscopy, XPS X-ray photoelectron spectroscopy,

RBS Rutherford backscattering spectrometry, ERD elastic recoil detection,

SIMS secondary ion mass spectrometry, SNMS secondary neutral mass spectroscopy,

SALI surface analysis by laser ionization, LAMMA laser microprobe mass analysis

many cases the data given in Table 1 represent, as is frequently done in such comparisons, the ultimate performance with respect to a particular parameter. In practice, the ultimate performance depends on the kind of material and can not be obtained for several parameters simultaneously.

From the point of view of depth resolution, EPMA, to be described as electron beam excited X-ray spectrometry, just fills the gap between surface analytical techniques (e.g., AES, XPS, SIMS) and a typical "bulk" technique like XRF (the meaning of the acronyms is given in Table 1). It should be mentioned that for some particular applications XRF can be applied to the determination of film thickness and composition [9], an interesting approach with respect to production control using a relatively simple apparatus (no vacuum required). The depth resolution of EPMA is adjustable over a range of about 50–1000 $\mu g/cm^2$(0.1–2 μm for a density of $\rho = 5$ g/cm^3), which fits the thickness range of many thin film applications. In the case of homogeneous films, EPMA provides an overall chemical analysis over the entire film thickness, without significant influence of the typically non-specific composition of the surface region. Surface analytical techniques, when applied to thin film characterization, have to be combined with time-consuming ion-sputtering procedures to eliminate the influence of surface contamination. Moreover, ion-sputtering may cause particular problems with respect to accurate quantitative analysis.

The accuracy of quantitative analysis (=possible range of systematic errors) provided by EPMA is in the range of a few percent relative for major elements, based on calibration by the pure element or an arbitrary compound standard in combination with first-principles matrix correction procedures. This is comparable to the performance of "bulk" chemical analysis by XRF and considerably superior to the accuracy of surface analytical techniques. Accurate quantitative analysis is also provided by RBS, although the technique is difficult to apply to certain analytical situations, e.g., low atomic number elements in a high atomic number matrix, low atomic number films on a high atomic number substrate, and the separation of heavy elements of similar atomic number. ERD is of interest as a unique technique of quantitative hydrogen analysis and hydrogen depth-profiling. Quantitative mass spectrometry is expected from the further development of post-ionization techniques, i.e., SNMS [10] and SALI [11]. Some interesting aspects of non-destructive in-depth analysis by EPMA will be discussed later. However, it should be noted at this point, that EPMA enables one to establish fairly rough in-depth analysis only, when compared to the depth profiles established by RBS or surface analytical techniques combined with ion beam sputtering.

EPMA of "Thick" Films

Provided that the surface film is sufficiently thick, it is possible to reduce the primary electron energy (E_0) so that the depth range of X-ray emission is entirely included inside the film, apart from possible effects of fluorescence excitation. Analysis is performed without influence of the substrate, which may be unknown or may contain elements of the film. The depth of analysis is adjusted to the provided film thickness by a suitable choice of E_0, in practice in the range of 100–1000 $\mu g/cm^2$ (0.2 to 2 μm for $\rho = 5$ g/cm^3). In favourable situations "bulk" analysis of films with a geometrical film thickness of 0.1 μm or even less is possible, e.g., when a high density material is studied by use of low energy X-rays. Although EPMA of "thick" films may be regarded as an extension of conventional "bulk" analysis, some particular aspects have to be considered.

Depth Resolution and Sensitivity

The ultimate geometrical depth of X-ray generation (z_{max}) as given by the equation of Castaing [12]

$$z_{max}[\mu m] = 0.033(E_0^{1.7} - E_c^{1.7})\, A/Z\; 1/\rho \qquad (1)$$

depends on the primary electron energy (E_0), the critical excitation energy of the X-ray line (E_c), the atomic number (Z), the atomic weight (A), and the density (ρ) of the material. In cases of strong matrix absorption effects it is useful to calculate the ultimate depth of X-ray emission as a more realistic estimation of the depth of analysis. The ultimate depth of X-ray emission can be derived from the corresponding $\Phi(\rho z)$ depth distribution function of the emitted X-rays. The example of SiC (Fig. 1) demonstrates that the depth of analysis, i.e., the ultimate depth of X-ray emission (150 μg/cm^2, $E_0 = 10$ keV), may be only a fraction of the ultimate depth of X-ray generation (350 μg/cm^2), particulary in the case of ultralight elements, where X-ray absorption plays an important role. The ultimate depth of X-ray generation (350 μg/cm^2, 1.1 μm for $\rho = 3.2$ g/cm^3) as given in Fig. 1 is in good agreement with the value of $z_{max} = 1.05$ μm ($\rho = 3.2$), which one would obtain using eq. (1). This supports the applicability of this expression to estimate the depth of analysis in cases where X-ray absorption is moderate.

Figure 2 shows an example for the depth of X-ray emission and the analytical sensitivity as a function of E_0. The lower limit of detection (LLD) [13] was calculated

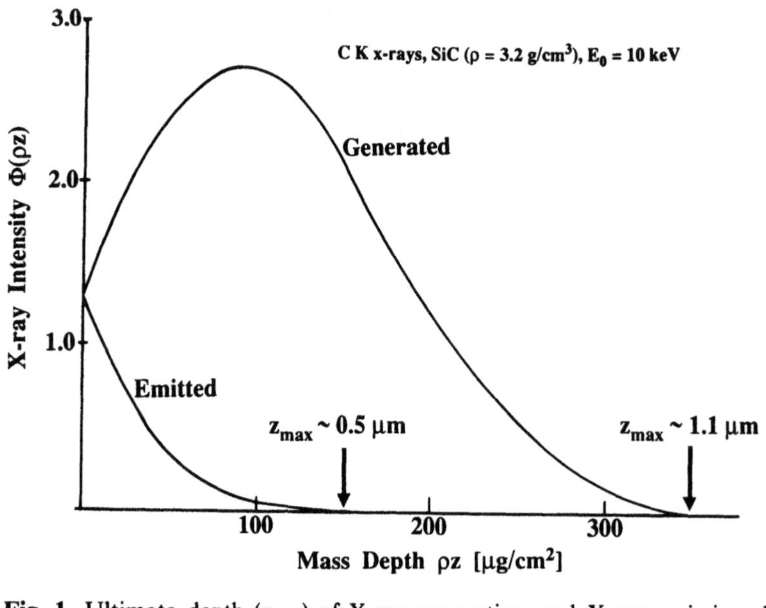

Fig. 1. Ultimate depth (z_{max}) of X-ray generation and X-ray emission derived from $\Phi(\rho z)$ depth distribution functions. Calculations performed by use of the PAP model [19]. X-ray take-off angle 40°

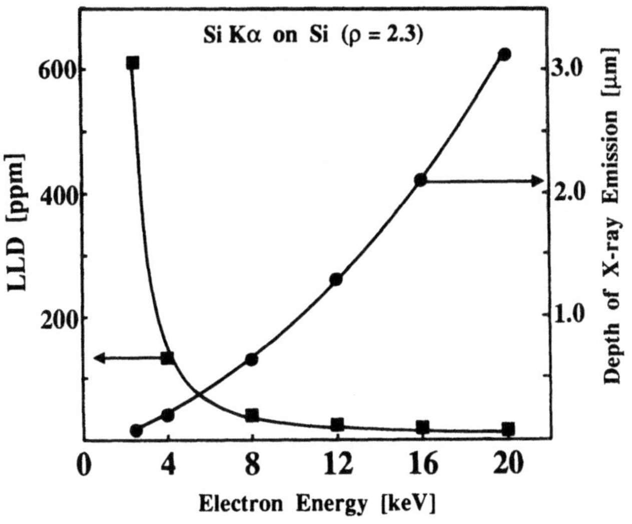

Fig. 2. Depth of X-ray emission and analytical sensitivity (lower limit of detection LLD) as a function of electron energy. LLD refers to a beam current of 100 nA and a counting time of 100 s (WDS, TAP crystal, 40° take-off angle)

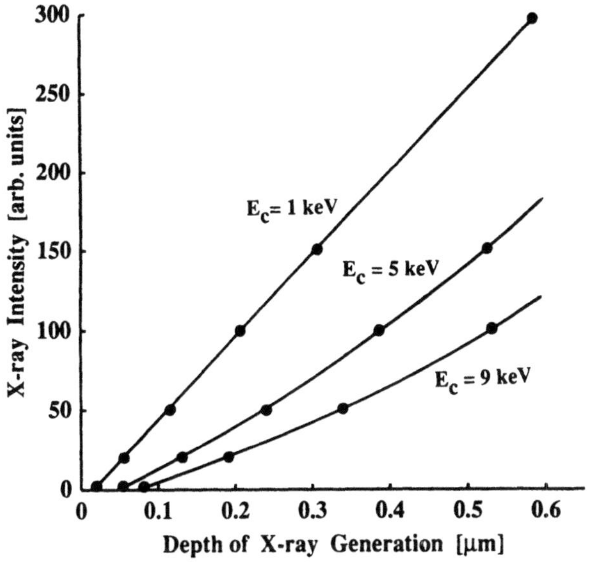

Fig. 3. X-ray intensity (WDS, 40° take-off angle) related to the depth of X-ray generation. X-ray lines of low (Cu Lα, TAP), medium (Ti Kα, PET) and high (Cu Kα, LIF) critical electron excitation energy (E_c) are considered. Geometrical depth of X-ray generation calculated by assuming $\rho = 8.9$ g/cm^3(Cu) and $\rho = 4.5$ g/cm^3(Ti)

from experimental Si Kα intensities and peak/background ratios in WDS. At $E_0 >$ 8 keV ($\rho z_{max} > 115$ μg/cm^2, $z_{max} > 0.5$ μm for $\rho = 2.3$ g/cm^3), Si can be determined with a sensitivity of 50 ppm or even less. The LLD of Si increases to about 600 μg/g at $E_0 = 2.5$ keV, where the corresponding depth of analysis is in the order of 20 μg/cm^2 (~80 nm for $\rho = 2.3$ g/cm^3). Typically, experiments at conditions of $\rho z_{max} < 100$ μg/cm^2 are performed at high beam currents ($I > 100$ nA) and relatively long counting times ($T > 50$ s) to compensate to a certain extent the loss of X-ray intensity for measurements at a low overvoltage E_0/E_c. In some practical situations this loss of intensity can be minimized by selection of a low energy X-ray line (Fig. 3). The normalized intensities of Fig. 3 were calculated from intensity measurements (WDS) of Cu Kα ($E_c = 8.98$ keV, LiF crystal), Ti Kα ($E_c = 4.97$ keV, PET crystal), and Cu Lα ($E_c = 0.93$ keV, TAP crystal) on the pure elements at E_0 of 2 – 20 keV. For instance, for $\rho z_{max} = 180$ μg/cm^2($z_{max} = 0.2$ μm for $\rho = 8.9$ g/cm^3), the intensity provided by Cu Lα (E_0 ~ 6.5 keV) is about 5 times higher than the intensity of Cu Kα ($E_0 \sim 11.5$ keV). Using transition element L-lines experiments at a sufficient level of sensitivity are possible even at conditions of z_{max} in the range of less than 100 nm [14]. However, EPMA based on transition element L-lines is not always recommended in view of particular problems of quantification, which will be discussed later. Sometimes analysis has to be performed at two or even more electron energies to get a similar depth of analysis for the different elements of a compound film, e.g., a Cr–Si alloy should be determined at $E_0 = 8$ keV for Cr Kα and at $E_0 = 5$ keV for Si Kα to get a similar depth of analysis of about 80 μg/cm^2(0.2 μm, $\rho = 4$ g/cm^3) for Cr and Si.

Influence of Surface Contamination

Carbon contamination and the existence of oxide films on the surface may seriously influence the accuracy of EPMA carried out at conditions where the depth of analysis is less than 250 μg/cm^2(0.5 μm for $\rho = 5$ g/cm^3). Typically the film thickness of contamination is $1 - 10$ μg/cm^2 (4–40 nm for $\rho = 2.5$ g/cm^3), which means a considerable fraction of the total depth of analysis. Carbon contamination at the point of electron impact is a severe problem due to thin film analysis being frequently performed at high beam currents in combination with long counting times. Liquid nitrogen cold traps and air-jet devices have proved to be an efficient means with respect to carbon contamination [6]. Air-jet operation even enables the almost complete removal of carbonaceous deposits.

Oxide contamination is an important source of errors in cases where highly oxidable materials are used as standards, e.g., Mg, Al, Ti. Surface oxidation may cause errors up to 10%, particularly when analysis of freshly prepared "clean" specimens is calibrated by metals which have been used as standard over a long period of time. Provided that a material contains no oxygen, the intensity of the O Kα signal may be regarded as a measure of the oxide thickness, which can be calculated by a thin film program [14, 15]. The oxide film on the surface reduces the intensity of the (metal) X-ray line as compared to the "clean" material. The calculated loss of intensity for the X-ray line of the metal is represented as a correction factor; Figure 4 shows the example of Ge. The typical GeO_2 oxide thickness, as it was established by use of the O Kα signal, is in the order of 2 μg/cm^2(5 nm for $\rho = 4$ g/cm^3). Consequently, the intensity of Ge Lα on Ge has to be multiplied by 1.04 ($E_0 = 3$ keV) to compensate for the effect of surface oxidation (Fig. 4). In a similar way the C K intensity can be used to handle the problem of carbon contamination or to determine the influence of carbon coating deposited on non-conducting materials [16].

Accuracy of Matrix Correction

The limitations of the conventional ZAF method, e.g., absorption correction for low energy X-rays, high atomic number correction at low voltages or overvoltages, have stimulated to elaborate improved models of data reduction [17, 18]. Most of these models are based on a more or less realistic description of $\Phi(\rho z)$ depth distribution functions. The following examples of quantitative analysis have been established by

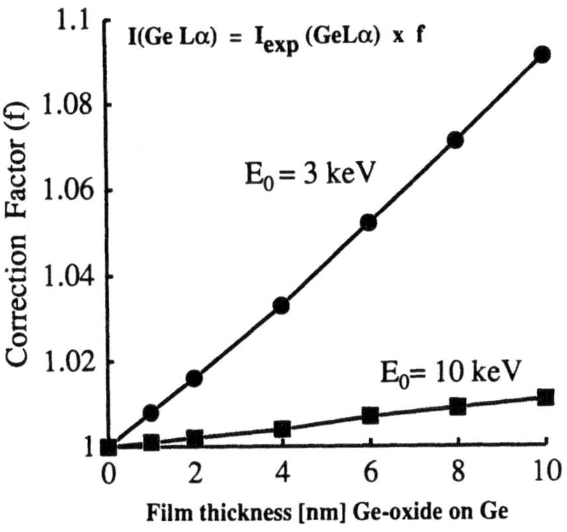

Fig. 4. Correction functions for the intensity of Ge Lα to compensate for the influence of surface oxidation on Ge. The oxide thickness (assuming GeO_2 with $\rho = 4.2$ g/cm^3) can be derived from the intensity of O Kα. X-ray take-off angle 40°

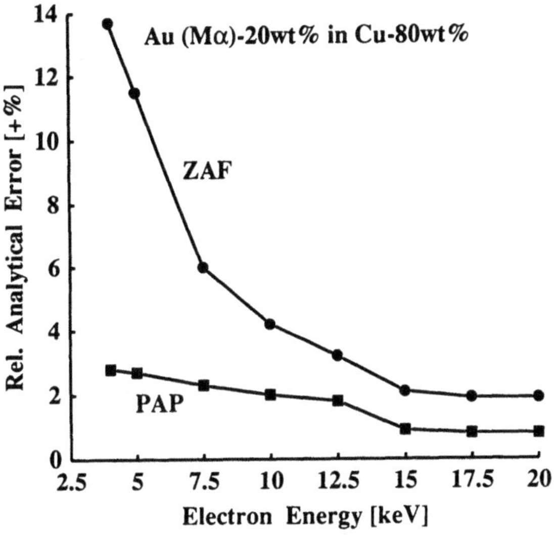

Fig. 5. Accuracy of matrix correction models as a function of electron energy. The Au–Cu alloy represents an example of high atomic number correction for Au Mα (standard used is Au)

use of the PAP procedure [19] as a representative of the new generation of EPMA data reduction.

Figure 5 shows the situation of extremely high atomic number correction. The "classical" ZAF–COR2 [20] model leads to large over-correction at low electron energies, whereas the PAP model results in an accuracy of within 3%, even at $E_0 = 4$ keV, where the depth of analysis for Au Mα is in the order of 50 μg/cm^2 (50 nm for $\rho = 10$ g/cm^3). Examples of quantitative EPMA at a very low electron energy of $E_0 = 3$ keV ($\rho z_{max} \sim 30$ μg/cm^2, $z_{max} \sim 0.1$ μm for $\rho = 3$ g/cm^3) are given in Table 2. Analysis of the relatively thick (~ 540 μg/cm^2, ~ 1.8 μm for $\rho = 3$ g/cm^3) films was also performed at more conventional experimental conditions of $E_0 =$ 10 keV ($\rho z_{max} \sim 300$ μg/cm^2, $z_{max} \sim 1$ μm for $\rho = 3$ g/cm^3). The results are sufficiently consistent. The slightly higher concentrations for carbon and oxygen at 3 keV may be due to the influence of surface contamination. EPMA by use of L-lines for the elements of the 4th period is very attractive from the point of view of sensitivity at conditions of $\rho z_{max} < 150$ μg/cm^2($z_{max} < 0.3$ μm for $\rho = 5$ g/cm^3). However, the accuracy of quantitative analysis may be limited due to anomalous

Table 2. EPMA of coatings (film thickness ~540 $\mu g/cm^2$, ~1.8 μm for density $\rho = 3\ g/cm^3$) at "conventional" conditions ($E_0 = 10$ keV, depth of analysis ~300 $\mu g/cm^2$), and under conditions of "near-surface" analysis ($E_0 = 3$ keV, depth of analysis ~30 $\mu g/cm^2$)

		Composition [wt%]					
Sample	E_0[keV]	Ge	C	N	O	Ar	Total
#1	3	83.8	14.8	<0.1	1.9	0.34	100.8
	10	84.3	14.2	<0.1	1.7	0.31	100.5
#2	3	43.0	48.6	<0.1	10.4	<0.01	102.0
	10	43.2	47.9	<0.1	9.8	<0.01	100.9
#3	3	68.7	0.9	<0.1	31.4	<0.01	100.9
	10	68.5	0.3	<0.1	30.9	<0.01	99.7
#4	3	77.3	1.2	18.7	0.8	0.11	98.1
	10	78.5	0.6	19.3	0.5	0.09	99.0
#5	3	67.4	16.4	14.0	3.4	<0.01	101.2
	10	67.8	15.7	14.3	3.1	<0.01	100.9

Standards: Ge metal (Ge Lα), C(vitreous carbon), O($Y_3Fe_5O_{12}$), N(Cr_2N), Ar(5.2 wt% Ar in Si)

mass absorption coefficients as a consequence of alloying [21], particularly for the elements of Sc to Ni, of which the 3d-band is partly empty.

EPMA of ultralight elements (B–O) may be regarded as a favourable situation with respect to a low depth of X-ray emission (Fig. 1). Typically, experiments performed at $E_0 = 3 - 5$ keV correspond to a depth of analysis in the range of 20–60 $\mu g/cm^2$(0.1–0.3 μm for $\rho = 2\ g/cm^3$). The decrease of the X-ray intensity is moderate, the use of synthetic multilayer monochromators is of additional advantage. The various aspects (experimental, matrix correction, mass absorption coefficients) of EPMA applied to quantitative analysis of ultralight elements have been recently reviewed by Bastin and Heijligers [7]. In summary, it can be concluded that an accuracy of at least 5% is feasible by use of the new correction models, e.g., PROZA [7], PAP [22] or XPP [23], particularly when analysis is performed at low electron energies of 5–10 keV, where X-ray absorption is reduced and uncertainties in the mass absorption coefficients are less important.

The influence of chemical shift (shift of the peak position, peak shape alterations) is of great importance with respect to ultralight elements, especially for boron and carbon. A general solution of this problem is the determination of integral peak areas as a measure of X-ray intensity, as it was done to establish the k-ratios of C K X-rays the examples of Table 2. Table 3 shows the accuracy for the determination of boron in a situation of extreme X-ray absorption (mass absorption coefficient of 84000 for B K X-rays in Si). The corresponding size of the matrix correction is in the order of 5 (weight fraction/k-ratio). Peak area measurements, correction of carbon coating, and the use of the PAP model lead to an accuracy of about 4% [16] for the determination of boron in coatings of borophosphosilicate glass (Table 3), in a comparison with results obtained by chemical analysis (ICP-OES, inductively coupled plasma–optical emission spectrometry) as a reference method.

Table 3. EPMA of borophosphosilicate (BPSG) glass (film thickness > 70 $\mu g/cm^2$, >0.3 μm for density $\rho = 2.2$ g/cm^3, Si ~ 40 wt%, O ~ 54 wt%). The results for B and P are compared to those obtained by chemical analysis (ICP-OES: inductively coupled plasma–optical emission spectrometry)

	Composition [wt%]			
	Boron		Phosphorus	
Sample	EPMA	ICP-OES	EPMA	ICP-OES
#1	0.07	0.10	0.54	0.52
#2	0.92	0.99	5.58	5.50
#3	2.21	2.18	4.86	4.90
#4	2.48	2.33	5.12	5.10
#5	3.48	3.50	2.17	2.20
#6	4.25	4.20	3.20	3.19

$E_0 = 6$ keV, standards: B metal, $Ca_5F(PO_4)_3$

EPMA of "Thin" Films

The term "EPMA of thin films" is used to describe the situation that the film thickness is less than the ultimate depth of X-ray emission. The X-ray intensity of the elements of the film depends on the film composition, the film (mass) thickness and the mean atomic number of the substrate. For simple structures information about film thickness and composition can be deduced from "conventional" experiments performed at a single electron energy. In this context the following preconditions have to be considered:

- The substrate is defined with respect to its mean atomic number.
- The elements of the film are not present in the substrate.
- Each element of a multilayer structure is a component of only one layer or the substrate.
- Each element is assigned to a single layer of the layered structure. The sequence of the single layers is known.
- All elements of the layer(s) have to be measured, with the exception of an element calculated by stoichiometry. Determination of an element by difference is not possible.

In practice, the above mentioned basic requirements are met in many relevant analytical situations. The procedures of data acquisition are identical to those of conventional bulk analysis. Results with respect to film thickness(es) and composition(s) can be computed directly from a set of k-ratios. The principle of fast thin film analysis is of particular interest to samples, of which the lateral distribution of film thickness and composition has to be studied at many points of analysis.

Some Basic Principles of Data Reduction

In the most simple case, where a single film of element A having the mass thickness ρz_A is deposited on a substrate, one gets:

$$k_A = \frac{\int_0^{\rho z_A} \Phi_A(\rho z) \exp(-\chi_A \rho z)\, d\rho z}{\int_0^{\infty} \Phi_{AP}(\rho z) \exp(-\chi_{AP} \rho z)\, d\rho z} \tag{2}$$

The denominator of Eq. (2) represents the intensity of the pure element bulk standard. χ_A and χ_{AP} are the absorption coefficients of the film and the standard. Provided that algorithms to compute $\Phi(\rho z)$ functions are available, analytical or numerical integration leads to the relative intensity k_A corresponding to the assumed mass thickness ρz_A. An iterative procedure, by modification of ρz_A, enables one to compute the final value of ρz_A, for which k_A is identical to the experimental k-ratio.

In general the atomic number of the film (Z_A) is different from the atomic number of the substrate (Z_S). This situation may cause fairly complicated shapes of $\Phi_A(\rho z)$, the depth distribution function of the substrate-film combination [24, 25]. Only for very thin films, as compared to the ultimate depth of X-ray generation, it may be justified to use the substrate atomic number (Z_S) for the calculation of $\Phi_A(\rho z)$. In practice, $\Phi_A(\rho z)$ is regarded as a continuous function corresponding to a fictitious homogeneous sample. Typically $\Phi_A(\rho z)$ is computed using e.g.

$$Z_{AS} = Z_s + (Z_A - Z_S)(\rho z_A / 2\rho z_{A\max})^{0.65} \tag{3}$$

to obtain Z_{AS} as an effective atomic number of the substrate-film combination [26]. In equation (3) ρz_A is the estimated mass thickness obtained by assuming $Z_{AS} = Z_S$ in the first iteration, and $\rho_{A\max}$ represents the ultimate depth of X-ray generation. Equation (3) is an example of a rather crude weighting law. More sophisticated approximations have been proposed [2, 3, 14], where the influence of Z_A and Z_S on the various $\Phi(\rho z)$ shape parameters is considered more individually. This enables one to handle complicated layered structures based on experiments over a wide range of electron energies, including the intensity of the substrate as a measure of film thickness [2].

In the case of multiconstituent films, in a first step, the k-ratios are converted, using the substrate atomic number, into a set of "elemental" mass thicknesses ρz_i. For an alloy film consisting of n elements, the approximated weight fraction C_i of element i is obtained as follows:

$$C_i = \rho z_i \Big/ \sum_{i=1}^{i=n} \rho z_i \tag{4}$$

Further iterations, based on the information about the film thickness and the film composition, include the calculation of the corresponding effective atomic numbers. Moreover, X-ray absorption by the companion elements is taken into account. The determination of multilayers requires the qualitative description of the layered structure as an additional input, i.e., the sequence of the layers and which element is a component of which layer. X-ray absorption by the top layers and the contribution of buried layers to the effective atomic number lead to fairly complex iterative computations [26]. Possible effects of characteristic and continuous secondary fluorescence in layered specimens have to be kept in mind. Analytical expressions

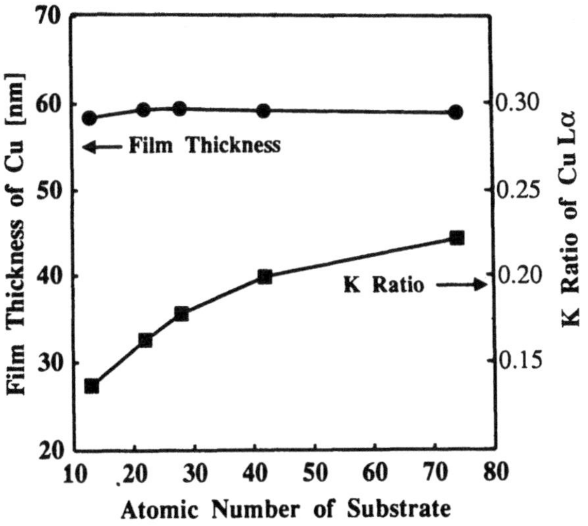

Fig. 6. K ratio ($E_0 = 15$ keV, X-ray take-off angle 40°) and film thickness of pure Cu deposited on various substrates. In all cases the film thickness was identical: 58 ± 2 nm as defined by mechanical stylus experiments (TALYSTEP). Nominal density of 8.96 g/cm^3 for Cu was assumed

to describe the influence of fluorescence excitation have been developed [14]. Taking the effect of secondary fluorescence into consideration will certainly improve the accuracy of thin film data reduction [23].

Application and Accuracy of Thin Film Analysis

The following examples of thin film analysis based on experiments at a single electron energy have been established by our own thin film program [26]. This program uses the formalism of the PAP model [19] to compute $\Phi(\rho z)$ functions. The program includes Eq. (3) as a weighting law and the previously discussed iterative procedures to handle multiconstituent films and multilayers. Figure 6 shows the determination of the film thickness for identical Cu films ($\rho = 8.9$ g/cm^3) deposited on substrates consisting of elements with atomic number 13 to 73. The results (Cu Lα, $E_0 = 15$ keV) agree within 4% to the nominal film thickness provided by mechanical stylus experiments (TALYSTEP). EPMA of the Cu films by use of Cu Kα ($E_0 = 25 - 35$ keV) leads to an accuracy of 3.5%. Additionally, pure element films of Au (Mα, Lα) and Si deposited on various substrates ($Z_S = 13 - 73$), were studied at electron energies of 15 – 30 keV. From these data, i.e., films of Si, Cu, and Au deposited on various substrates with a film thickness in the range of 20–200 μg/cm^2, an average accuracy of 3.6% (RMS based on 60 thin film determinations) was calculated. This estimation of accuracy is valid for experiments at a sufficiently high electron energy, at conditions where the film thickness does not exceed about 20% of the ultimate depth of X-ray generation. Table 4 gives an example of EPMA applied to the characterization of double layer configurations. Film thickness and composition of the magneto-optic alloy (buried layer) and the protective coating (surface layer) are obtained simultaneously, based on the k-ratios of the elements determined at a single electron energy [26]. The results with respect to composition and film thickness agree within 5% of those established by RBS. RBS was performed on regions, prepared by use of a movable mask, where the surface layer and the buried layer were accessible as single layers on the Si substrate. Direct characterization of the double layers by RBS was difficult due to problems of mass interference. In practice, relatively fast (about 60 s for one point of analysis)

Table 4. Characterization of Ni–Cr on Gd–Fe–Pt double layers deposited on Si. Comparison of compositions and film thicknesses established by EPMA ($E_0 = 25$ keV) and Rutherford backscattering spectroscopy (RBS)

		Surface layer			Buried layer			
		Composition [wt%]		Film thickness [nm][1]	Composition [wt%]			Film thickness [nm][2]
Sample	Analysis by	Ni	Cr		Gd	Fe	Pt	
#1	EPMA	14.2	85.8	69.1	29.4	50.7	19.9	23.6
	RBS	14.4	85.6	68.3	28.6	51.4	20.0	24.6
#2	EPMA	15.0	85.0	52.0	28.7	51.7	19.6	57.5
	RBS	15.4	84.6	51.2	28.3	52.9	18.8	55.8
#3	EPMA	14.0	86.0	69.5	28.0	51.8	20.2	101.9
	RBS	14.4	85.6	68.0	27.7	52.9	19.4	96.3

[1] $\rho = 7.4$ g/cm^3, [2] $\rho = 10.6$ g/cm^3

and local analysis provided by EPMA performed at a single electron energy enables one to study the lateral homogeneity of stratified materials by use of automated data acquisition procedures, e.g., electron beam or specimen stage "step scan" operation.

Non-Destructive In-Depth Analysis

The operation modes of "Thick Film Analysis" and "Thin Film Analysis", as it has been discussed in the previous sections, may be regarded as an extension of conventional procedures, particularly from the view point of simple and fast data acquisition. Although these operation modes fit the demand of many practical analytical situations, some pre-conditions have to be considered always, which limit the general application to complex structures. This lack of general applicability was realized by Pouchou and Pichoir in 1984 [2], when they proposed a universal procedure of non-destructive in-depth analysis by EPMA. The basic principle of this approach is the determination of a specimen by combining experiments at different electron energies, for which the depth of X-ray emission varies in the range of about 50 μg/cm^2(0.1 μm for $\rho = 5$ g/cm^3) to the ultimate depth of 1500–2500 μg/cm^2(3–5 μm for $\rho = 5$ g/cm^3) that can be reached by electrons having an energy of about 40 keV, which is the technical limit of most commercial EPMA instruments.

As a possible alternative to the variation of the electron energy one can consider the technique of determinations at various tilt angles of the specimen [27]. Grazing incidence of the impinging electrons considerably reduces the mean penetration depth, but not the ultimate depth of X-ray generation [5]. The experiments can be easily performed in most SEM/EDS instruments. The advantages and limitations of this principle with respect to non-destructive in-depth analysis have to be studied, particulary in view of practical applications.

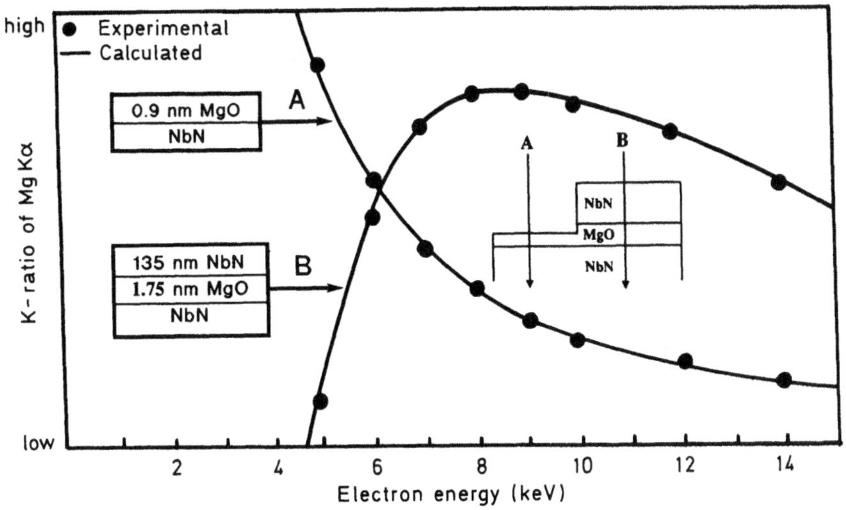

Fig. 7. K-ratio versus E_0 display to determine the thickness (curve A) of MgO ($\rho = 3.6$ g/cm^3) on NbN. Curve B is used to determine the depth and thickness of MgO covered by NbN ($\rho = 8.4$ g/cm^3). The models giving the best fit to the experimental data are included. K-ratios ($\sim$0–0.02) are relative to Mg. X-ray take-off angle is 40°

The Procedure of Data Reduction

The following discussion describes the procedure of commercial [28, 29] thin film programs, which are closely related to the basic concept of Pouchou and Pichoir [2, 14]. Approaches covering a similar range of application, but which are based on different algorithms to compute $\Phi(\rho z)$ functions, have been proposed recently by August [30] and Bastin et al. [31] (the latter is also available as a commercial program [32]).

As a first step of data deconvolution, the k-ratios of the element(s) of the layer(s) and the substrate are displayed as k-ratio versus E_0 plots. In general, the determination of all elements of the layered structure and the substrate is recommended, and, if possible, by using different X-ray lines for one element. For relatively simple structures, as given in Fig. 7, the consideration of only one or a selected number of elements may be sufficient. In the example of Fig. 7, it was required (position A) to determine the thickness of Mg-oxide (to be assumed as MgO, $\rho = 3.6$ g/cm^3) on Nb-nitride (to be assumed as NbN) and (position B) the thickness and depth of a similar MgO layer buried under NbN($\rho = 8.4$ g/cm^3). After display of the experimental (Mg Kα) k-ratios, the configuration of the layered structure has to be estimated and presented as a model, which consisted of the composition of the substrate (NbN), the composition and thickness of MgO, and the composition and thickness of NbN covering MgO. Based on the assumed model, the corresponding k-ratio versus E_0 data are computed and displayed. A procedure of "trial and error", in the examples of Fig. 7 the parameters "MgO film thickness" and "surface NbN thickness" were manipulated, led to a final model of the layered structure, for which the calculated k-ratios are in agreement with the experimental data.

Previous knowledge of some parameters of the layered structure is not generally necessary, although this knowledge is available in many practical situations. However, it can imply a considerable saving of time with respect to computation and, what is more important, the range of the required experimental data can be limited. Frequently it is advantageous to start with a simulation to find out the most informative range of electron energies, e.g., 5–12 keV for the examples of Fig. 7. After some practical experience, the operator is able to deduce some qualitative information from the display of the experimental data without any computation, e.g., curve A of Fig. 7 is typical for a thin film on the surface, whereas curve B is characteristic for a buried layer.

The accuracy of the results obtained by modelling the k-ratio versus E_0 plots, can be estimated by considering a limited number of relatively simple analytical situations. For example, the double layers of Table 4 were studied at electron energies of 9, 10, 12, 15, 20, 25, 30 keV. The experimental k-ratios were modelled using the configurations established by RBS. For all electron energies the agreement between experimental and calculated k-ratios was found to be within 5%. A similar degree of accuracy was obtained for the composition and film thickness ($\sim$100 nm) of Ta-oxide ($\sim Ta_2O_5$, $\rho = 8.2$ g/cm^3) on Ta [15]. In principle, it should be interesting to compare EPMA of stratified materials with the detailed information provided by conventional depth profiling techniques, e.g., AES, XPS, SIMS. However, with the exception of RBS, accurate quantitative results, which can be used to test the performance of EPMA applied to non-destructive in-depth anaysis, are not generally available.

The practical limits of non-destructive in-depth analysis have been achieved when different sets of parameters lead to a similar degree of fitting. In this case, the parameters of the final model are poorly defined. The level of precision, with respect to the parameters describing the final model, depends on the complexity of the layered structure, the range of film thickness, the range of electron energies, and the precision of the experimental data. For the example of Ta-oxide ($\sim$30 nm, ρ = 8.2 g/cm^3) on Ta studied at $E_0 = 4 - 12$ keV [15], the parameters "film thickness" and "oxygen concentration" can be varied within 20% without any significant ($<3\%$) change of the computed k-ratios, which are always in agreement with the experimental data, of which the precision is about 3%.

Examples of Application

Figure 8 shows an in-depth analysis of a Ta:C:(O:Ar) double layer structure deposited on Si. Surface and buried layer are identical with respect to their qualitative composition. "Bulk" EPMA at $E_0 = 4$ keV of corresponding single layers, prepared by use of a movable mask, gave the following results: Surface layer 58.5 wt% Ta, 39.3 wt% C, buried layer 90.2 wt% Ta, 6.8 wt% C. The results of in-depth analysis (Fig. 8), when compared to "bulk" EPMA of the single layers, are fairly consistent. The most significant deviation concerns the concentration of carbon in the buried layer (8 wt%), where a value of 6.8 wt% was established on the corresponding single layer. The discrepancy may be explained by the limited sensitivity for the determination of a low concentration buried under a high concentration of the same element. Moreover, data reduction is critical because of large variations of the mean atomic number and the high magnitude of the absorption correction for C Kα. The film thicknesses, as given in Fig. 8 were calculated by

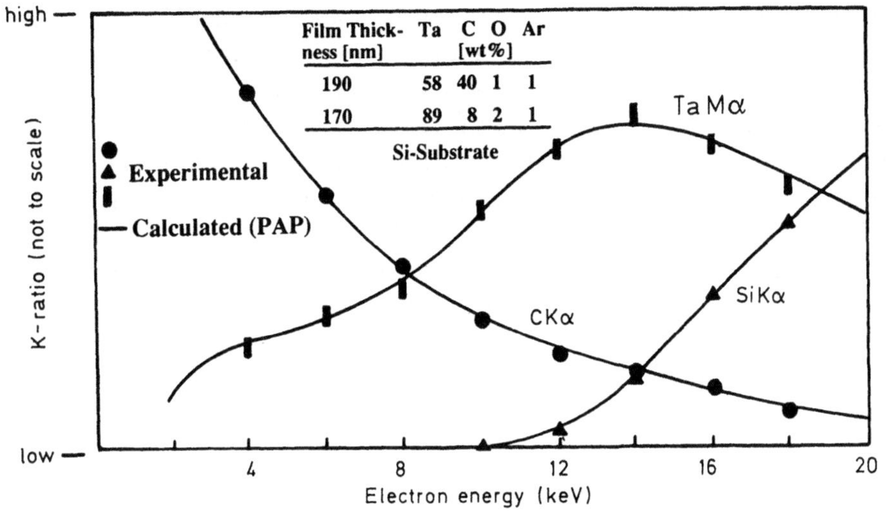

Fig. 8. K-ratio versus E_0 display for a Ta–C double layer deposited on Si (X-ray take-off angle 40°, standards: Ta, Si, vitreous carbon). The model giving the best fit to the experimental k-ratios is included, assuming densities of $\rho = 10.4$ g/cm^3 (top layer) and $\rho = 14.9$ g/cm^3 (buried layer)

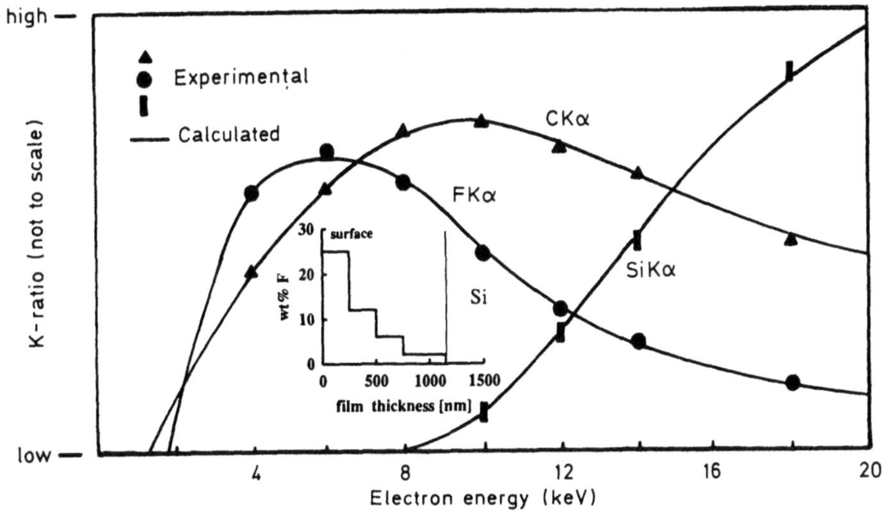

Fig. 9. K-ratio versus E_0 display of a non-homogeneous C–F film deposited on Si. The best fit to the experimental data refers to the given four layer "depth profile", assuming a constant density of $\rho = 1.5$ g/cm^3. X-ray take-off angle is 40°. Standards: Si, vitreous carbon, LiF

weight-averaging the nominal densities of tantalum ($\rho = 16.6.$ g/cm^3) and carbon ($\rho = 2.0$ g/cm^3). Based on this procedure, which may be rather crude with respect to possible structural changes, the film thicknesses provided by in-depth analysis agree within 10% of the results established by mechanical stylus experiments (TALYSTEP).

The example of Fig. 9 concerns the investigation of a continuous compositional variation in the near-surface region. A polymer-like film of C:F:(O:H) was deposited on Si. From the parameters of preparation a variation of the C/F ratio as a function of depth was expected. EPMA was carried out after deposition of Au as a conductive coating. The influence of Au on the surface was considered by calculating the Au film thickness ($\sim 7.5\ \mu g/cm^2$, 4 nm for $\rho = 19\ g/cm^3$) from experimental Au Mα intensities [33]. The display of the experimental k-ratio data (Fig. 9) already points out an enrichment of fluorine in the surface region; a homogeneous material would exhibit an almost identical distribution for C K and F Kα. Quantitative modelling was started by assuming a configuration of four layers and the model representing the best fit is given in Fig. 9. No significantly better quality of fitting was obtained by further refinements, e.g., by increasing the number of layers. The "steplike" model of Fig. 9 represents a typical result one can expect from non-destructive in-depth analysis by EPMA. Nevertheless, the schematic of Fig. 9 contains some interesting quantitative information, which is difficult to obtain by other techniques of depth profiling.

Conclusions

As summarized in Table 5, EPMA offers three principal modes of operation to characterize thin films and layered structures with respect to composition and film thickness. In combination with a lateral resolution of about 1 μm and the possibility of X-ray/SEM imaging, EPMA has to be regarded as a versatile tool to study solids in the near-surface region. In principle the various modes of operation can be applied by use of SEM-EDS or EPMA-WDS instruments, the latter is to be preferred in view of its superior sensitivity, particularly for the ultralight elements. Quantitative analysis is possible with a high degree of accuracy by use of pure element or simple compound standards in combination with fully developed procedures of data reduction. The aspect of quantitative analysis enables one to consider EPMA as an interesting supplement to the frequently used procedures of depth profiling based on surface analytical techniques.

Table 5. Some characteristics of operation modes to investigate thin films and layered structures by use of EPMA

Operation mode	Film thickness (typical)	Parameters	Accuracy	Advantages	Disadvantages
Modified "bulk" $z > R_x$	$z > 0.2\ \mu m$	composition	<5%	independent of substrate	reduced sensitivity, contamination
"Thin Film" $z < R_x$ single E_0	$z < 1\ \mu m$	film thickness(es) composition(s)	<5%	fast, multilayers	element present in (one) layer or substrate only
"Thin Film" $z < R_x$ multi E_0	$z < 1\ \mu m$	film thickness(es) composition(s)	5–20%	universal, non-destructive in-depth analysis	time consuming, limited accuracy and precision

R_x: Ultimate depth of X-ray emission

References

[1] K. F. J. Heinrich, *Electron Beam X-ray Microanalysis*, Van Nostrand Reinhold, 1981, p. 430.
[2] J. L. Pouchou, F. Pichoir, *La Recherche Aérospatiale* **1984**, *5*, 47.
[3] R. Packwood, G. Remond, J. D. Brown, *Proc. 11th ICXOM*, London, Canada, 1987, p. 274.
[4] G. F. Bastin, H. J. M. Heijligers, J. M. Dijkstra, *Microbeam Analysis* **1990**, 159.
[5] J. L. Pouchou, F. Pichoir, D. Boivin, *Proc. 12th ICXOM*, Krakow, 1990, p. 52.
[6] G. F. Bastin, H. J. M. Heijligers, *Microbeam Analysis* **1988**, 325.
[7] G. F. Bastin, H. J. M. Heijligers, *Scanning* **1990**, *12*, 225.
[8] H. W. Werner, A. Torrisi, *Fresenius Z. Anal. Chem.* **1990**, *337*, 594.
[9] N. S. Parekh, K. Nieuwenhuizen, J. J. M. Borstrok, O. Elgersma, *J. Electrochem. Soc.* (*to be published*).
[10] K. H. Müller, H. Oechsner, *Mikrochimica Acta* [*Wien*] **1983**, [Suppl.] *10*, 51.
[11] C. H. Becker, K. T. Gillen, *Anal. Chem.* **1984**, *56*, 1671.
[12] R. Castaing, *Advances in Electronics and Electron Physics, Vol. 13* (L. Marton, ed.), Academic, New York, 1960, p. 317.
[13] T. O. Ziebold, *Anal. Chem.* **1967**, *39*, 858.
[14] J. L. Pouchou, F. Pichoir, *Scanning* **1990**, *12*, 212.
[15] P. Willich, K. Schiffmann, *Mikrochimica Acta* [*Wien*] **1992** [Suppl.] *12*, 221.
[16] P. Willich, K. Schiffmann, *Microbeam Analysis* **1990**, 177.
[17] V. D. Scott, G. Love, *Scanning* **1990**, *12*, 193.
[18] G. F. Bastin, F. J. J. van Loo, H. J. M. Heijligers, *X-ray Spectrum.* **1984**, *13*, 91.
[19] J. L. Pouchou, F. Pichoir, *Proc. 11th ICXOM*, London, Canada, 1987, p. 249.
[20] J. Henoc, M. Tong, *Proc. 12th MAS Conference*, Boston, 1977, p. 46A.
[21] J. L. Pouchou, F. Pichoir, *Microbeam Analysis* **1988**, 319.
[22] J. L. Pouchou, F. Pichoir, *Microbeam Analysis* **1988**, 315.
[23] J. L. Pouchou, F. Pichoir, D. Boivin, *Microbeam Analysis* **1990**, 120.
[24] P. Karduck, N. Amman, W. P. Rehbach, *Microbeam Analysis* **1990**, 21.
[25] N. Amman, P. Karduck, *Microbeam Analysis* **1990**, 150.
[26] P. Willich, D. Obertop, *Surf. Interf. Anal.* **1988**, *13*, 20.
[27] M. Wendt, *Fresenius Z. Anal. Chem.* **1991**, *340*, 193.
[28] *TFCD-Thin film concentration display*, CAMECA, Courbevoie, France, 1988.
[29] *STRATA-X-ray microanalysis software*, SAMx, Guyancourt, France, 1991.
[30] H. J. August, *Mikrochimica Acta* [*Wien*] **1992** [Suppl.] *12*, 131 and 139.
[31] G. F. Bastin, J. M. Dijkstra, H. J. M. Heijligers, D. Klepper, *Mikrochimica Acta* [*Wien*] **1992** [Suppl.] *12*, 93.
[32] *TFA/MLA-Thin film analysis program*, TECHAN, Eindhoven, The Netherlands, 1991
[33] P. Willich, D. Obertop, *Proc. 12th ICXOM* Krakow, 1990, p. 100.

Mikrochim. Acta (1992) [Suppl.] 12: 19–36

Quantitative EPMA of the Ultra-Light Elements Boron Through Oxygen

Guillaume F. Bastin* and Hans J. M. Heijligers

Laboratory of Solid State Chemistry and Materials Science, Centre for Technical Ceramics, University of Technology, P.O. Box 513, NL-5600 MB Eindhoven, The Netherlands

Abstract. The practical problems in quantitative electron probe microanalysis of the ultra-light elements B, C, N, and O are discussed and solutions to these problems are given. It is shown that the basic requirement for accurate intensity measurements is the integral recording of the light element emission peaks in order to deal with the strong alterations in the peak shapes that sometimes occur. It is demonstrated that when these effects are properly accounted for and a good matrix correction program in conjunction with consistent mass absorption coefficients is used good quantitative results can be obtained.

Key words: Quantitative electron probe microanalysis, ultra-light elements.

The need for accurate quantitative electron probe microanalysis of ultra-light elements is the direct result of the strongly grown interest in new classes of materials with extremes in properties like hardness, toughness, wear-resistance, high-temperature behaviour, etc. Quite frequently these specific properties can only be achieved by the application or incorporation of significant amounts of compounds containing ultra-light elements, like B, C, N, and O. Since the properties of almost any materials depend largely also on their microstructures, microanalytical techniques will be necessary for the characterization of the material. In this respect the electron probe microanalyzer with its ability to analyze volumes down to 1 μm^3 could play a major role, provided that the specific problems connected with the detection and quantification of these elements can be overcome.

Basically, the procedure for the analysis of ultra-light elements is the same as the one applied in the analysis of medium to high atomic number elements and consists of three steps: 1) The specimen preparation, 2) The intensity measurement, 3) Matrix correction. The only difference is that specific items in each step become more important in the case of ultra-light elements, so that they require much more attention and care than is necessary for heavier elements. Therefore, the analysis of ultra-light elements is bound to be more time-consuming than the average operator

* To whom correspondence should be addressed

is used to. Nevertheless, it must be clear from the beginning that a good quantitative result can only be expected from a good matrix correction program working with the correct input: the proper k-ratios. The latter, in turn, can only be measured using well-defined procedures in which the operator is fully aware of all the problems that might interfere with the measurements and knows how to deal with them. Since many of these problems have to do with the condition of the specimen, and especially with its surface, it is inevitable that the first concern should be with specimen preparation. The reason why the surface conditions play such a large role in ultra-light element analysis is that the emission of X-rays of these elements usually takes place only in extremely shallow volumes and this has a number of far-reaching consequences.

In the first place the presence of contamination layers (whether carbonaceous or oxidic [1] in nature) can have a disproportionally large effect on the intensity measurements. Likewise, local variations in the X-ray take-off angle as a result of surface roughness can lead to surprisingly large variations in X-ray intensity. Not in the last place, however, the shallow emission of X-rays forces attention to be focused on the very first part of the X-ray distribution function with depth, the so-called $\Phi(\rho z)$ curve. Especially the surface ionization value $\Phi(0)$ and the variation of the $\Phi(\rho z)$ curve in the near-surface region play a crucial role in the calculation of the amount of X-rays that is finally emitted from the specimen. In these calculations, which are typically carried out by a matrix correction program, a major role is also played by the magnitude of the mass absorption coefficient (mac) of the particular wavelength analyzed in the specific compound examined. We have argued on several occasions [2–5] that a variation of 1% in the magnitude of a mac can lead to a 1% variation in the concentration for ultra-light elements.

It will be obvious, however, that the required 1% relative accuracy in the mac's cannot be guaranteed for ultra-light element radiations by any of the existing sets of mac's in literature [6, 7]; this is not even the case for radiations like Al–Kα or Si–Kα. Fortunately, the strong improvements in matrix correction procedures in the past decade [5, 8–10] have now made it possible to use these programs in order to test existing sets of mac's on correctness and internal consistency; in some cases new and better values can even be produced. Of course, such work can only be done if large data bases containing measurements on a wide variety of compounds and a wide range in accelerating voltages are available. Our laboratory has been active in this respect from 1983 on and the collection of data on B–Kα, C–Kα, N–Kα, and O–Kα intensity measurements has now been successfully completed.

Apart from the problems related to the surface condition of the specimens there are several more problems connected with the actual intensity measurements themselves. Among these are the (frequently) low count rates and low peak to background ratios for ultra-light element radiations, especially when a conventional lead-stearate crystal is used. The use of the newer synthetic multilayer crystals (like W/Si and Mo/B4C) causes important improvements [5, 10], not only in terms of improved count rates and peak to background ratios but also because of the strong suppression of higher order reflections such crystals sometimes exhibit. The latter effect is extremely beneficial for an accurate background determination.

Rather than discussing each specific problem in great detail, which has been done on several previous occasions already [2–5], it is more useful at present to follow the complete experimental procedure for the analysis of ultra-light elements from beginning to end and to discuss relevant problems where necessary.

Specimen Preparation

Although the *generation* of ultra-light element radiations may not take place in unduely shallow volumes their *emission* can sometimes be confined to extremely shallow regions. This is the result of the huge absorption which frequently takes place in the specimen itself. This is especially the case when the wavelength of the analyzed element falls right into an absorption edge of an accompanying element in the matrix. One of the most notorious examples that we know of in this respect is found in the combination carbon and boron where C–Kα radiation falls exactly into the B–K edge, which results in a huge mac for C–Kα X-rays: 37,000 cm^2/g [6]. The impact this high mac value has on the differences between generated and emitted intensity in a specimen like B_4C is shown in Fig. 1. Whereas the intensity is being generated up to depths of more than 1.5 μm the large absorption in the sample allows the generated X-rays only to escape from an average depth of .15 μm, i.e. 1500 Å. This explains why the presence of surface contaminants can have such a disastrous influence on the measurement: a thickness of even a 100 Å of contaminant can easily occupy a significant fraction of the total emission volume available for the intensity measurement. Since this part of the emission volume contains the surface region where absorption is normally low the effect of a contamination layer can be disproportionally large and can easily account for something like 10% of the total emitted intensity.

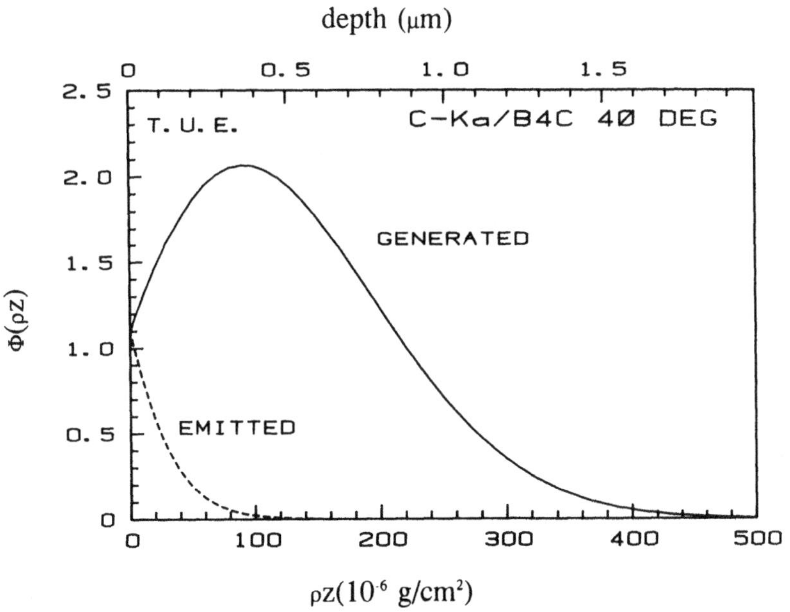

Fig. 1. Influence of a large mac value on the emission of C–Kα X-rays in B_4C. Solid curve represents the *generated* intensity, while the dashed curve shows the *emitted* intensity after correction for absorption. Accelerating voltage 10 kV; take-off angle 40°. Calculations according to the "PROZA" program [5] with mac's: C–Kα in C 2373 and in B 37,000 cm^2/g

Table 1. Influence of the X-ray take-off angle on the calculated *k*-ratios of *C*–Kα and B–Kα in B_4C for an accelerating voltage of 10 kV. Mac's used: B–Kα in B 3400, B–Kα in C 6500, C–Kα in B 37,000, and C–Kα in C 2373 cm^2/g. Calculations by the "PROZA" program [5]

Take-off angle (deg)	*k*-ratio C–Kα	*k*-ratio B–Kα
25	0.0181	0.6869
30	0.0201	0.6952
35	0.0222	0.7019
40	0.0242	0.7074
45	0.0261	0.7118
50	0.0279	0.7154
55	0.0295	0.7184

The example discussed here is not even the worst; the absorption of B–Kα X-rays in Si is even much stronger with a mac of 84,000 [6]. In such cases the emphasis is so heavy on the surface regions that attempts to do quantitative EPMA are almost like trying to do surface analysis.

Likewise, small variations in the X-ray take-off angle can have an exaggerated effect on the intensity measurements in cases of severe absorption. This is demonstrated in Table 1 where the results of some calculations by the "PROZA" program [5] for B_4C at 10 kV are given for a number of take-off angles. It is clear that a variation of only 5 degrees can produce a variation of 10% in the *k*-ratios for C–Kα. The results for B–Kα are relatively insensitive due to the much lower absorption. Because most ultra-light element compounds are extremely hard it is not always so easy to produce completely flat cross-sections during metallographic polishing, especially at interfaces with softer compounds or at edges bordering on the mounting resin. The resulting topography, which is usually present to some degree, may easily lead to apparent variations in concentrations which may, in fact, quite often be the result of a variation in the local geometry. Such cases can be discovered by rotating the specimen 180 degrees. A previous increase in intensity will now be transformed into a decrease or vice versa.

Summarizing this paragraph it can be stated that utmost care has to be exercised in the preparation of flat cross-sections which are free of contamination layers before attempts are being made for quantitative EPMA of ultra-light elements.

Intensity Measurements

Contamination Phenomena

Because the intensity measurements for ultra-light elements are usually much more time-consuming than one is used to for heavier elements contamination effects under the electron beam can play a more important role than they normally do. While the contamination with e.g., carbon can safely be ignored during a measurement of Cu–Kα during 10 s it is no longer safe to do this during the integral recording of a

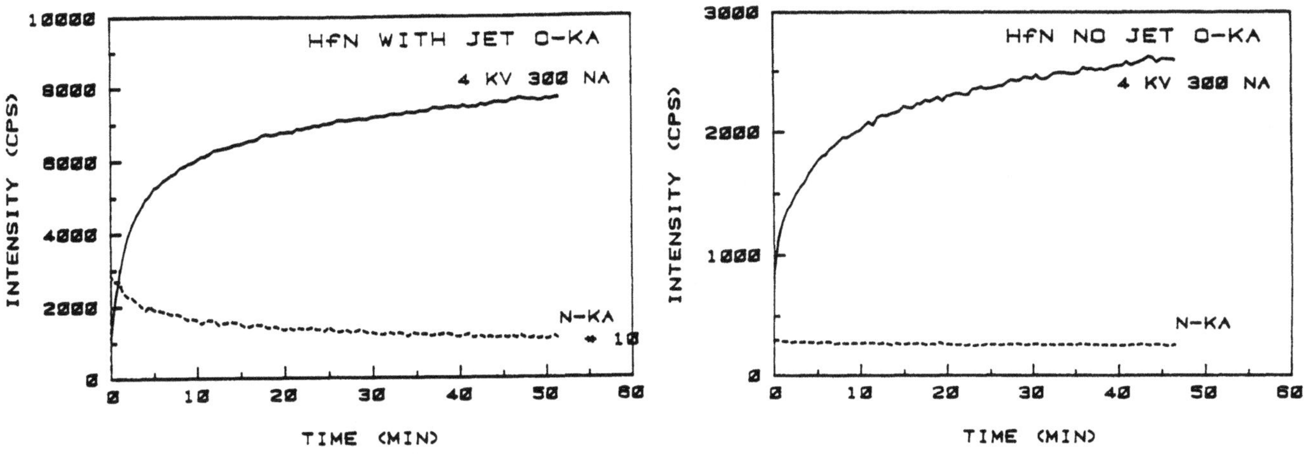

Fig. 2. In situ oxidation phenomena on a HfN specimen at 4 kV and 300 nA beam current. Left: With the use of an air jet; Right: without air jet. The use of the air jet leads to a rapid increase in O–Kα signal (top) and a gradual decrease in N–Kα signal (bottom)

C–Kα or N–Kα emission peak which can take up to 1 h on the same location of the specimen. The effects of carbon contamination can be reduced quite effectively by using a simple device like an air jet [1–5] which can bring down carbon contamination to an absolute minimum level, which is, however, unfortunately never quite equal to zero. On the other hand, the application of an air jet can lead to dramatic oxidation phenomena on certain sensitive specimens. This is demonstrated in Fig. 2 for a HfN specimen which can apparently be oxidized extremely rapidly when an air jet is used. Even without air jet, oxidation can set in immediately after the positioning of the beam although the process then takes place at a much lower rate and no noticeable losses in N–Kα intensity occur. It is very difficult to give strict rules on how the effects of various contamination phenomena should be diminished other than the general advice to watch closely for adverse effects. The best thing to do is probably to monitor the intensity of the element to be measured for a prolonged period of time, together with the intensities of possibly contaminating elements. Then it can be decided how to act in the most appropriate way in order to keep contamination phenomena at a minimum level and yet have a stable and correct signal with time.

Peak Shape Alterations

After the brief discussion of the obvious effects which might lead to erroneous measurements we now proceed to the more fundamental problem: how the actual intensity measurements should be carried out. In routine EPMA it is common practice to perform a peak search on standard and specimen and to compare the net peak intensities in order to obtain their ratio: the well-known k-ratio. From a fundamental point of view one could argue that it would be more appropriate to compare the integrals of the emitted intensities over the relevant wavelength range, in other words: the total emitted intensities associated with the specific electronic transition under consideration. If this is the correct thing to do then peak intensity measurements can only lead to correct k-ratios if it can be assumed that the peak

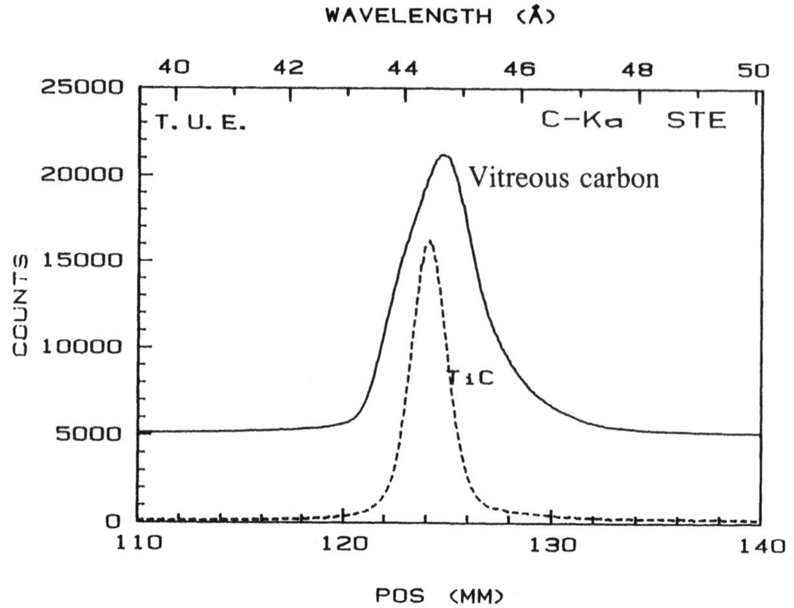

Fig. 3. Carbon-Kα emission profiles recorded from vitreous carbon and TiC at 10 kV and 300 nA (stearate crystal). Spectra have been scaled to yield the same net peak count rates. Note the peak shift and the extreme peak shape alteration

intensity is proportional to the integral emitted intensity. Although this assumption appears to be correct for the K- and L-radiations of medium–to–high Z-elements it no longer holds for ultra-light element radiations. Since in the latter cases the emission of X-rays is the result of transitions of the electrons involved in the chemical bond it must be expected that the X-ray emission spectra reveal to some extent the nature of that bond. Hence, it is quite natural to expect peak shifts and peak shape alterations in ultra-light element X-ray emission spectra. The effects of peak shifts do not really present a big problem in a wavelength-dispersive spectrometer because they can easily be accounted for by performing a new peak search procedure when moving from standard to specimen. Peak shape alterations, on the other hand, are a real nuisance because they are not so easy to cope with. Yet, it has been shown on several occasions [1–5] that serious errors of 30% or more can easily be made if such alterations in peak shapes are ignored.

Why it is necessary to perform the intensity measurements in an integral fashion will be clear upon inspection of Fig. 3 where the peak shape of C–Kα emitted from vitreous carbon is compared to that of TiC. The conspicuous difference in peak width makes it unfair to simply compare peak heights. The intensity from TiC would be highly overestimated at the expense of that from vitreous carbon. Similar observations can be made for B–Kα; the presence of two additional peaks in the spectrum from hexagonal BN (Fig. 4), which belong to the B–Kα emission band, makes it obvious that the intensities from both peaks should be included in the comparison with the spectrum from elemental boron. The latter also shows signs of the presence of several distinct components which add up to one peak.

The two examples discussed here imply that it is imperative to perform the intensity measurements for ultra-light element radiations in an integral manner.

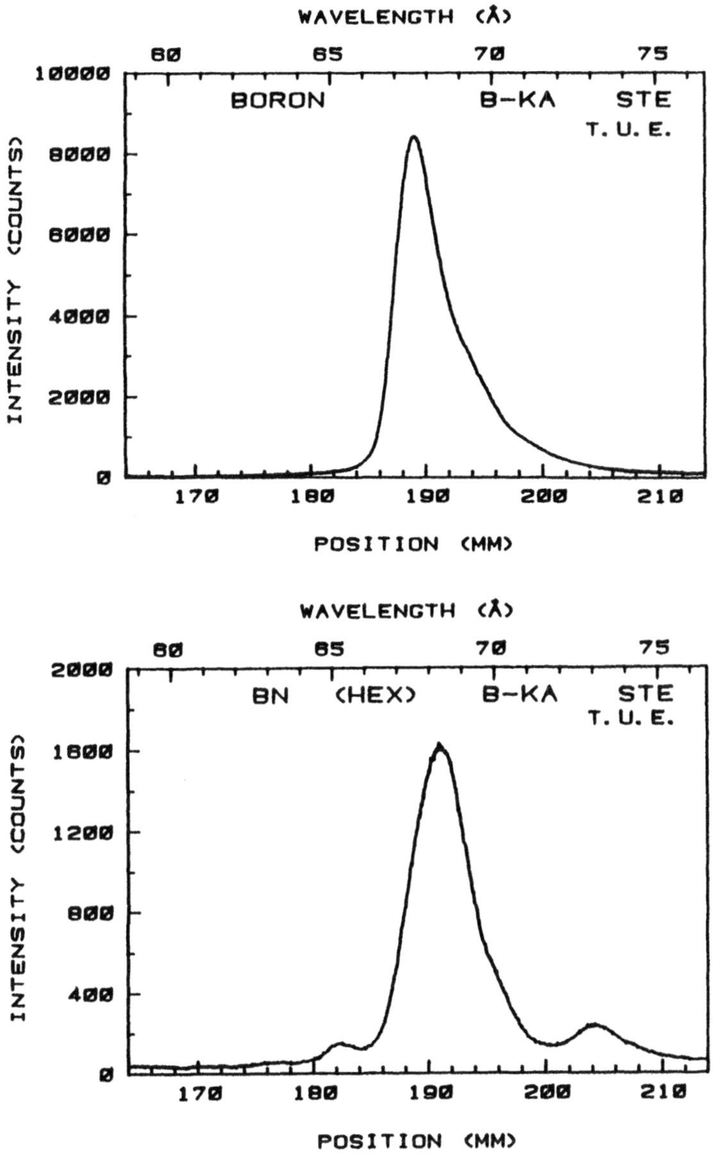

Fig 4. Peak shape alterations in the B–Kα peak between boron standard and hexagonal BN. The two satellites in the spectrum from BN actually belong to the B–Kα emission band. Experimental conditions: 10 kV, 300 nA, stearate crystal

Since this requires tedious and time-consuming procedures with a wavelength-dispersive spectrometer it is not a very attractive prospect for on-line analysis. Unfortunately, if only a few measurements in a particular system are required, there is no alternative to the full integral recording of the peaks in standard and specimen. When a large series of measurements in the same system are expected or when measurements have to be performed over a wide range in accelerating voltages it is advantageous to use the "area/peak factor" concept. This concept is based on the following considerations:

It has been demonstrated before [1–5] that there is a fixed ratio between the correct integral (or area) k-ratio and the usually incorrect but easy to measure peak k-ratio. This fixed ratio is called the area/peak factor (APF). In cases that repeated measurements in the same system are required it can save time to establish once the value of the APF for the particular compound relative to the selected standard from integral recordings of the peaks. Any further measurements, even at different accelerating voltages, can then be carried out on the peak and the peak k-ratios can easily be converted into more meaningful integral k-ratios by simple multiplication with the APF.

In many respects the APF can be regarded as a relative width-to-height ratio (relative to the standard) or also as a kind of weight factor which has to be assigned to a certain peak intensity. It must not be confused, however, with the full width at half maximum (FWHM) of the peak because in the vast majority of cases peak broadening takes place at the foot of the peak and, as a consequence, the FWHM value is usually insensitive to peak shape alterations. This is observed clearly in the B–Kα spectrum from BN in Fig. 4 with its two additional peaks.

Furthermore, it is important to keep in mind that APF values measured in one particular microprobe are strictly specific to that instrument, because they are strongly determined by the resolution of the analyzer crystal and spectrometer used.

The only quantity that should not be dependent on the particular instrument or crystal used, is the integral k-ratio for a given accelerating voltage and take-off angle.

In Table 2 a survey is given of APF's for C–Kα radiation, measured with a lead stearate crystal, in a variety of binary carbide specimens relative to an Fe_3C (cementite) standard. The large variation in peak shapes is reflected in the variation of APF, with extremes as low as 0.715 for ZrC and as high as 1.379 for vitreous carbon. A most conspicuous feature in Table 2 is the saw-tooth like variation of

Table 2. Area/Peak Factors (APF's) for carbon-Kα radiation in binary carbides with respect to Fe_3C on the stearate crystal (2d-spacing 100.23 Å)

Carbide	APF	Peak pos. (mm)	Wavelength (Å)	Crystal structure
Vitr. Carb.	1.379	124.82	44.700	Amorphous
B_4C	1.048	124.34	44.528	Rhombohed.
α-SiC (0001)	0.861	124.28	44.507	Hexagonal
TiC	0.723	124.16	44.464	Cubic
V_2C	0.775	124.40	44.550	Hexagonal
VC	0.773	124.34	44.528	Cubic
$Cr_{23}C_6$	0.801	124.34	44.528	Cubic
Cr_7C_3	0.803	124.28	44.507	Hexagonal
Cr_3C_2	0.825	124.16	44.464	Ort. rhomb.
Fe_3C	1.000	124.34	44.528	Ort. rhomb.
ZrC	0.715	123.98	44.399	Cubic
NbC	0.787	124.34	44.528	Cubic
Mo_2C	0.822	124.58	44.614	Ort. rhomb.
HfC	0.831	124.10	44.442	Cubic
Ta_2C	—	124.46	44.571	Hexagonal
TaC	0.968	124.52	44.593	Cubic
W_2C	1.021	124.76	44.679	Hexagonal
WC	0.974	124.04	44.421	Hexagonal

Table 3. Area/peak factors for carbon-Kα radiation in a number of binary carbides with respect to Fe_3C on the W/Si crystal (2*d*-spacing 59.8 Å)

Carbide	APF	Peak pos. (mm)
B_4C	1.010	208.50
SiC	0.933	208.30
TiC	0.868	208.00
V_2C	0.873	208.30
VC	0.873	208.30
ZrC	0.880	207.80

APF with atomic number of the metal partner. It is evident that strong carbide-forming elements like Ti and Zr hold the lowest APF values, which means that their carbides have the narrowest peaks compared to Fe_3C. The influence of the resolution of the analyzer crystal is shown in Table 3, where a number of APF's measured on the W/Si multilayer crystal are given. The observation that now all APF's are closer to unity means that the W/Si crystal simply has a somewhat poorer spectral resolution. Although the latter property is usually considered a disadvantage it is actually an advantage in light element work because it makes the measurements less sensitive to peak shape alterations.

For the lighter element boron the situation is the same to start with: Strong peak shifts and pronounced peak shape alterations relative to elemental boron, at least with a stearate crystal. Of the 27 binary boride specimens investigated so far approximately 50% can be measured in an identical manner as the carbides. That means that in the first place peak shifts should be closely watched and in the second place the APF should be accurately known.

In the remaining 50% two additional complications were observed:

1. Peak positions in these borides can vary strongly from one crystal in the specimen to another; apparently the peak position is strongly dependent on the crystallographic orientation of the specimen with respect to the spectrometer.
2. Emitted peaks vary their shapes simultaneously with the shifts in the peak positions. This can either be observed during rotation of the specimen under the electron beam or when moving from one grain in the specimen to another.

Typical peak shifts on stearate can be of the order of 1 mm (0.357 Å) and the associated variations in peak shapes, expressed in APF values, can be seen in Table 4. The broadest peak with the lowest peak intensity is always found at the longest wavelength and the narrowest one with the highest peak intensity at the shortest wavelength. It seems, therefore, that the loss in peak intensity is compensated by a broadening of the peak in order to yield approximately the same integral intensity. Upon rotation of the specimen in its own plane under the electron beam the boron peak can be made to shift back and forth in a 4 times 90 degrees cycle and at the same time it will change its shape continuously.

Table 4. Area/peak factors for boron-Kα radiation in binary borides with respect to elemental boron on the stearate crystal ($2d$-spacing 100.23 Å). Those cases where a range in parameters is reported represent the compounds where a crystallographic dependence of peak shape and position exists

Boride	APF	Peak pos. (mm)	Wavelength (Å)
B	1.000	189.00	67.60
B_4C	1.014	189.00	67.60
BN	1.198	191.05–191.60	68.33–68.53
AlB_2	1.152–1.095	189.50–189.90	67.78–67.92
AlB_{12}	1.008	189.03	67.61
SiB_3	1.003	188.85	67.55
SiB_6	0.922	189.00	67.60
TiB	0.690–0.835	188.80–189.45	67.53–67.76
TiB_2	0.799–0.945	189.35–190.25	67.73–68.05
VB_2	0.950–1.045	189.85–190.70	67.91–68.21
CrB	0.825–1.000	189.25–190.25	67.69–68.05
CrB_2	0.950–1.090	189.90–190.90	67.92–68.28
Fe_2B	1.242	188.90	67.57
FeB	0.985–1.160	189.55–190.00	67.80–67.96
Co_2B	1.015	189.10	67.64
CoB	1.186	190.40	68.10
Ni_3B	1.020–0.935	189.40–189.85	67.74–67.91
Ni_2B	1.059	189.10	67.64
NiB	1.060–1.310	189.80–190.45	67.89–68.12
ZrB_2	0.665–0.915	188.90–189.90	67.57–67.92
NbB	0.775–0.855	189.05–189.55	67.62–67.80
NbB_2	0.810–1.025	189.50–190.55	67.78–68.16
MoB	0.936	189.70	67.85
LaB_6	0.898	189.55	67.80
TaB	0.853–0.881	189.55–189.85	67.80–67.91
TaB_2	0.990–1.120	189.95–190.60	67.94–68.16
WB	0.978	189.80	67.89
UB_4	1.028	189.95	67.94

The origin of these peculiar effects is found in polarization phenomena [11] in the emitted B–Kα radiation. Under certain conditions the Pb-stearate crystal can act as a polarization filter and the conditions are optimal when the angle of incidence of the X-rays on the analyzer crystal is 45 degrees. In our spectrometer this angle is 42.5 degrees, which is very close to the optimum angle. It could be expected that spectra recorded with an Mo/B_4C multilayer, with a $2d$-spacing of 144.8 Å, and hence, a correspondingly much lower angle of incidence of the X-rays on the crystal, would be less sensitive to polarization effects. Our preliminary experiments with this crystal show that this is the case, although the phenomena can still be observed and peak shape alterations still have to be accounted for. The occurrence of polarization phenomena is strongly related to the crystal symmetry [5, 11] of the particular compound examined: all compounds with a symmetry lower than cubic and higher than triclinic may exhibit this behaviour. Quite conform to the expectations the relatively few cubic borides did not show an orientation-dependence of peak shape and peak position. Neither did elemental boron with its rhombohedral symmetry and this property made it suitable as a convenient standard for the boron measurements.

On cubic BN a peculiar observation was made later: It turned out that the left-hand additional peak, which was rather prominent in hexagonal BN (Fig. 4) was almost gone; the integral intensity, however, as well as the APF were exactly the same.

It can hardly be overemphasized how important the effects of shape alterations can be on the intensity measurements because it is clear from Tables 2 and 4 that very large errors can easily be made when these alterations are ignored. The magnitude of these errors is simply dictated by the choice of the standard in relation to the compound examined and can easily amount to more than 10%. In those cases with a crystallographic dependence of peak shape and position, it is not even safe to use the same compound as a standard. Variations in peak intensity, even when the peak position is continuously monitored, might easily be misinterpreted as variations in concentration while they are, in fact, entirely the result of shape alterations. What can be done in these particular cases in order to reduce the measuring efforts is to measure the APF as a function of peak position which requires the integral recordings of the B–Kα peaks in a number of different grains (orientations) in the specimen and to make a plot of APF vs. observed peak position [5]. Future measurements on the same compound can then again be carried out at the peak and from the known correlation between peak position and APF it is possible to correct the peak k-ratios into integral k-ratios.

Our experience shows that the shape alteration effects (on stearate) are most severe for the lightest element boron while they are much less severe for carbon because the crystallographic effects and polarization phenomena are absent. The latter applies also to the elements nitrogen and oxygen, where, in addition, the effects of shape alterations are very much smaller than for boron and carbon. For nitrogen, measured relative to Cr_2N, APF values hardly deviate 5% from unity on stearate. With the W/Si crystal the results for nitrogen are even less pronounced and the same is true for oxygen.

The Collection of Data Bases and Their Use

Meaningful tests on the performances of the various matrix correction programs currently available can only be carried out on large data bases containing many light element measurements taken over a wide range in accelerating voltages and including a large variety of systems. Because in the end the performance of any correction program depends largely on the quality and consistency of the available mass absorption coefficients it is inevitable that the assessment is iterative in nature: No program can function well without proper mass absorption coefficients and good m.a.c's will be of no use when used in a poor program. It will be clear, therefore, that the availability of large data bases is crucial in the process of evaluation. Not only can these data bases help in the improvement of existing correction procedures, they can also be of great value in establishing more reliable and consistent mac's.

Therefore, we will give a brief summary of the experimental outlines along which the data bases were collected.

In general the measurements for the elements boron, carbon, and nitrogen were carried out in the voltage range between 4 and 30 keV, with an exception for some of the most heavily absorbing borides which were measured only up to 15 kV. The relatively simple oxygen measurements were even extended to voltages of up to 40 kV. Boron and carbon were exclusively measured with the stearate crystal. For oxygen only the W/Si multilayer was used because of the problems we encountered with stearate in the background determination on the short-wavelength side of the oxygen peak as a result of the proximity of the mechanical limit of the spectrometer. Extensive use was made of the APF's which had previously been established and which enabled us to carry out large series of measurements on the peaks quite rapidly, with good statistics as a result. Unfortunately, this approach could not be followed for the nitrogen measurements because of the strong curvatures which sometimes occur in backgrounds on stearate and, to a lesser extent, on the W/Si multilayer. In this case the measurements were carried out in an integral manner on both crystals simultaneously over the full voltage range. After completion the results were averaged in a 2 : 1 weight ratio in favour of the W/Si multilayer because of the much better count rates on the latter crystal.

For carbon a data file of 117 measurements relative to Fe_3C on 13 binary carbides between 4 and 30 kV could be collected. For boron a total number of 180 measurements, relative to elemental boron, on 27 binary borides between 4 and 15 kV (in some cases up to 30 kV) have been realized. The data base for nitrogen contains at present 144 measurements on 17 binary nitrides, relative to Cr_2N between 4 and 30 kV.

The work on oxygen has resulted in a compilation of 344 measurements (on the W/Si multilayer), relative to Fe_2O_3 as a standard, for accelerating voltages between 4 and 40 kV. These measurements have been carried out on more than 30 oxides, carefully selected for sufficient electrical conductivity, just like the standard used. Non-conducting oxides, which present a number of specific problems, difficult to deal with, are at the moment the subject of a separate investigation. More details of the work on nitrogen and oxygen will be published in the near future.

Finally, it must be mentioned that apart from the light elements also the X-ray lines of the metal partners have always been measured over the full range in voltages. These data can also serve to test existing and future matrix correction programs on their performance. The full numerical details have been published in a series of internal reports [3, 4, 12, 13].

Matrix Correction

In the introduction it has already been emphasized that mac's play a crucial role in the process of matrix correction in which measured k-ratios are finally converted into concentration units. In our opinion this role is of equal importance as the performance of the particular matrix correction program used and it is not well possible to test both items separately.

It is obvious that, as far as the correction procedure is concerned, the absorption correction part in the program deserves most of the attention in the present case of ultra-light elements because the corrections for absorption can easily be an order of magnitude larger than for atomic number effects. Since the emission of ultra-light element X-radiation frequently takes place only in the near-surface layers of the specimen it is essential not only to have the proper X-ray distribution functions with depth ($\Phi(\rho z)$ curves) but especially to have the correct $\Phi(0)$ value. As a matter of fact the complete very first part of the $\Phi(\rho z)$ curve, close to the specimen surface, plays a vital role. It must, therefore, be expected that only those programs based on realistic descriptions of $\Phi(\rho z)$ curves can be successful in coping with the heavy demands being made on the absorption correction in ultra-light element work. Fortunately, a number of such programs have become available in the eighties [5, 8, 9] and they have proved successful in these difficult circumstances.

As argued before, matrix correction programs can only be tested on large data bases containing measurements on a wide variety of compounds over a wide range in accelerating voltages. Performance tests are usually carried out by comparing the k-ratio calculated by a specific program (k') for the reported composition of a compound in the data base to the measured k-ratio (k). The ratio k'/k is commonly displayed in a histogram showing the number of analyses vs. k'/k.

The success of a particular program is measured in terms of the narrowness of the histogram obtained, expressed in the relative root-mean-square deviation (r.m.s. in %) and the final average k'/k value.

It has been pointed out before [5] that with the various sets of published mac's for light element radiations [6, 7] it is virtually impossible for any of the currently used correction programs to obtain r.m.s. values below 10%; a conclusion which applies to all our data bases for the elements boron through oxygen. We have always attributed these poor results to inconsistencies in the published mac's and experimental evidence for this statement is mounting.

The most prominent observations concern the inexplicable discrepancies between the results obtained in sequences of similar compounds of neighbouring elements in the periodic system, like e.g. the carbides ZrC, NbC and Mo_2C. The use of one particular set of mac's may lead to satisfactory results only for NbC and while much too low k-ratios can be calculated for ZrC the calculations for Mo_2C may come out much too high. If systematic differences in X-ray emission can be ignored these peculiar observations must be regarded as a serious indication that the mac for C–Kα X-rays is much too high in Zr and much too low in Mo. Since we are dealing here with compounds of neighbouring elements the discrepancies can never be explained by improper functioning of the correction program used because the equations upon which all programs are based are invariably smooth functions of atomic number Z and atomic weight A, either in an explicit or implicit way.

Similar experiences occurred with oxygen measurements in a sequence of neighbouring oxides: while excellent results were obtained up to 40 kV on ZrO_2, Nb_2O_5, MoO_3 and RuO_2 with the "PROZA" program using Henke et al.'s mac's for O–Kα, in spite of the very large absorption effects, the results on SnO_2 were suddenly 30% off. It turned out that in this particular case the mac has inadvertently been extrapolated across an absorption edge. The correct value must be much lower (presumably approx 15,050 cm^2/g) than the published [6] value of 23,090 cm^2/g. This was later corroborated by measurements on In_2O_3 films on Si substrates [14]. Our calculations on these data show that here too the published mac for O–Kα in In (22,200 cm^2/g) must be much too high; a value around 15,000 cm^2/g gives an important improvement in the results.

During the evaluation of the results on boron and nitrogen often similar experiences were made and quite frequently major adjustments were necessary, especially for mac's in elements like Zr, Nb, and Mo. This can easily be explained by the close vicinity of the M_5 edge in these elements to the wavelengths of the ultra-light element radiation. This usually produces regions with strong non-linear variations of mac with wavelength where it is very difficult to inter- and extrapolate.

Nevertheless it will be clear that such systematic evaluations can only be carried out when large data bases of accurate measurements are available. The use of these data bases makes it possible to test any correction program in conjunction with any set of mac's. Ultimately, it is even possible to come up with much more consistent sets of mac's with a relative accuracy which may be better than can ever be expected from direct physical measurements.

Finally, it is very important to point out that the improvements that can be achieved by the use of the improved sets of mac's are not confined only to the particular program used. Our experiences [2–4] as well as other's [15] indicate that the use of these mac's will produce most significant improvements in virtually all current correction programs in which suitable absorption correction models are being used.

A brief discussion is given of what can be achieved with strongly improved matrix correction procedures in combination with consistent sets of mac's. Figure 5 shows

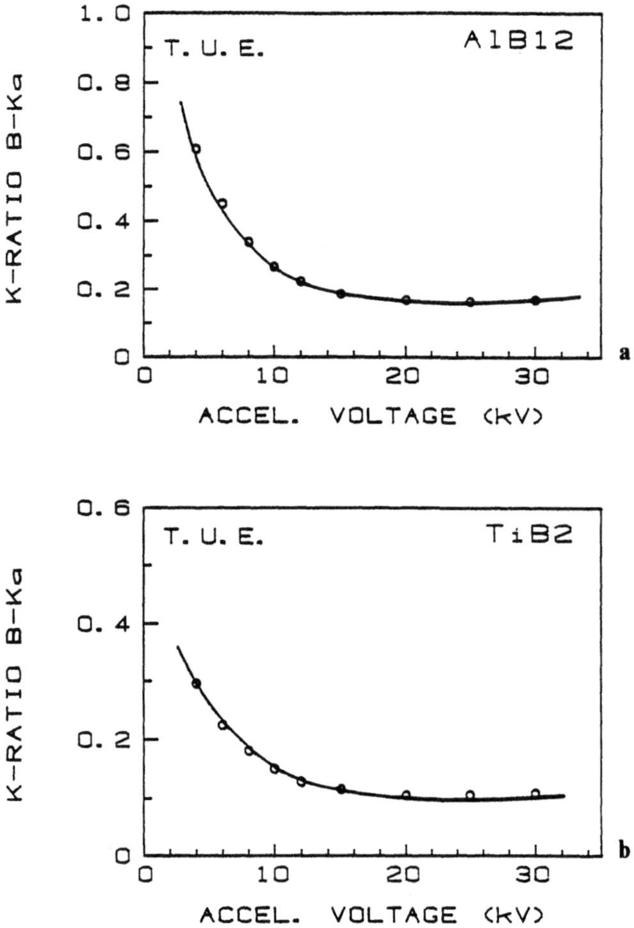

Fig. 5. Comparison between the experimental integral *k*-ratios for B–Kα relative to boron measured on stearate and the predictions of the PROZA [5] program (solid curve). **a** AlB_{12} (82.78 wt % B); **b** TiB_2 (30.07 wt % B). Mac's used: B–Kα in B: 3400, in Al: 65,000, and in Ti: 14,700 cm^2/g

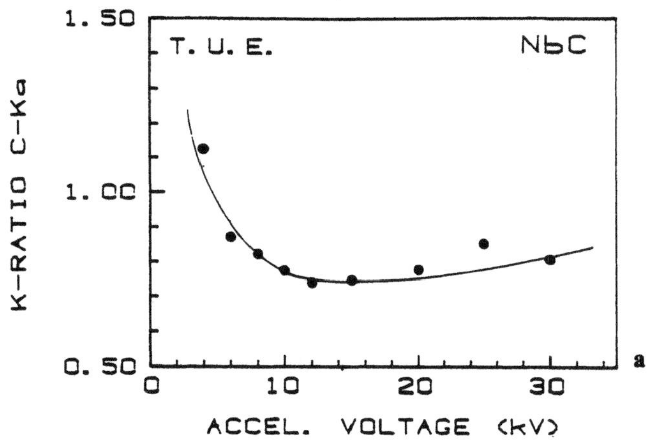

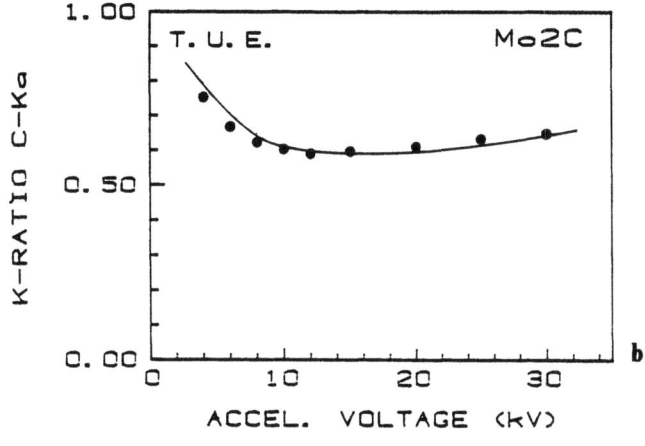

Fig. 6. Comparison between the experimental integral k-ratios for C–Kα relative to Fe_3C measured on stearate and the predictions of the PROZA [5] program (solid curve). **a** NbC (8.55 wt % C); **b** Mo_2C (5.58 wt % C). Mac's used: C–Kα in C: 2373, in Nb: 24,800, in Mo: 20,600 and in Fe: 13,500 cm^2/g

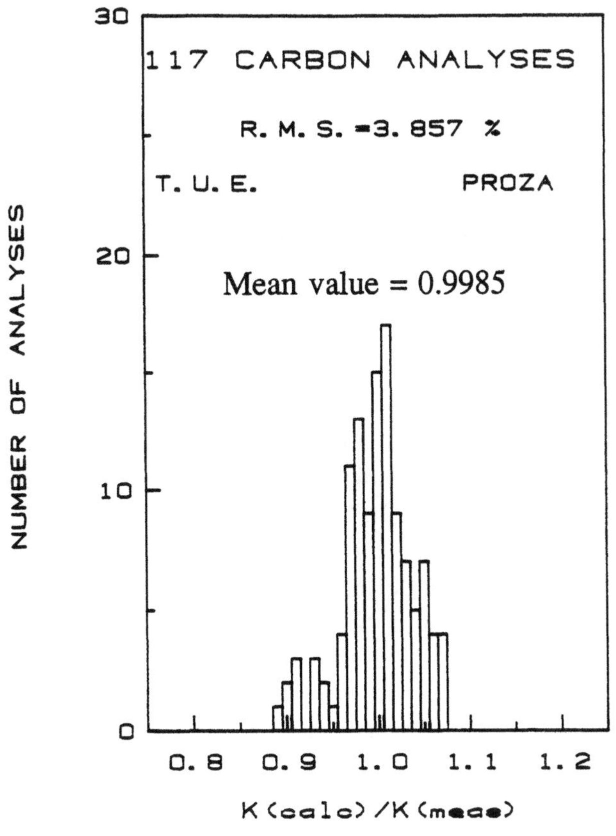

Fig. 7. Histogram showing the results obtained with the PROZA program on 117 carbon analyses between 4 and 30 kV

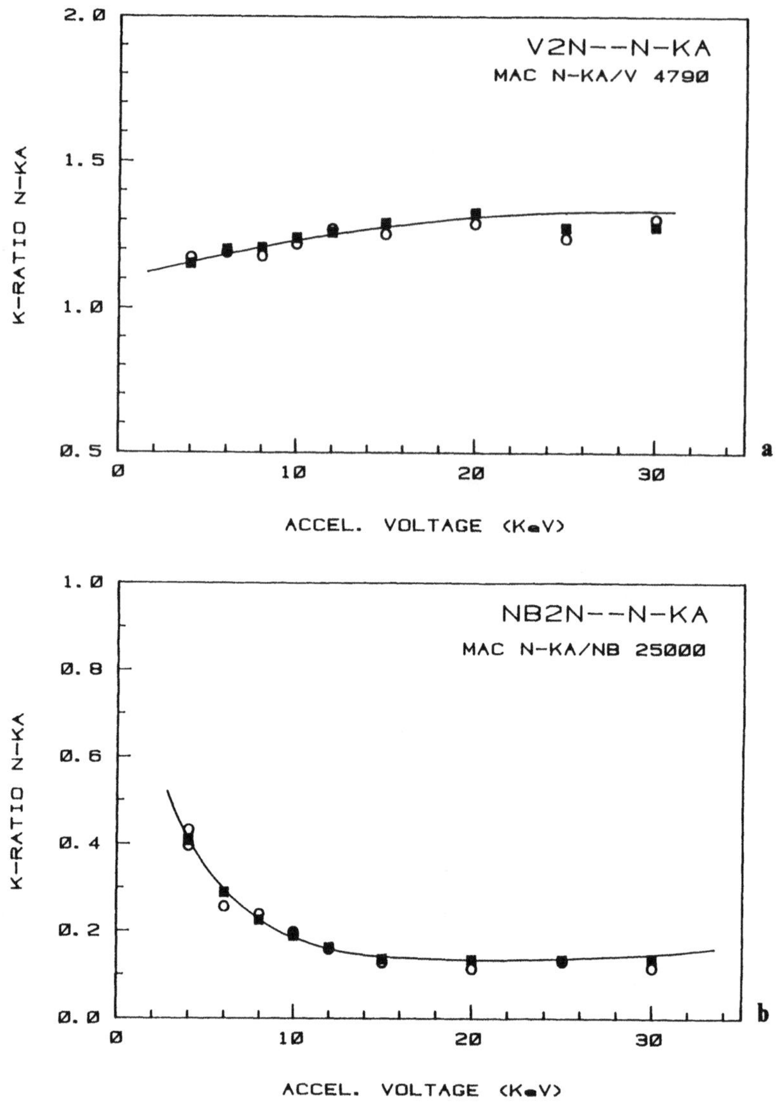

Fig. 8. Comparison between the experimental integral k-ratios for N–Kα relative to Cr_2N (10.70 wt % N, 0.48 wt % O, balance Cr) measured with stearate (open circles) as well as with a W/Si multilayer crystal (solid squares) and the predictions of the PROZA program [5] (solid curve). **a** V_2N (11.99 wt % N, 0.78 wt % O); **b** Nb_2N (6.55 wt % N, 0.46 wt % O). Mac's used: N–Kα in N: 1810, in O: 2530, in Cr: 5630, in V: 4790, and in Nb: 25,000 cm^2/g

two examples of boron measurements in comparison to the predictions made by the PROZA program. The agreement is good, certainly if one takes into consideration the very large range in accelerating voltages. On the complete data file of 180 measurements an average value of $C_{calc.}/C_{measured}$ of 0.9603 and an r.m.s. value of 6–7% was obtained. The lowest r.m.s. figure and a better average would be achieved if the data for the three Ni-borides Ni_3B, Ni_2B and NiB, with their systematically too low (±15%) emission of B–Kα X-rays, would be left out.

Some examples of our carbon measurements are represented in Fig. 6 where they are compared with the calculations of the PROZA program. Again the agreement is satisfactory. The overall results for the complete data base of 117 measurements are given in a histogram in Fig. 7.

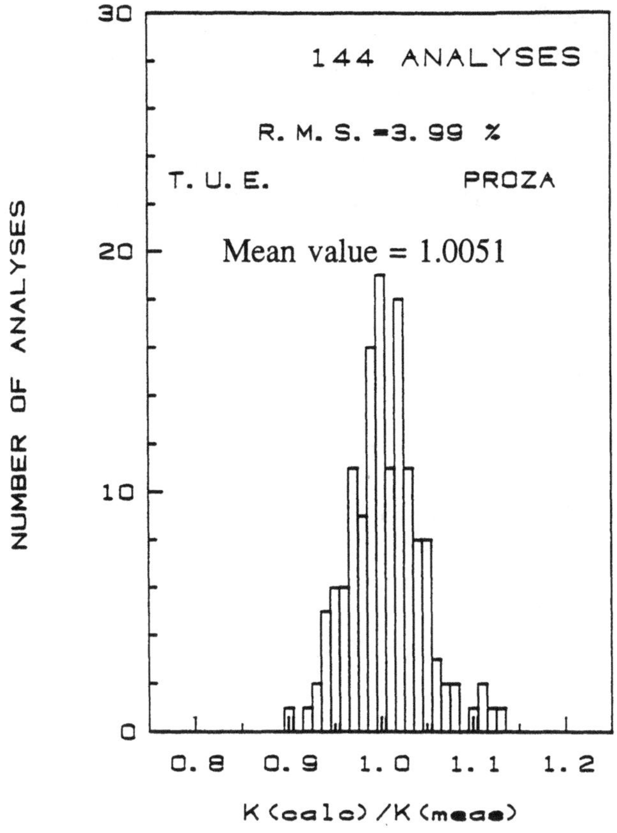

Fig. 9. Histogram showing the results obtained with the PROZA program on 144 nitrogen analyses between 4 and 30 kV

Figure 8 shows some of the results of the nitrogen measurements, in a relatively low-absorbing system (V_2N) and in a heavily-absorbing system (Nb_2N). The agreement between predictions and measurements is again good. Also the agreement between the *k*-ratios measured on two different crystals is satisfactory. The overall results for the complete data file of 144 measurements can be found in Fig. 9.

The results on our oxygen data file of 344 measurements relative to hematite (Fe_2O_3) as a standard can be summarized as follows: Average k'/k value of 0.9992 and an r.m.s. figure of 2.48%. This compares favourably to the results which are usually claimed for data bases containing measurements on medium-to-high Z elements, certainly if one takes into consideration that our oxygen measurements have been carried out up to 40 kV.

Conclusions

The general conclusion that can be drawn from the present work is simple: If one is prepared to exercise sufficient care in the measurements and keeps an open eye for effects like peak shifts and peak shape alterations then surprisingly accurate and reproducible intensity measurements are possible for ultra-light element radiations. From then on it is only a question of selecting a good matrix correction program in combination with consistent mac's before finally reliable quantitative results are obtained.

It has to be emphasized that in the present work the experimental conditions for our measurements have been stretched far beyond the point one would normally consider useful and this has been done for academic purposes: testing of programs and mac's. Because the results are still remarkably good it must be expected that in more moderate conditions, e.g. in the range of 8–20 kV, the results would be even better.

One has to accept the fact that the actual procedures involved in the collection of light-element data will always be more time-consuming than those for medium-to-high Z-elements.

References

[1] G. F. Bastin, H. J. M. Heijligers, in: *Microbeam Analysis*, San Francisco Press, San Francisco, 1988, p. 325.

[2] G. F. Bastin, H. J. M. Heijligers, *X-ray Spectrom.* **1986**, *15*, p. 135.

[3] G. F. Bastin, H. J. M. Heijligers, *Quantitative Electon-Probe Microanalysis of Carbon in Binary Carbides, Internal. Report*, Eindhoven, University of Technology, March, Feb. 1990.

[4] G. F. Bastin, H. J. M. Heijligers, *Quantitative Electon-Probe Microanalysis of Boron in Binary Borides, Internal Report*, Eindhoven, University of Technology, March 1986.

[5] G. F. Bastin, H. J. M. Heijligers, *Scanning*, **1990**, *12*, p. 225.

[6] B. L. Henke, P. Lee, T. J. Tanaka, R. L. Shimabukuro, B. K. Fujikawa, *At. Data Nucl. Data Tables* **1982**, *27*, 1.

[7] B. L. Henke, E. S. Ebisu, *Adv. X-ray Anal.* **1974**, *17*, 150.

[8] R. H. Packwood, J. D. Brown, *X-ray Spectrom.* **1981**, *10*, 138.

[9] J. L. Pouchou, F. Pichoir, *Réch. Aérospat.* **1984**, *3*, 13.

[10] G. F. Bastin, H. J. M. Heijligers in: *Proc. 11th ICXOM Conference*, London, Canada (J. D. Brown, R. H. Packwood, eds.), 1987, p. 257.

[11] G. Wiech, in: *Emission and Scattering Techniques, Ser. C* (P. Day, ed.), Nato Adv. Study Inst., 1981, p. 103.

[12] G. F. Bastin, H. J. M. Heijligers, *Quantitative Electon-Probe Microanalysis of Nitrogen, Internal Report*, Eindhoven, University of Technology, September 1988.

[13] G. F. Bastin, H. J. M. Heijligers, *Quantitative Electon-Probe Microanalysis of Oxygen, Internal Report*, Eindhoven, University of Technology, December 1989.

[14] A. McKenzie, personal communication, Cavendish Laboratory, Cambridge, U.K., 1990.

[15] V. D. Scott, G. Love, in: *Electron Probe Quantitation* (K. F. J. Heinrich, D. E. Newbury, eds.), Plenum, New York, 1991, p. 19.

Mikrochim. Acta (1992) [Suppl.] 12: 37–52

Auger Microscopy and Electron Probe Microanalysis

J. Cazaux

LASSI/GRSM, BP 347, Faculté des Sciences, F-51062 Reims Cédex, France

Abstract. The parameters, characterizing the performance (elemental information, sensitivity, lateral resolution, minimum detectable mass), and the trends of Auger microscopy, are analyzed in detail by referring, at each step, to the corresponding parameters of electron probe microanalysis (EPMA). Special attention is paid to the surface sensitivity of the two techniques, to the EPMA analysis of a thin coating and to the quantification problems associated with the determination of the Auger backscattering factor or the $\Phi(0)$ function in EPMA.

Some commonly admitted opinions are reconsidered. The main result is that, in spite of some differences, each technique may benefit from the progress established in the other technique, instead of ignoring each other.

Key words: Auger microscopy, electron probe microanalysis.

When a solid specimen is submitted to the bombardment of incident electrons (with primary energies E_0 in the 10-keV range), it is well known that the core-ionization of atoms is followed by two complementary de-excitation processes: X-ray emission and Auger electron emission.

When these processes are applied to the elemental characterization of specimens by detecting either the emitted X-ray photons in EPMA (electron probe microanalysis) or the emitted electrons in AES (Auger electron spectroscopy), one may expect a close analogy between the two techniques concerning for instance the interactions of incident electrons inside the specimen and the ionization of atoms.

Despite this obvious remark, it is surprising to observe that the formalisms used for these spectroscopies have been developed quite independently of each other. One possible explanation (among many others) of this situation may be found in the fact that EPMA became rapidly mature after the pioneering work of Castaing in the Fifties [1], while AES, which was operated initially in a LEED (low electron energy diffraction)-instrument with primary electrons in the 1–3 keV range [2, 3], took longer to become mature.

During the last decade, however, there was a trend to increase the primary energy of AES to 20 keV or more and therefore the situation has to be reconsidered. The present paper is devoted mainly to AES and SAM (scanning Auger microscopy), the information it yields, its intensity, surface sensitivity, and quantification problems, the lateral resolution of SAM and imaging properties, as well as the trends of this technique. Each step will be discussed by referring to the corresponding performance and procedure of EPMA in order not only to indicate the differences between the two approaches but also to outline the benefit that can be implemented in one technique, looking at the progress of the other.

Elemental and Chemical Information

In principle, all the elements composing the specimen, except H and He, can be identified by both techniques but practically, the detection of Li and Be and the quantification of light elements are very difficult in EPMA [4], whereas there is no specific difficulty in detecting and measuring elements ranging from Li to Na by AES. As an illustration, the paper of Kruger et al. [5], may be referred in which AES is used to analyse qualitatively the change of the oxygen concentration on the grain surface and at the grain boundaries of a high-Tc superconductor.

Both spectroscopies may also provide additional information related to the chemical state of the element of interest through the shape and the shift of the lines if the emission processes involve the valence band. Nevertheless the exact shape of the valence band density of state is more difficult to extract from the shape of a CVV (core valence valence) Auger line [6] than from the shape of a CV (core valence) transition in EPMA. On the other hand the chemical shift of the Auger lines, which can be sometimes very large, are extensively exploited in AES with a special reference to the SiO_2/Si system where it reaches around 15 eV for both the LVV (L-shell valence valence) and the KLL (K-shell L-shell L-shell) Si Auger lines [7].

Signal Intensity, Sensitivity and Quantification

Signal Intensity

In Auger spectroscopy, the detected Auger signal intensity, $dI_A(\mathrm{A})$, associated with an element A with atomic concentration, C_A, and located very close to the surface between depths z and $z + dz$, can be written (at normal incidence) as [8]:

$$dI_A(\mathrm{A}) = I_0 \cdot [N^\circ C_A(1 + R_{A/S})Q_A\, dz]a_{ijk}e^{-z/\lambda \sin\theta}T_A. \qquad (1)$$

The expression between brackets corresponds to the number of ionized atoms (electronic level i) per incident electron, where Q_A is the ionization cross-section; $R_{A/S}$ is the Auger backscattering coefficient, describing the reinforcement of the Auger signal (of element A) by the electrons backscattered by the substrate S; N° is the atomic density ($\sim 5 \cdot 10^{22}$ atoms/cm^3). Further, T_A is the collection efficiency of the analyser, a_{ijk} is the Auger yield and the exponential term describes the attenuation of Auger electrons generated in the specimen at depth z when going towards the analyser (λ = attenuation length and θ = take off angle).

The detected X-ray signal intensity $dI_X(\mathrm{A})$, related to the same layer dz very close to the surface of the same specimen submitted to the same experimental conditions has obviously a similar form, with analogous terms between brackets:

$$dI_X(\mathrm{A}) = I_0 \cdot [N^\circ C_A(1 + R_{A/S})Q_A\, dz]\omega_{ij} \cdot e^{-\mu z/\sin\theta}T_X. \qquad (2)$$

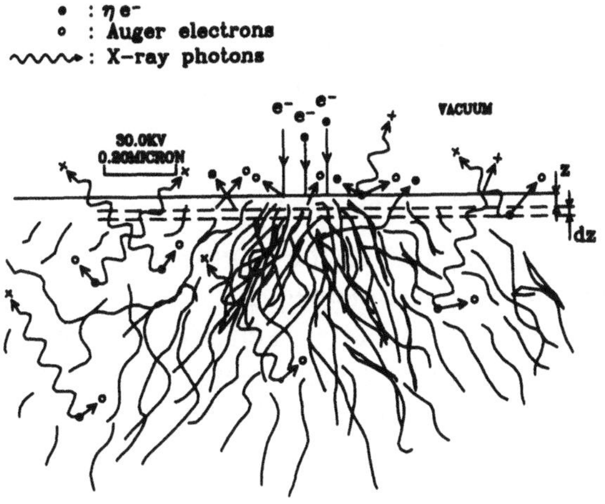

Fig. 1. Schematic diagram of the Auger electron and the X-ray photon production. Due to their very short attenuation length only the Auger electrons created very close to the surface can escape into the vacuum to be measured

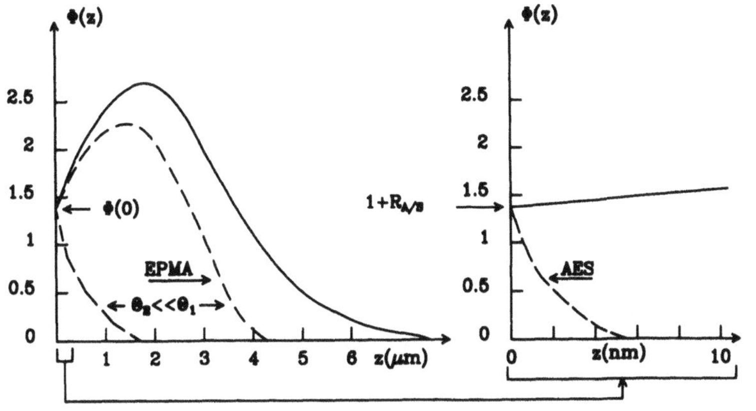

Fig. 2. The $\Phi(\rho z)$ function (solid line) is the same for EPMA (left) and AES (right) when applied to the same specimen (Al) under the same conditions ($E_0 = 30$ keV), but they differ considerably when they are weighted by the corresponding attenuation function (dashed lines). This difference may be partly reduced by a decrease of the take off angle (θ) in EPMA (left insert)

By observing that X-ray photons are also emitted (see Fig. 1), but with a yield ω_{ij}, eq. (2) has been derived from eq. (1) by replacing a_{ijk} with ω_{ij}, λ with μ^{-1} (μ: linear absorption coefficient) and T_A with T_X (collection efficiency of the X-ray analyser).

From the definition of the $\Phi(\rho z)$ function, used in EPMA, it is important to point out that:

$$1 + R_{A/S} = \Phi(0). \tag{3}$$

Quantitatively, the main difference between eq. (1) and (2) results from the fact that, being in the nanometer range, λ is several orders of magnitude less than μ^{-1}: the interaction of electrons (of kinetic energy in the 50 eV–2 keV range) with matter is stronger than that of X-ray photons. Consequently the $\Phi(\rho z)$ function is the same for the two spectroscopies (when both the primary beam energy, E_0, and the specimen are the same); the information depths, however, are strongly different from each other when this common function is weighted by the attenuation term (Fig. 2).

Surface Sensitivity

The order of magnitude of the minimum detectable concentration, C_m, of AES, using pulse counting, can be evaluated, in atomic fraction:

$$C_m \simeq 3\sqrt{\mathrm{BG}}/S_{\mathrm{A}}^{\infty}, \tag{4}$$

where BG is the number of counts of the background in the sample under study and S_{A}^{∞} is the number of counts of the signal for a pure element A.

In AES, most of the background is due to the backscattered electrons contribution, η, (mainly at high primary energies and for high kinetic energy Auger lines) and η increases with the mean atomic number of the atoms in the specimen (up to the depth of complete diffusion). The background decreases when the primary beam energy E_0 is increased above 10 keV (because η remains quite constant and is spread over a wider energy interval). In X-ray spectroscopy, the background (associated with the bremsstrahlung generation) increases also with the mean atomic number of the specimen (in a way similar to that of AES) but it also increases with the increase of E_0. Despite the fact that the signal -to- background ratio in AES is very different from that of EPMA ($S^{\infty}/\mathrm{BG} \simeq 10^{-1} - 1$ in AES; 10–100 in EDS-EPMA), the sensitivity of the two techniques is in the same range (about 0.1%), whereas it is about 0.02% in WDS-EPMA, because of its better energy resolution (that reduces the background signal–relative to EDS-EPMA) [9].

These sensitivities lead to the same fraction of the corresponding information depths for the evaluation of the minimum detectable thickness (MDT). In AES, this MDT is a fraction of an atomic monolayer (λ being in the nanometer range), while it is in the nanometer range in EPMA because the information depth is in the micrometer range.

Nevertheless the surface sensitivity of EPMA can be improved by decreasing its effective information depth, for example by lowering the primary beam energy and by operating at grazing incidence (in order to reduce the generation depth of the characteristic X-rays). Inspired by AES, an alternative solution would be the decrease of the take-off angle θ_X ($\mu^{-1} \sin \theta_X \leq$ few nm).

In fact the sensitivity depends on the nature of the element to be detected and in order to compare, in detail, the surface sensitivities of AES and EPMA, one needs to evaluate the number of counts, $S_{\mathrm{A}}^1(\mathrm{A})$ and $S_X^1(\mathrm{A})$, given by the two techniques for a surface atomic monolayer of an element A on a substrate S of different composition. From eqs. (1) and (2) one obtains:

$$S_{\mathrm{A}}^1(\mathrm{A})/S_X^1(\mathrm{A}) = a_{ijk} \cdot T_{\mathrm{A}} \cdot \tau_{\mathrm{A}}/\omega_{ij} \cdot T_X \cdot \tau_X, \tag{5}$$

where the same electronic subshell i is ionized and $\tau_{\mathrm{A}}(\tau_X)$ is the time of measurement per channel for the Auger (X-ray) signal.

From eq. (5) the advantage of using AES over WDS-EPMA in the detection of light elements (up to Na) results from the larger value of the Auger yield and the collection efficiency of the electron analyser (relative to ω_{ij} and T_X), the times of the (sequential) measurement being assumed to be the same for a given duration of the experiments. Nevertheless the very large disadvantage of WDS-EPMA is reduced by the fact that the background level in an X-ray spectrum is far lower than that of Auger spectroscopy and its surface sensitivity allows to detect thin oxide layers [10].

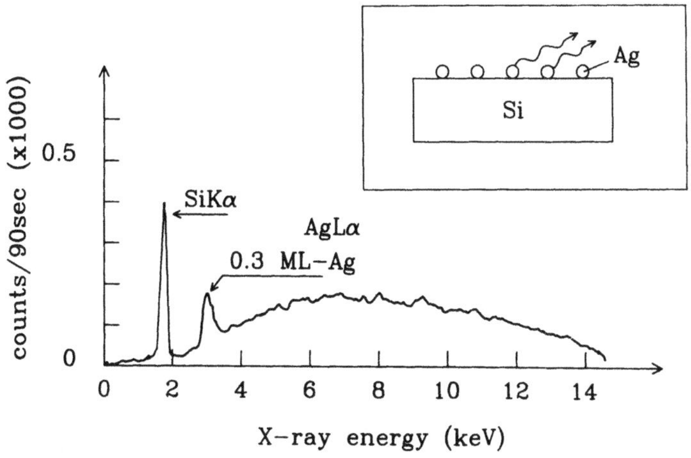

Fig. 3. X-ray spectrum of a 0.3 monolayer of Ag on a Si substrate at a small take-off angle (from [11])

In EDS-EPMA of electronic levels in the 1–2.5 keV range (i.e from Na to S for the K-lines), the advantage of AES over EPMA seems also less obvious because of the rapid increase of ω_{ij} when Z is increased and because of the fact that T_A/T_X is now in the 0.1–1 range (depending on the specific arrangements of the analysers in AES and EPMA). For a given duration of the experiments, τ, (i.e for a given dose on the specimen), the ratio of eq. 5 may be less than 10 (down to unity or even less), if one considers that the ED spectrometer is intrinsically a parallel detector ($\tau \simeq \tau_X$), while—up to now—almost all the spectra acquisitions in AES are performed in the sequential mode ($\tau_A \ll \tau$). The signal-to-background ratio remaining better in EDS-EPMA than in AES, which may result finally in better detection limits of EPMA than of Auger for some selected elements and specific samples. Such an advantage may also be extended to heavy elements by selecting characteristic X-ray lines related to the excitation of deeper core levels, i', involved in the Auger signals in order to improve $Q_A(i') \cdot \omega_{i'j'}$ relative to $Q_A(i) \cdot a_{ijk}$.

A good example of the surface sensitivity of EDS-EPMA, taken from [11], is shown in Fig. 3, where a fraction of a silver monolayer (0.3) on a silicon substrate gives rise to a signal which can be clearly detected from the statistical fluctuations of the background. In this example the detection limit is certainly in the 1/10–1/100 of a monolayer and the signal intensity has been improved by choosing a take-off angle around the critical angle for total reflection.

In summary, AES is of general use for surface identification of almost all the elements whereas EDS-EPMA is restricted to the surface elements being detectable ($Z > 10$), not present in the bulk and in rather simple samples (to avoid confusion between various elements because of the poor energy resolution of the ED system).

All these EDS experiments can be developed in an EDS standard vacuum instrument (instead of a UHV instrument in AES), because of the insensitivity to carbon contamination. From a practical point of view, an important application is the possibility to perform EDS-EPMA measurements in parallel with RHEED during molecular beam epitaxy (AES is also possible but is technically more difficult).

The general use of Auger electron allows this technique to be combined to ion erosion for z-profiling (see [12] for a recent example) and this important ability has to be compared to depth profiling by varying E_0 in EPMA with specific advantages for each of the two approaches.

Quantification Problems in AES

In AES, the various quantification procedures (see [3] chap 5, p. 205) are all derived from eq. (1) and the precision in the deduction of the concentrations of the elemental components of a surface C_A, C_B, is often very poor ($\pm 20\%$). To improve on this, an international cooperation program has been established (VAMAS Program Technical Working area 2, Surface Analysis [13]).

The goal of this program is to improve the precision of surface analytical techniques down to that of EPMA ($\sim 1\%$).

The main difficulty lies in an accurate knowledge of the key parameters λ and R, in the right hand side of eq. (1) (The use of Gryzinski's expression for the cross section has a large consensus among the AES community). In EPMA, if the knowledge of the absorption coefficient μ is required, this coefficient only appears as a correction term (absorption correction) whereas the Auger intensity is directly proportional to the attenuation length λ in homogeneous alloys (as it can be established by integrating eq. (1) from 0 to ∞). A consequence of the short length of λ is that an absolute error on λ of 0.2 nm would lead to a relative error, $\Delta\lambda/\lambda$ of 20% for λ in the nm range.

Like μ^{-1}, λ depends on the (kinetic) energy of the detected particles and on the (unknown) composition of the medium in which they propagate before reaching the detector [14]. Moreover the dominant interaction of X-ray photons with matter is absorption while an Auger electron may suffer elastic as well as inelastic interactions leading to differences between the calculated inelastic mean free path and the useful value of the attenuation length [15].

The Auger backscattering coefficient, R, depends also on matrix effects like the subsurface composition which cannot be deduced from a simple Auger experiment. For example the $(1 + R)$-factor of the same 10-nm Al film, changes from ~ 1.4 to ~ 2.2 ($E_0 = 15$ keV; normal incidence) when the silicon substrate is replaced by a gold substrate (the silicon or gold Auger lines being undetectable).

Various theoretical or semi-empirical expressions have been proposed to estimate the backscattering correction factor $R_{A/S}$ in AES. At normal incidence this factor is defined by [16]:

$$R_{A/S} = \frac{\int_{E_B}^{E_0} Q_A(E)\{\int_Q \sec\theta(\partial^2\eta/dEd\Omega)\, d\Omega\}\, dE}{Q_A(E_0)}, \tag{6}$$

where E_B is the binding energy of the ionization shell involved in the Auger transition process and $\partial^2\eta/\partial E\partial\Omega$ represents the angular energy distribution of backscattered electrons. Initiated by Bishop and Riviere [17], Monte Carlo simulations are frequently used to find a functional representation of R.

For example the expression deduced by Ichimura and Shimizu is [18]:

$$R_{A/S} = (2.34 - 2.1\, Z_S^{0.14})U_0^{-0.35} + (2.58\, Z_S^{0.14} - 2.98), \tag{7}$$

where $U_0 = E^0/E_B(\mathrm{A})$.

Due to the fact that $1 + R_{A/S} = \Phi(0)$ and despite the difference in the range of the primary energies it is surprising to observe that rather few authors [19, 8] involved in the determination of $R_{A/S}$ in AES have compared their results with one of the various functional expressions proposed for $\Phi(0)$ in EPMA and deduced from similar Monte Carlo calculations such as proposed in [20]. Conversely, hardly any

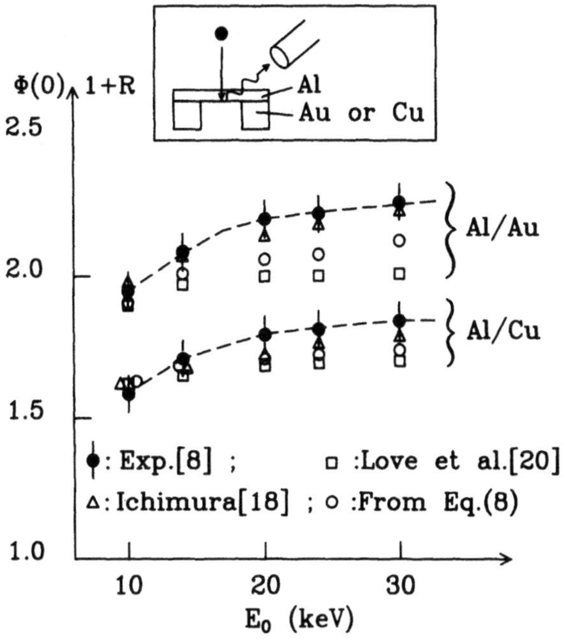

Fig. 4. Comparison between the experimental determination (•) of (1 + R) (or $\Phi(0)$) and various theoretical estimates— □: Love et al [20]; △: Ichimura and Shimizu [18]; ○: From eq. (9) using Hunger's expression for η [22]. The possible influence of characteristic (V) and continuous (W) radiations generated in the substrate has not been taken into account. The experimental arrangement is shown as insert [8]

EPMA scientist, has used the experimental or theoretical Auger values of $R_{A/S}$ for finding $\Phi(0)$.

From the experimental point of view, the identity (3) allows one to obtain the Auger backscattering factor from EPMA experiments by evaluating the ratio between the X-ray signal, emitted by a thin film (of element A) on a substrate S, to that of the same unsupported film [8].

Such a result is illustrated in Fig. 4 for Al/Au and Al/Cu systems in which the experimental results are compared to the calculated values deduced from Love et al., for $\Phi(0)$ [20], and Ichimura and Shimizu, for $1 + R_{Al/Au}$[18]. We have also added our own calculations based on the following simplifications:

i) The angular distribution of the backscattered electron follows a cosine-law and R is:

$$R = \frac{2}{Q_A(E_0)} \int_{E_B}^{E_0} Q(E) \cdot \frac{\partial \eta}{\partial E} \, dE \tag{8}$$

ii) The spectral distribution of the backscattered electrons is replaced by a δ-function located at $\bar{E} = (1 + \eta)E_0/2$.

This leads to $R = 2\eta_{eff} \cdot Q(\bar{E})/Q(E_0)$

iii) The ionization cross-section takes the Bethe form: $Q_A(E) \propto U^{-1} \log U$ and $\eta_{eff} = \eta(1 - U_0^{-1})$ to take into account the efficient backscattered electrons with energies greater than E_B only.

The result is:

$$R_{A/S} = \frac{4\eta}{1+\eta}(1 - U_0^{-1}) \left[1 + \log\left(\frac{1+\eta}{2}\right) \cdot \log^{-1} U_0 \right] \tag{9}$$

A comparison between the experimental results and the calculations in Fig. 4 leads to the following remarks:

i) It is surprising to see that the calculations deduced from simulations for which E_0 was 3 keV < E_0 < 10 keV [18], seem to work better than those deduced from simulations in the 20 keV range [20], while the reverse was claimed for Auger experiments performed below 10 keV [19].

ii) Despite its crude simplifications, eq. 9 seems to work better than the more sophisticated expression proposed by Love et al. [20]. Eq. 8 (and others) underlines the fact that the precision on R (or on $\Phi(0)$) is directly related to the precision in the knowledge of η.

For η, instead of the Reuter's expression [21], we prefer that of Hunger and Kuchler which includes the influence of E_0 on η [22].

iii) At rather high primary beam energies, E_0, some Auger electrons may also be emitted from the surface as a result of atomic excitations by X-ray photons generated in the bulk. The corresponding reinforcements of the Auger signal by characteristic (U) and continuous (V) radiations have been evaluated in [23], but ignoring the identity of this correction with the fluorescence correction for EPMA of thin coatings such as that proposed by Cox et al. [24].

. iv) Finally, when these fluorescence corrections will be added, it is not so sure that Ichimura's expression will remain better than Love's one. For evaluating $R_{A/S}$, an advantage of eq. (9) is to separate the influence of the element of interest, A, on the surface (through U_0) and the influence of the substrate (through η). Considering η, the main advantage of eq. 9 is that the Auger backscattering correction can be performed even when the substrate composition is unknown if η is measured, for instance, in the 2D-chemical mapping of a surface on a strongly heterogeneous substrate as well as during the in-depth profiling of heterogeneous specimens.

Like for EPMA, the other problems related to quantification in AES are:

i) The Accurate Determination of the Characteristic Signal Intensities

This determination requires the knowledge of the response function of the Auger instrument and overall the spectrometer transmission function and detector sensitivity [25]. It also requires the background removal and the corresponding error is in AES larger than in EPMA because of its large amplitude. Various procedures have been proposed (see [26] for a review), but it seems that only the use of coincidence techniques (between the incident electron, having ionized the atom and the corresponding Auger electron) enables the background suppression and the deduction of its exact shape in normal AES experiments [27].

ii) The Reduction of Topographic Effects

AES, like EPMA signal intensities are influenced by topographic effects, but the magnitude of these effects is much larger in AES because of its very high surface sensitivity. A step of a monoatomic layer is sufficient to induce these effects in AES and the surface cannot be polished (like in EPMA) to prevent this. To reduce the effects of the surface topography in SAM, Prutton et al. suggest to evaluate the S/(S + BG) or the S/(S + 2BG)-ratios [28]. An alternative solution is to record simultaneously two Auger spectra at the same point. Because of the change of the two take-off angles, the ratio of Auger intensities of the element A, obtained in the two channels will change when scanning the incident electron beam on a non-flat surface (see Fig. 5). The same experimental arrangement allows one to distinguish

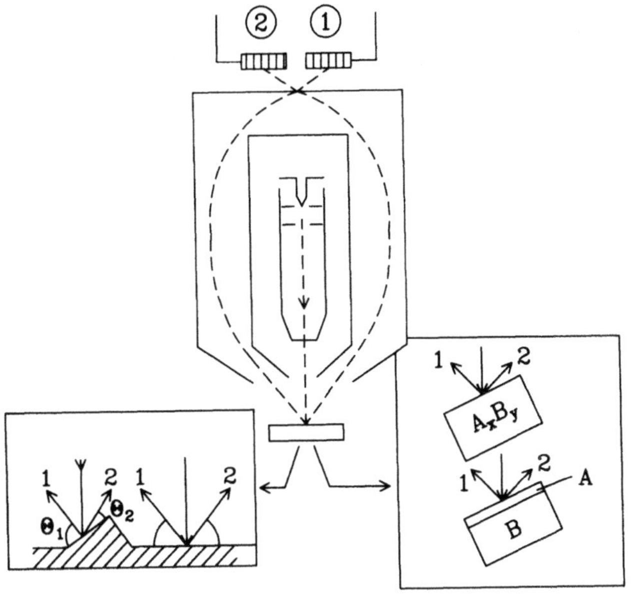

Fig. 5. Schematic AES experimental arrangement [29], to distinguish between homogeneous and stratified specimens (right insert) and to detect topographic effects (left insert). The operation consists in obtaining two Auger spectra at the same point (see text for further explanation)

between homogeneous alloys and stratified specimens [29]. If might be useful to apply a similar arrangement (with two ED detectors) in EPMA.

iii) The Crystalline Effects

Eqs. (1) and (2) are only valid when crystalline effects can be neglected (i.e for amorphous materials).

When single crystals are investigated by AES or EPMA, the channeling effects on the incident electron beam and backscattered electrons are obviously the same when EPMA and AES are carried out on the same sample under the same conditions [30]. In addition, like the diffraction of photoelectrons in XPS, the diffraction of Auger electrons in AES cannot be neglected entirely [31] as the diffraction of the emitted X-rays in EPMA.

iv) The Analysis of Insulators and Sensitive Materials

The difficulties of investigating insulators using incident charged particles is well known. Here again, the electric field built-up in the specimen during electron irradiation would be the same in AES and EPMA (under the same conditions). It would lead to the same spurious effects (such as ion migration), even if it seems different because of the difference in the surface sensitivity of the two spectroscopies [32]. In practice, the surface of the specimen is coated by a thin conducting layer in EPMA (while it is not in AES). This operation increases the comfort of the experiment because the electric field is nihil outside the specimen and the incident electron beam is not deflected. Unfortunately, the electric field inside the specimen is not suppressed (it is maximum at the metal coating/insulator interface) and most of the difficulties for getting quantitative results from these specimens remain (see ref [33] for a recent analysis of these effects).

We also have to mention the well known difficulty (of EPMA and AES) to analyze materials like polymers sensitive to electron beam irradiation. This difficulty

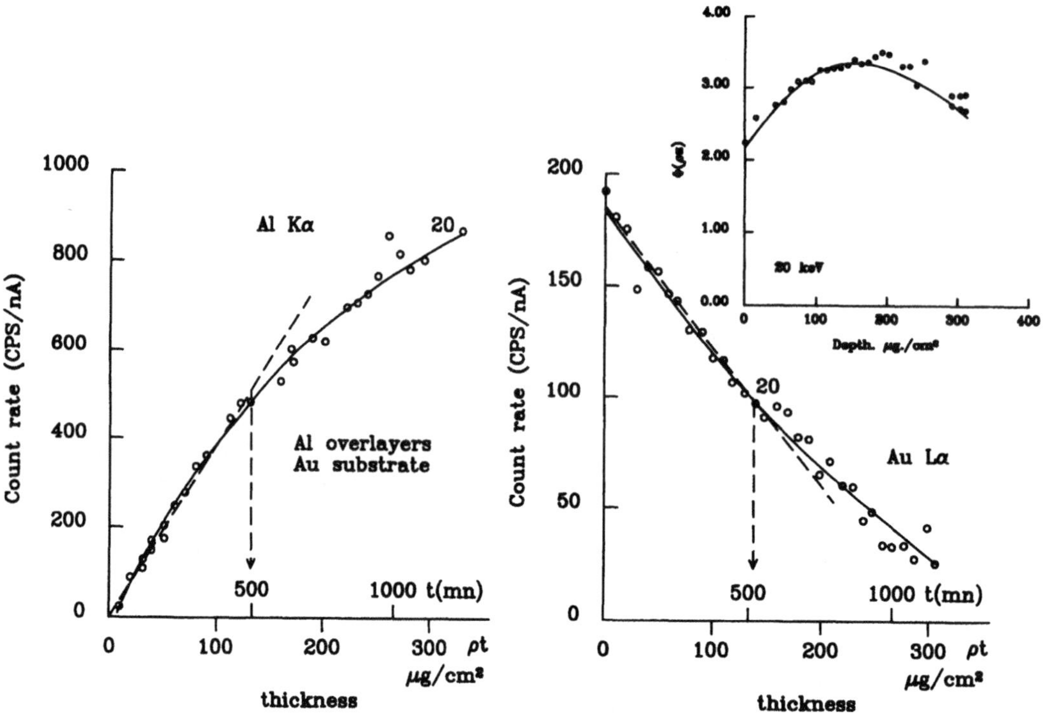

Fig. 6. Al coatings on a gold substrate. Left: measured Al–Kα (X-ray) intensities from the overlayers; right: measured Au–Lα X-ray intensities from the substrate (both as a function of the coating thickness) Insert: $\Phi(\rho z)$ function. The experimental points and solid lines are taken from [34]. The straight dotted lines have been added to underline the thickness range where the $I_X = f(t)$-relationship is linear. The upper limit corresponds to $t_{max} \simeq 0.5\ \mu m$ with a range Re of $\simeq 4\ \mu m$ (at $E_0 \simeq 20$ keV)

is mainly due to the large energy deposited into very small volumes of matter and the common but very poor solution consists in minimizing the dose received by the specimen by defocussing the incident electron beam.

Thin Coating Analysis by EPMA

By integrating the intensity, $dI_X(\mathrm{A})$, over the thickness of the coating, a practical consequence of eq. 2 is to expect a proportionality between the characteristic emitted intensity and thickness t. This linear relationship is only valid when the underlying assumptions of eq. 2 are satisfied i.e. for thin coatings. From many published experimental data, it has been deduced that the linear relation is very often satisfied if $0 < t \leq R_e^c/10$ (with R_e^c = calculated range of incident electrons in a bulk material having the same composition as the coating) [8]. Taken from the experimental data of Brown and Chan [34], Fig. 6 illustrates this result, which leads to a $\Phi(\rho z)$ function which is quite constant in the thickness range of interest and which seems to be conflicting with the common bell shape $\Phi(\rho z)$-function. In fact Brown and Chang have obtained a standard $\Phi(\rho z)$-function by differentiating their experimental data (see insert Fig. 6). As shown in Fig. 6, the explanation is that the position of the experimental points is submitted to statistical deviations (a few per cent) and it is possible to draw between them both a straight line as well as a more complex line (resulting from the integration of the $\Phi(\rho z)$-function). More precisely

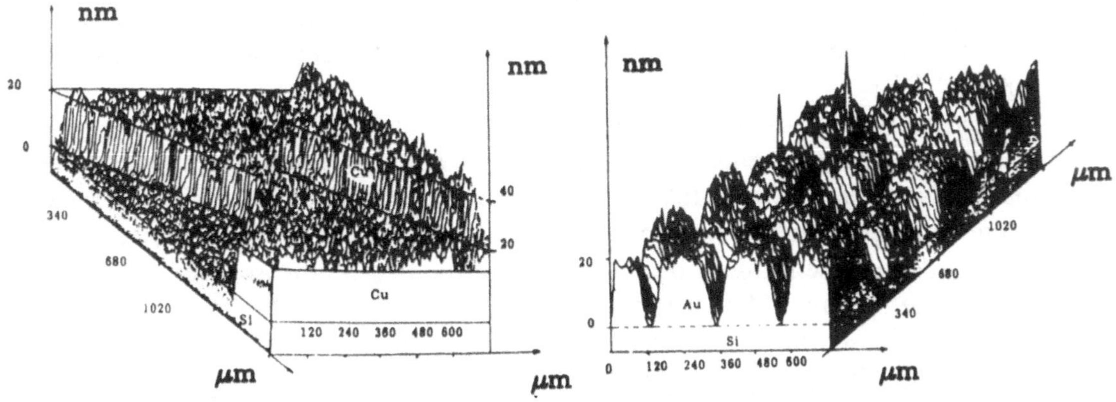

Fig. 7. Three dimensional profilometry of an Au/Cu coating on a silicon substrate using Cu Lα and Au Mα radiations [8]

if $\varphi(z)$ is of the form $\phi(0) + az - bz^2$ when t is $< R_e/10$, I_X is $\phi(0)t + at^2/2 - bt^3/3$ and the two last terms cancel each other more or less ($a, b > 0$).

The linearity between the signal and the thickness of the coatings leads to a very simple procedure for characterizing, in a non-destructive manner, unknown coatings with a precision of $\sim 5\%$ and using a reduced number of reference specimens. Furthermore, by surface scanning with the incident electron beam, it is possible to obtain x profiles or $x - y$ images leading to a visualization of the coatings in two or three dimensions. An example is shown in Fig. 7 (see ref [8] for further details).

Lateral Resolution, Minimum Detectable Mass and Trends in SAM

For characterizing the localization properties of microanalytical techniques such as AES and EPMA, two different parameters may be considered: the lateral resolution and the minimum detectable mass (or the minimum number of detectable atoms, or a minimum dimension of a detail to be detected) [25, 35].

Lateral Resolution in SAM

In the past, much attention has been given to the lateral resolution problem of scanning Auger microscopy. This was due to the fact that there is not a single criterion for the evaluation of the lateral resolution (point-to-point resolution, edge profile, spot size, transfer function, cross correlations) and the choice of the criterion was important to see whether the spatial resolution limits are determined almost entirely by the incident electron beam diameter or by the lateral extent of the backscattered contribution. Using an analytical expression for describing the radial distribution function of the emitted Auger electrons, it has been established that the point to point resolution of Auger microscopy is governed by the incident probe size if this probe size is at least 4 to 5 times smaller than the backscattering halo dimension [36]. In commercially available instruments operated at around 20–30 keV, this probe diameter is presently in the 20–50 nm range using mainly a field emission gun (the incident beam intensity being of few nA). Using a modified scanning transmission electron microscope, the best resolution (8 nm) has also been obtained with a commercial instrument but operated at the unusual primary energy of 100 keV [37].

The same point-to-point resolution criterion, applied to EPMA of bulk specimens, leads to an ambiguous answer: the lateral resolution will be governed by the diffusion of incident electrons into the specimen, if the two points are located in the plane of complete diffusion (leading to a resolution in the micrometer range). It is only governed by the incident beam diameter if the two points are located on the surface of the specimen. For a more practical point of view, this means that the point-to-point resolution of EPMA, applied to two small details on a substrate of different composition, is the same as in X-ray microanalysis of thin foils in analytical microscopy. It is governed by the incident spot size together with the beam broadening and the X-rays generated by the electrons backscattered by the substrate have no effect [8].

Minimum Detectable Mass in SAM

All the criteria used for the evaluation of the lateral resolution deal with relative intensities. They characterize the response function of the whole instrument but, dealing with points, they are independent on the nature of these points [35].

The minimum detectable mass (MDM) combines the localization character of a microanalytical technique and its sensitivity (C_m). It can be deduced from the minimum number of detectable atoms, y_m, by:

$$y_m = N \cdot C_m, \tag{10}$$

where N is the total number of the analyzed atoms in the analyzed volume V (with $N = nV$ and n: atomic density $\simeq 5 \cdot 10^{22}$ atoms/cm^3), C_m is given by eq. (4) and depends on the signal-to-noise ratio. Due to its very short information depth, the analyzed volume in Auger microscopy is restricted to a disc having approximately the diameter of the incident beam and a height in the nanometer range. Using a probe of 8 nm in diameter, a few thousands of Pd atoms (MDM $\simeq 10^{-18} - 10^{-19}$ g) on a Al_2O_3 substrate have been detected [38].

Because of the larger analyzed volume, this limit seems not to be reached in EPMA on bulk specimens. Nevertheless, combining the remarks on the sensitivity of EDS-EPMA with that on the lateral dimension of the analyzed volume (restricted to the incident probe diameter: see above), this is not impossible for a specific specimen for which the bremsstrahlung level is very low. At a first glance, there is no fundamental reason (but many practical ones) to reach in EPMA of bulk specimens, the minimum detectable mass of X-ray microanalysis of thin foils in analytical electron microscopy (where a few thousands of atoms are also detected [39]): for example Ag or Pd atoms on carbon (or Be) substrates.

Trends in Scanning Auger Microscopy

Added to the continuing effort of the scientific community to improve the quantification procedure in AES, the trends for future instrumental development, in SAM, fulfill two major requirements:

i) improve the performance with a special attention to spatial resolution.
ii) increase the amount of information extracted from a given specimen.

Requirement i) consists of obtaining the most intense probe in the smallest diameter and in optimizing the collection of the signals. The increase of the primary beam energy, E_0, increases the brightness of the electron gun β_S, because:

$$\beta_S = \beta_E\left(\frac{V_A^X}{V_E}\right), \tag{11}$$

where V_A^X is the relativistic accelerating voltage: $V_A^X = V_0(1 + eV_0/2m_0c^2)$ and β_E is the parameter characterizing the brightness of the electron source which gives a strong advantage to the use of field emission guns.

The next step is putting the specimen at the optimal position, i.e., where the spot size is minimum (between the pole pieces of a magnetic lens). The difficulty is to extract the Auger signal but the use of a "parallelizer" not only allows to overcome this difficulty but even to use it; the collection efficiency approaches 100% [40].

It is next possible to analyze the emitted electrons using a parallel detector, Kruit at Delft University and Venables at Arizona State University have built instruments following this consistent analysis. Recently, excellent Auger spectra have been obtained with a probe size of 5 nm, and the last goal would be to lower the spatial resolution under the nanometer level [41, 42]. An elementary evaluation shows that the ultimate detection limit of a single atom by Auger microscopy will be obtained in the next few years.

When the primary beam energy increases, the signal decreases (through the decrease of the cross section $\propto U^{-1} \log U$) but in Auger microscopy the background is also decreased (BG $\propto E_0^{-1}$) and the result is an increase of the signal-to-background ratio and a slight decrease of the S/N ratio (like $E_0^{-1/2} \log U$). At $E_0 = 100$ keV, in the experiments above described, this disadvantage is overcompensated by the increase of the brightness of the source and of the collection efficiency of the "parallelizer" and the question is to know whether such advantages can be extrapolated to higher primary energies. In this respect the work of Zaluzec (Argonne, USA) is very interesting, because there are recent attempts to obtain Auger spectra at $E_0 = 300$ keV (private communication and ref. [43]).

Another question concerns the possibility to transfer this obvious progress of SAM instrumentation to EPMA by using higher primary beam energies. If all the positive points concerning the incident probe are the same, it remains to increase the collection efficiency of the solid state detectors. But the main negative point is in this case, the increase of the background with E_0 leading to a large deterioration of the S/N ratio. Nevertheless the analysis with high lateral resolution of small details on specific (light) substrates seems interesting with also the extension of the linear range in EPMA of thin films on substrates of different composition. Instead of using sophisticated but complicated procedures for the characterization and imaging of coatings in the micron range for the thickness, it seems easier to increase E_0 in order to apply the simple procedure described in section thin coating analysis in the EPMA.

Concerning point ii) related to the increase of the information extracted from a given specimen, its usefulness in AES is obvious if the increase of information leads to the advantage of knowing the subsurface composition, of reducing the effects on Auger yields of variations in backscattering factor due to the changes in subsurface composition and of reducing the topographic effects. All these requirements are satisfied in the third generation of Auger microscopes constructed by Prutton et al. in York (GB) [44] with a parallel detector to acquire simultaneously the Auger spectra, a set of quadrants acting as backscattered electron detectors, various channeltrons for the secondary electron signals and a bakable Si(Li) X-ray detector. Thus last attachment suggests that the correlation between Auger images and X-ray images will be obtained in the near future.

The various images thus obtained require the use of modern arrangements of electronics for data acquisition and the use of sophisticated image processing facilities.

The generalization of the use of computer facilities is a common point of instrumental development of almost all the experimental techniques and, concerning microanalytical techniques, a specific spectrum or image processing system developed for one technique can be easily transposed to the others. It is the case in factor analysis for XPS and AES spectra [26]. It is also, for example, the case in cross-correlation processes called "scatter diagrams", initiated by Browning for SAM [45] and widely popularized in the Auger community by Prutton et al. [44]. It was easily adapted to EPMA of bulk specimen [46] and thin coatings [8]; in the near future the unconventional methods of EELS elemental mapping, developed by Trebbia and Bonnet [47] will be certainly transposed to SAM and EPMA.

In this field, the developments in progress in our laboratory concern the acquisition of a two dimensional histogram correlating the two Auger images of the same element using two different angular channels (see Fig. 5) and in EPMA of thin films, the correlation of the characteristic image of a coating and the specimen current image. In EPMA, our final goal is to correct the thickness profilometry of coatings from the backscattering artefacts associated to heterogeneities in the composition of the substrate (correction through eq. (9)). This approach to obtain $\Phi(0)$ and its change when the incident beam is scanned may also be useful, for EPMA of bulk materials.

Conclusion

AES and EPMA are presently two well-established techniques widely applied for elementary characterization of solids in materials science. In the past, they have been developed while quite ignoring each other. Keeping their own specificity, we believe in the advantages of each of the two when looking at the progress in methodology, quantification procedures and instrumental aspects of the other, before a cross-correlation of the two techniques for characterizing accurately the material in the first atomic layers of a surface down to the bulk. Illustrated by examples, chosen arbitrarily, this was the goal of this paper.

Acknowledgements. I wish to thank my colleagues of the laboratory, O. Jbara, S. Rondot and X. Thomas for allowing me to use some of their results prior to publication.

References

[1] R. Castaing, *Thèse*, Paris, 1951 and *Advances Electron. Electr. Phys.* **1960**, *13*, 317.

[2] L. A. Harris, *J. Appl. Phys.* **1968**, *39*, 1419.

[3] M. P. Seah, D. Briggs, in: *Practical Surface Analysis* Vol 1: Auger and XPS (D. Briggs, M. P. Seah, eds.), Wiley, Chichester, USA, 1990, 2nd ed.

[4] G. F. Bastin, and M. J. M. Heijligers, *J. Micros. Spectros. Electron.* **1986**, *11*, 215.

[5] D. M. Kroeger, A. Choudhary, J. Brynestad, R. K. Williams, R. A Padgett, W. A. Coghlar, *J. Appl. Phys.* **1988**, *64*, 33.

[6] D. E. Ramaker, *Scanning Microscopy* **1990**, *4*, [Suppl.] 207.

[7] L. E. Davies, N. C. Mac Donald, P. W. Palmberg, G. E. Riach, R. E. Weber, *Handbook of Auger Electron Spectroscopy*, Physical Electronics Division, Perkin Elmer, U.S.A., 1978.

[8] J. Cazaux, O. Jbara, and X. Thomas, *Surf. Interf. Anal.* **1990**, *15*, 567.

[9] W. A. Van Borm, F. C. Adams, *Anal. Chim. Acta* **1989**, *218*, 185.

[10] M. J. Romand, F. Gaillard, M. Charbonnier, D. S. Urch, *Advances in X-Ray Anal.* **1991**, *34*, (in press).

[11] S. Hasegawa, H. Daimon, S. Ino, *Surf. Sci* **1987**, *186*, 138.

[12] P. Etienne, J. Chazelas, G. Creuzet, A. Friederich, J. Massies, F. Nguyen-Van-Dau, A. Fert, *J. Crystal Growth* **1989**, *95*, 410.

[13] C. J. Powell, M. P. Seah. *Surf. Interf. Anal.* **1986**, *9*, 79.

[14] C. J. Powell, *J. Electron Spectrosc. Rel. Phenom.* **1988**, *47*, 197.

[15] A. Jablonski, *Surf. Interf. Anal.* **1990**, *15*, 559.

[16] D. Ze-Jun, R. Shimizu, S. Ichimura, *Surf. Int. Anal.* **1987**, *10*, 253.

[17] H. E. Bishop, J. C. Riviere, *J. Appl. Phys.* **1969**, *40*, 1740.

[18] S. Ichimura, S. Shimizu, *Surf. Sci.* **1981**, *112*, 368.

[19] T. Sato, Y. Nagasawa, T. Sekine, Y. Sakai, A. D. Buonaquisti, *Surf. Interf. Anal.* **1989**, *14*, 787.

[20] G. Love, M. G. Cox, D. Scott *J. Phys. D. Appl. Phys.* **1978**, *11*, 23.

[21] W. Reuter, *X-Ray Optics and Microanalysis* (G. Shinoda, K. Kohra, T. Ichinokawa, eds.), Tokyo University Press, Tokyo 1978, p. 121.

[22] H. J. Hunger, L. Kuchler, *Phys. Stat. Sol. A* **1979**, *56*, K 45.

[23] J. Cazaux, S. Moutou *Surf. Interf. Anal.* **1984**, *6*, 62.

[24] M. G. C. Cox, G. Love, V. D. Scott, *J. Phys. D. Appl. Phys.* **1979**, *12*, 1441.

[25] M. P. Seah, G. C. Smith, *Surf. Interf. Anal.* **1990**, *15*, 751.

[26] J. T. Grant, *Surf. Interf. Anal.* **1989**, *14*, 271.

[27] J. Cazaux, O. Jbara, K. H. Kim, *Surf. Sci.* **1991**, *247*, 360.

[28] M. Prutton, L. A. Larson, H. J. Popa, *J. Appl. Phys.* **1983**, *54*, 374.

[29] J. Cazaux, T. Bardoux, D. Mouze, J. M. Patat, G. Salace, X. Thomas, J. Toth, *Proc. ECASIA 91, Budapest* (in press).

[30] B. Akamatsu, P. Henoc, F. Maurice, C. LE Gressus, K. Raoudi, T. Sekine, T. Sakai, *Surf. Interf. Anal.* **1990**, *15*, 7.

[31] H. E. Bishop, *Surf. Interf. Anal.* **1990**, *15*, 27.

[32] J. Cazaux, *J. Micros. Spectr. Electron.* **1986**, *11*, 293.

[33] J. Cazaux, P. Lehuede, *J. Electr. Spectros. Rel. Phenom.* (in press).

[34] J. D. Brown, A. Chan, *Proc. 12th Conf. X-Ray Optics and Microanalysis, Vol. 1* Krakow (S. Jasienska, L. M. Maksymovicz, eds.), 1989, p. 60.

[35] J. Cazaux, C. Colliex, *J. Electr. Spectros. Rel. Phen.* **1990**, *52*, 837.

[36] J. Cazaux, *Surf. Interf. Anal.* **1989**, *14*, 354.

[37] J. Cazaux, J. Chazelas, M. N. Charasse, J. P. Hirtz, *Ultramicroscopy* **1988**, *25*, 31.

[38] J. Chazelas, J. Cazaux, G. Gilmann, J. Lynch, R. Szymanski, *Surf. Interf. Anal.* **1988**, *12*, 45.

[39] J. J. Goldstein, C. E. Lyman, J. Zhang, *Proc. 12th Int. Congr. Electron Microscopy, Vol. 2*, Seattle, San Francisco USA 1990, p. 450.

[40] P. Kruit, J. A. Venables. *Ultramicroscopy* **1988**, *25*, 183.

[41] G. G. Hembree, F. Ch. Luo, J. A. Venables, *Proc. 12th Int. Congr. Electron Microscopy, Vol. 2*, Seattle, San Francisco Press, San Francisco, 1990, p. 382 and p. 378.
[42] A. J. Bleeker, P. Kruit, *Proc. 12th Int. Congr. Electron Microscopy, Seattle, Vol. 2*, San Francisco Press, San Francisco 1990, p. 380.
[43] H: S. von Harach, J. A. Colling, R. Keyse, J. Morphew, *Proc. 12th Int. Congr. Electron Microscopy, Vol. 2*, Seattle, San Francisco Press, San Francisco 1990, p. 136.
[44] M. Prutton, C. G. H. Walker, J. C. Greenwood, P. J. Kenny, J. C. Dee, I. R. Barkshire, R. H. Roberts, M. M. El Gomati, *Surf. Interf. Anal.* **1991**, *17*, 71.
[45] R. Browning, *J. Vac. Sci. Technol.* **1985**, *A3*, 1959.
[46] D. S. Bright, D. E. Newbury, R. B. Marenko, *Microbeam Analysis* **1988**, 18.
[47] P. Trebbia, N. Bonnet, *Ultramicroscopy* **1990**, *34*, 165.

Mikrochim. Acta (1992) [Suppl.] 12: 53–74

Quantitative X-Ray Microanalysis of Ultra-Thin Resin-Embedded Biological Samples

Hugh Y. Elder*, Stuart M. Wilson, W. A. Patrick Nicholson[1], John D. Pediani, Scott A. McWilliams, D. McEwan Jenkinson, and Christopher J. Kenyon[2]

Institute of Physiology, [1] Department of Physics and Astronomy, and [2] MRC Blood Pressure Unit, University of Glasgow, Glasgow G12 8QQ Scotland, U.K.

Abstract. For X-ray microanalysis, unlike materials samples, most biological specimens present a complex of characteristics which prescribe the preparative procedures, determine the microanalytical configurations and limit the achievable statistical accuracy. The five principal limitations are the high water content, the ionised state and high mobility of the majority of the inorganic elements of interest, the low, sometimes trace concentrations of these elements, the ubiquity and diversity of the organic molecular and structural components of the cellular and extracellular matrix, often rendering quantification of their component elements redundant, and the sensitivity of biological material to radiation damage and beam-induced mass loss, particularly in hydrated samples. For mobile elements cryofixation is the only acceptable initial processing step. Spatial resolution requirements and beam sensitivity restrictions usually indicate microanalysis of thin section (≤ 1 μm) by energy dispersive detectors. Amongst briefly reviewed options for production of thin sections, the merits of low temperature freeze-drying, vacuum resin embedding and dry sectioning are presented, particularly for the location of targets, which are difficult to locate deep within tissues. Evidence is presented that plastic embedding need not cause elemental translocation and that intracellular gradients and physiological changes in diffusible elements are preserved by this preparative route. Prospects for improved accuracy in quantification are considered.

Key words: X-ray microanalysis, quantitative biological microprobe, low temperature freeze drying, resin section analysis.

In materials science the majority of microanalytical applications involve bulk samples. Thin specimens, which can often be produced by erosion-thinning of solid samples and sometimes by evaporative deposition, give good X-ray yields, while offering greatly improved spatial resolution in comparison with bulk samples.

* To whom correspondence should be addressed

Biological tissues are comprised of very large numbers of small living units, the cells, each usually only about 10–15 μm in diameter, within a variable amount of extracellular matrix of very different composition. Further, both cells and matrix are highly inhomogeneous and the whole is of low average atomic number. From these considerations biological microanalysts are constrained to adopt thin section microanalysis to achieve the spatial resolution of 0.1 to 0.2 μm necessary for most applications. Further, the sensitivity of biological material to beam-induced mass loss argues in favour of the use of energy dispersive (ED) detectors. Because of their ability to collect X-ray photons over the whole spectral range simultaneously and the larger solid angle which they subtend to the specimen, compared to wavelength dispersive (WD) detectors, the total electron dose which the specimens receive during analysis using ED detectors can be about a tenth of that necessary to achieve comparable analytical data with WD detectors. Sensitivity to radiation damage varies greatly amongst biological specimens and with the preparative method followed. Where damage or mass loss become significant, analysis at low temperature becomes mandatory. The topic has recently been reviewed by Echlin [13] and Lamvik [53].

Microanalysis of biological thin film samples differs from materials thin film analysis in many important respects. In comparison to most materials science applications, five major problems are associated with biological samples. These are firstly, the high water content of most biological tissues, secondly, the high mobility of many of the small, soluble molecules, and particularly the inorganic ions, thirdly, the ubiquity and diversity of the organic molecules which usually renders measurement of C, O and N relatively meaningless, even if it can be performed but increases background, fourthly, the very low mass fractions of the inorganic elements which are often of greatest interest and fifthly, the sensitivity of most biological material to radiation damage, already referred to above. It is the purpose of this paper to consider some of these problems, created by the characteristics of biological material and to illustrate ways in which the problems may be circumvented.

Biological Microanalytical Parameters

Pre-eminently, amongst the problems of biological specimen microanalysis, is the high water content, usually about 75–80% of their mass in the native state. The water in liquid phase is incompatible with the vacuum of the microscope and must be removed, replaced or stabilised in the thin specimen by one of the ways outlined below.

Many of the elements of greatest interest in biological specimens are partly ionised and maintained in an inhomogeneous distribution by affinity with specific tissue components or maintained, often against their electrochemical gradients, by the energy dependent metabolism of living cells. Therefore their retention in the exact locations and concentrations, which they had in the intact living tissue, poses special problems; cryopreparation techniques have proved to be the only feasible answer. The initial stage is cryofixation, which lies outwith the present consideration, and has been extensively reviewed in recent years [70, 75, 23, 61, 84, 2, 18].

The major part of living material, with the important exception of mineralised tissues, is comprised of organic molecules. The principal elements of this organic matrix, C, H, N and O either cannot be detected by ED detectors, or can be detected only by the ultrathin-windowed or windowless types. Even then the data are usually of little value because of the ubiquity and diversity of their organic molecules. Fortuitously, therefore, the unwanted elemental peaks from the organic matrix are completely suppressed by using detectors with Be windows, which effectively allow detection of elements only of atomic number $\geqslant 11$ (Na). However, the concentrations of the elements usually of greatest interest (atomic numbers 11–20) are often so low in the hydrated organic matrix that the Bremsstrahlung created by the matrix reduces the signal-to-noise ratio of the elements of interest to close to their minimum detectable mass fractions and below. Biological microanalysts are thus usually in the position of performing what is effectively trace element analysis and have to be resigned to larger error limits than in most materials analysis.

Further consequences flow from the fact that biological material is live and therefore continually changing, interacting with, and dependent upon, its material environment. Many of the questions which physiologists, pathologists and cell biologists ask, therefore relate to the way in which the elemental composition of cells and tissues has *changed* with time or in response to some environmental factor. Fortunately, *relative quantities* are often more important than absolute ones in biological microanalysis and this tends to reduce some of the very large technical difficulties in working with biological material.

Biological Elemental Stabilization

The problems which the dynamic nature of the interaction between cells and their environment raises for specimen preparation are often underestimated by biologists. Undoubtedly one of the major factors in any cryopreparation protocol is ionic redistribution which can occur immediately when the tissues are separated from the body and its natural support systems. Such changes are particularly likely to occur following interruption of the vascular supply to tissues, such as brain, liver or heart, which are particularly sensitive to oxygen and carbon dioxide tension, or occur with changes in the tissue's normal environmental conditions of temperature, hydrostatic pressure or interstitial fluid composition. Anaesthesia and even stress in an individual or animal are known to alter the physiology of many tissues. However, tissues vary greatly in their susceptibility to fluctuation in their environmental conditions and some, such as skin and the sweat glands described here, are remarkably resilient to prolonged periods of abnormal temperature and lowered oxygen tension. The changes which can occur upon tissue excision and prior to cryofixation are now well documented [46, 104, 91, 103].

Biologists therefore prefer to work with simple organisms and cell cultures which are amenable to rapid cryofixation but this, depending upon the nature of the question to be addressed, is not always possible [103]. The dangers of error arising from preparatory changes such as in environmental pH and O_2 or CO_2 tensions and of osmotic pressure due to evaporative dehydration, are considerable when tiny biological specimens are isolated. If the project objectives dictate that tissues sampled from larger masses or whole organisms are required, one approach has been to develop comprehensive artificial environmental support systems in which the cryofixation can be performed [3, 4]. However, when samples have to be taken directly from surface-accessible tissue locations in the body, methods for *in situ* cryosampling are usually preferable to excision followed by cryofixation [103]. Numerous devices for *in situ* cryosampling, such as cryo-pliers [28, 46], the "popsicle" freezer [86], "cryosnappers" [29] and a cryo-needle [93] have been employed.

Biological Cryopreparative Options

A cryofixation protocol, as mentioned above, is the only recognised route for the retention of diffusible substances. There are four principal options for further cryopreparation of thin sections suitable for microanalysis from frozen samples. The first two involve cryoultramicrotomy with the production of sections directly, either in (i) the frozen-hydrated state or (ii) freeze-dried. In the second two, water is removed from the bulk tissue, either by (iii) low temperature freeze-drying or (iv) by low temperature freeze-substitution. This is followed by embedding in non-polar resin and dry sectioning. In Table 1, microanalytical objectives are combined with tissue factors in helping to provide a guide to the cryopreparative route which is primarily applicable to a proposed study.

Cryoultramicrotomy has been well described [1, 27, 82, 100, 101, 71] and the advantages and disadvantages of the frozen-hydrated and freeze-dried routes are well appreciated. In summary, it is generally agreed that sectioning of frozen-hydrated tissue should be performed at temperatures below $-120°C$ (at which ice vapour pressure is about 10^{-5} Pa) to avoid sublimation and that facilities for

Table 1. Suggested guide for choice of cryopreparative route

Microanalytical objectives for diffusible elements:
Extracellular fluids, localisation/measurement
Cellular, localisation/measurement
Subcellular organelles, localisation/measurement

Tissue factors:
Target cell/tissue is—Surface (say <50 μm deep)
Deep (say >200 μm deep)
Target cell/tissue is—Easy to locate
Hard (rare, small, no contrast, etc.)

Section preparation route:
Freeze-dried cryosection (FDC)
Block freeze-dried, vacuum resin embedded, dry-cut (FDR)
Frozen-hydrated cryosection (FHC)

		Microanalytical objective		
Tissue	Factors	Extracellular	Cellular	Subcellular
Surface	Easy	FHC	FDC or FDR	FDC
Surface	Hard	FHC?	FDC or FDR	FDC
Deep	Easy	FHC?	FDR or FDC	FDC?
Deep	Hard	Forget it!	FDR or FDC	FDC?

cryotransfer and microanalysis (in a cold stage) at temperatures lower than about −140°C are necessary. Strategies for target location, particularly in low magnification sections, such as partial freeze drying, dark field imaging, partial dark field contrast and STEM imaging [27, 42, 54] are required, since the scattering cross-section of water and organic matrix are very similar, causing extremely poor imaging contrast. Further, frozen-hydrated samples are subject to severe beam-induced mass loss due to radiolysis [90, 13], hydrated sections being about 100 times more sensitive than completely dry cryosections [102]. Some reduction of mass loss is afforded by a protective carbon coat. Since radiolysis is a surface process, the beam-induced mass loss becomes relatively less, for a given electron dose, as section thickness increases. Therefore, although microanalysis is precluded by radiolysis in ultrathin hydrated sections, it is manageable in sections of about 1 μm in thickness [26, 27, 32, 102].

Preparation by the second cryoultramicrotomy route is similar except that the sections are freeze-dried on support grids before examination and analysis. Earlier studies involved drying the sections (at ∼ −80°C) in dry nitrogen atmosphere within the cryoultramicrotome chamber or in the vacuum of a coating unit, after cryotransfer under liquid nitrogen. When performed anhydrously, these procedures produce excellent results [87, 85]. Freeze-dried sections, which are hygroscopic, are normally given a protective carbon-coating, as for frozen-hydrated sections, prior to transfer to the microscope. The advent of better, integrated cryotransfer coldstages has led some microanalysts to advocate direct cryotransfer of the frozen hydrated sections, in a carbon-coated sandwich grid, to the electron microscope for freeze drying in the column at −80°C [28, 80]. Although reliable extracellular microanalysis can not be performed with freeze-dried sections, contrast is much improved and beam-induced mass loss is much reduced. Loss of some 80% of the mass during drying, results in a much improved signal to noise ratio with correspondingly improved statistical accuracy of elemental measurement. Although freeze-drying inevitably involves tissue shrinkage, the mass fraction data derived by the continuum normalisation method discussed below are independent of shrinkage and the majority of microanalytical studies of intracellular elements have been made using this route.

Freeze-substitution and freeze-drying cryopreparative options are similar except that specimen dehydration in the former is achieved by low temperature substitution using an appropriate medium. Freeze-substitution has been well reviewed in recent years [40, 83, 41, 89, 16, 10] and will not be considered separately in the present discussion.

Freeze-Dried, Resin Embedded Tissues

This discourse focusses on the cryopreparative route involving cryofixation followed by low temperature freeze-drying, vacuum embedding with non-polar resin and dry-sectioning, which we have found to be of value in the microanalysis of small biological targets, which are difficult to locate in larger tissue masses. This latter point can be very important in specific studies, such as those on individual mammalian sweat glands detailed below, where the advantages of durability of the resin-embedded tissue and the facility to carry out repeated dry sectioning and light microscopical examination of stained sections until the target cells are located, outweighs potential problems of diffusible element translocation induced by resin infiltration. Greater than 90% targetting success is achieved with the resin embedded sections whereas, in our early attempts by cryoultramicrotomy to perform microanalysis on sweat gland profiles (~200 μm diameter), the location rate in the mass of surrounding connective tissue of more than 100 times the volume, was only 3%.

Comparative Sweat Gland Studies

The use of rapid cryofixation techniques to preserve electrolyte distributions, followed by low temperature freeze-drying of the tissue blocks, vacuum embedding in non-polar epoxy resin and dry cutting, is illustrated below by data from studies of the sweat glands from several mammalian species which we have recently undertaken to ascertain the cellular levels of the principle elements involved in formation of the primary secretion. The species from which these results are taken represent physiologically different gland types. Stimulation of human glands is principally by cholinergic activation and the formed sweat is a dilute solution mainly of NaCl [95]. In the horse adrenergic agonists predominate and the sweat is rich in K. Man and the horse both utilise sweating in thermoregulation and the glands can be thermally activated [94]. Rat footpad glands are cholinergically activated, the sweat is potassium rich and the glands cannot be thermally activated [59].

Materials and Methods

Tissue Samples

In man and the horse, a control sample of skin was obtained by high speed punch biopsy (4 mm diameter), without anaesthetic, at ambient conditions. Individuals were then put into a warm environment in a climatic chamber (40°C and 50% relative humidity). Water was supplied ad libitum and sweat output was monitored [94, 95]. Sweat glands of rats (200–250 g) were stimulated to secrete by subcutaneous injection of 0.3 ml pilocarpine (8 mg/ml in 0.9% NaCl) [59]. These procedures were in accord with animal legislation (horse and rat) and had the approval of the local medical ethical committee (human). Further biopsy samples were taken at the monitored onset of sweating and after 2–3 hours of continuous sweat output. The biopsies were rapidly cut into strips less than one mm thick and, within one minute of excision, cryofixed by plunging into 50 mm depth of mechanically stirred liquid cryogens under conditions for optimised quenching [11, 20, 75]. The essential features of this are that the samples were rapidly cut into the smallest pieces compatible with maintenance of structural and physiological integrity of the target glands. Freon 22, cooled to just above its freezing point (~ −160°C) in liquid nitrogen, was used for the human samples taken in hospitals and a propane/isopentane mixture, 70/30%, cooled to −196°C in liquid nitrogen, was employed for the horse and rat samples.

The data in Figs. 7–9 were obtained from bovine adrenal cells, obtained immediately post mortem, by collagenase digestion of the outermost region of 6–7 glands (predominantly the zona glomerulosa cells) and incubated in modified Medium 199. The cell suspensions were divided into control and experimental batches, with the latter being stimulated by incubation in medium plus Angiotensin II (1 μmol/1), for 5 minutes, immediately before further processing. The batches were centrifuged for 30 s at 12,000 G and small cell-pellets were cryofixed by optimised plunging into propane/isopentane at—196°C [49].

Specimen Cryoprocessing Protocol

The small tissue blocks (or cell pellets) were loaded under liquid nitrogen into a low temperature freeze drier [21] and held at $-80°C$ for three or four days under a vacuum of $\sim 10^{-3}$ Pa, in the presence of molecular sieve previously activated for 48h in a slow flow of dry nitrogen gas at 300°C. A copper block cryotrap, cooled by liquid nitrogen, was placed at about 10 mm from the specimens for the first 24h during drying of some batches. Omission of this cryotrapping did not adversely affect the quality of freeze-drying provided that the vacuum and molecular sieve were present. During the fourth day the specimens were slowly warmed, still under vacuum, at an approximate rate of 0.167° per minute until $-30°C$ was reached, after which a more rapid warming regime of about 1° per minute to 30°C was applied. Specimens were then vacuum embedded in degassed Araldite, gently taken up to atmospheric pressure with dry nitrogen gas and removed from the drier. After an overnight soak in fresh resin at room temperature, the blocks were cured at 60°C and stored in sealed plastic boxes with silica gel desiccant. Sections were dry-cut with a diamond knife at ~ 150 nm (LKB IV ultramicrotome), placed in the centre of nylon films (~ 50 nm) spanning single hole [63] titanium foil mounts, using eye lashes mounted on orange sticks, carbon coated (~ 20 nm) in a carbon filament coating unit (Emscope, Temcarb 500) and stored in sealed plastic boxes with silica gel desiccant until analysis. Comparably thick aminoplastic sections [79], doped with known concentrations of the elements of interest, were used as standards. The appearance of a section through a human sweat gland secretory coil, processed by this technique, is shown in Fig. 1. Although contrast is poor in these non-chemically fixed and unstained resin sections and resolution is reduced due to the thickness, enough detail can be discerned to enable placement of the ~ 0.2 μm diameter probe in various cytoplasmic locations well clear of the major folds in the section.

Microanalytical System

Spectra were obtained with a Link Systems 290 energy dispersive microanalyser using a horizontally mounted Kevex 30 mm^2 142 eV Si(Li) detector, placed at ~ 30 mm from the specimen and fitted with a colimator designed to stop hard X-rays from the specimen environment [67]. The microscope (JEOL 100 C), fitted with a free lens control for condenser 1, was modified as described by Nicholson et al. [67] to minimise extraneous instrumental X-ray contribution to the spectra. The column contained a large cryotrap at liquid nitrogen temperature, with a specimen anticontaminator at $-185°C$, designed to minimise extraneous radiation [66]. Single hole specimen mounts were held in low background specimen rods tilted at 30° towards the detector [66]. Accelerating voltage was 80 kV and the beam current ($3–4 \times 10^{-10}$ A) was monitored during specimen analyses with a "spade" detector mounted at viewing chamber level [64]. Spectra were collected from ~ 200 nm static probes, individually placed over selected cytoplasmic areas, for 100 s live time. The zero strobe and the Ti Kα peak were used to check spectra for zero drift and gain in the 0–20 keV range and spectra had 20 ev channel width. Quantification by the continuum normalization method [30] was performed by custom software using characteristic and background windows and a 5 keV continuum window centered on 12.5 keV, set according to the protocol of Nicholson and Dempster [65]. Ti-foil spectra and support film spectra taken from the carbon coated nylon support films adjacent to each analysed section, were used for calculation of specimen Bremsstrahlung.

Results and Discussion

There have been many microanalytical applications utilising the block freeze-drying and plastic embedding preparation route [39, 47, 48, 43, 44, 46, 9, 7, 52, 80, 81, 22, 98, 97, 15, 60, 59, 94, 95, 96, 78, 49]. This preparation route shares the criticism with the cryosectioned, freeze-dried preparation route, that the drying process causes tissue volume shrinkage of about 20% [5, 46] and introduces dangers of electrolyte

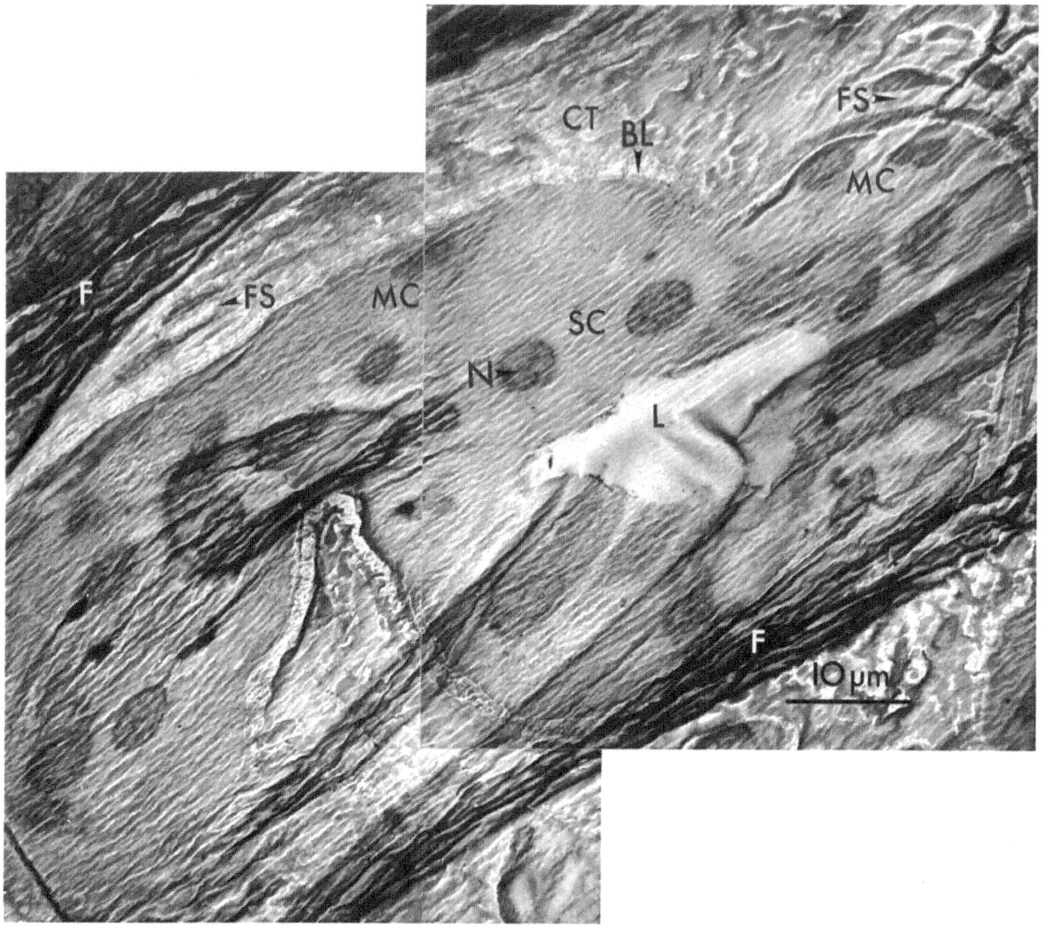

Fig. 1. Electron micrograph of a low temperature freeze-dried, vacuum-embedded Araldite resin section, dry-cut at 150 nm, of the secretory coil of a human sweat gland. BL, basal lamina; CT, connective tissue; F, section fold; FS, fibrocyte sheath; L, gland lumen; MC, myoepithelial cell; N, nucleus; SC, secretory cell

redistribution. However, it is generally agreed that, although diffusible substances may not be faithfully retained in extracellular spaces, particularly where there is little organic matrix, the drying process, when properly performed, does not cause elemental redistribution in intracellular sites at the spatial resolution of microanalysis [44, 38].

Data from glandular secretory epithelia of unstimulated and activated sweat glands from the horse, human and rat foot pad are presented in Fig. 2 and show that consistent changes in intracellular elemental quantities are measurable by this technique.

Tissue Temperature During Freeze-Drying

Small differences between freeze-drying protocols can apparently produce very different results and, as Condron and Marshall [10] point out, failure to obtain

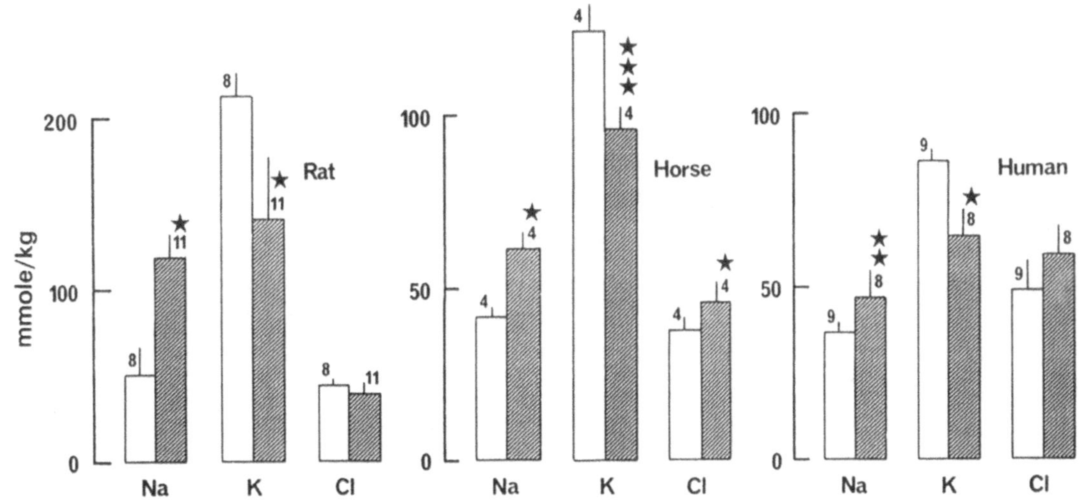

Fig. 2. Comparison of the measured intracellular quantities of Na, K and Cl, in human, horse and rat foot pad sweat gland secretory cells, between unstimulated (open columns) and activated glands (crosshatched columns). Significant differences between control and activated values; ★, $P < 0\ 05$; ★★, $P < 0.02$; ★★★, $P < 0.01$. Data replotted from: Rat, McWilliams et al. [59]; Horse, Wilson et al. [94]; Human, Wilson et al. [95]

good results is not a reason to condemn a method. The vexed question of how to "properly perform" low temperature freeze-drying probably underlies some of the criticism which the freeze-drying route has attracted in the literature [62, 55, 88] and the topic has been fully discussed in Ingram and Ingram [44, 46], and Robards and Sleytr [75]. Even if, as Livesey et al. [56] claim, vitreous ice starts to sublime from temperatures above $-160°C$, most multicellular cryofixed samples contain only crystalline ice or have, in the best cryofixed samples, a thin superficial layer of vitreous ice. Most authors, therefore, agree that low temperature freeze-drying must be performed in relation to the vapour pressures of crystalline ice, within the temperature range of $\sim -75°C$ and $\sim -100°C$. Warmer than this leads to a rapid increase in the rate of ice crystal growth. At colder temperatures the rate of sublimation of the crystalline ice phases becomes prohibitively slow [92].

One major difficulty is measurement of the temperature which the drying specimen actually experiences. Subliming water molecules are removed by one or more of three ways, by evacuation, by adsorption in desiccants such as a molecular sieve or by cryosorption pumping. The first two may easily become saturated. Although it is normal to employ two or more of these mechanisms together, cryosorption, using liquid nitrogen cooled surfaces, is the favoured method [14, 46]. Herein, however, lies the danger because, as the surface of the specimen dries, its thermal conductivity falls; it becomes more influenced by radiant heat transfer and the conductive flux through the stage becomes reduced. Thus, if its environment is at a temperature markedly different from that of the temperature controlled stage on which it lies, the specimen will adopt a temperature intermediate between the two. The problem can be illustrated by reference to Fig. 3. in which a hypothetical spherical sample of soft biological tissue, 1 mm diameter resting upon a freeze-drier

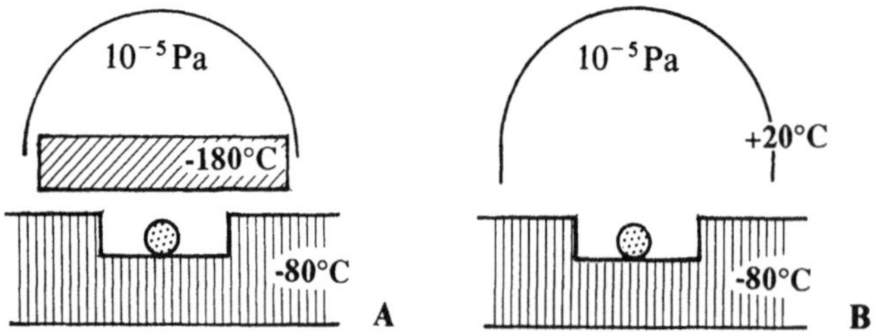

Fig. 3. Diagram of a theoretical 1 mm diameter biological sample (stippled circle) in a low temperature freeze-drying apparatus with the specimen stage (verticle hatch) set at −80°C. In **A** the specimen is shrouded by a cryotrap (diagonal hatch) at −180°C. In **B** the specimen stage environment is at 20°C. See text for explanation

stage controlled at −80°C, is alternatively exposed (a) to an "environment" one hundred degrees colder (liquid nitrogen cooled cryotrap at −180°C) or (b) to one a hundred degrees warmer than the stage (20°C room temperature surfaces). If we assume that the thermal properties of the hydrated specimen initially approximate to those of frozen water, calculation predicts that the temperature of the top surface of the specimen would be ~ −82°C in (a) and ~ −78°C in (b). Sublimation drying would commence, given adequate reduction of the partial pressure of water vapour in both situations. As drying proceeds, the surface layers dry first and the thermal properties will approximate to those of a dry, carbon-based matrix. Calculations, modelled upon thermal properties of plastic, suggest that the top surface of the specimen will now be ~ −120°C in (a) and ~ −40°C in (b). Under the conditions of (a) the drying times would be slowed by about two orders of magnitude [92], while in the situation of (b) ice crystal growth in the hydrated centre of the tissue would become rapid, leading to specimen collapse as documented by MacKenzie [57]. Remedies may include the construction of freeze-driers in which the specimens are not in "line-of-sight" of cryosorption surfaces [21]. Additionally, they probably include adoption of drying schedules which progress slowly through successively warmer temperatures [14, 46]. Slow warm up may be even more important than previously recognised and indeed, may be the only step of some protocols where the effective freeze-drying occurs. Slow warm up schedules have also been advocated for the different reason that components of tissue water do not sublime at the temperatures predicted from properties of pure ice [58].

Elemental Location During Embedding

Vacuum embedding with plastic introduces a further step at which, potentially, there could be translocation of mobile elements and components of the specimen. Even although non-polar resins are employed, the possibility that elements may be physically translocated within compartments of the specimen during the vacuum embedding stage has been extensively discussed [55, 44, 15, 19].

An experiment to determine if elemental translocation could be detected in our microanalytical system was briefly reported in Elder et al. [19]. Droplets of various salt standards (500 mmoles/l) dissolved in previously dialysed 400 kDa dextran

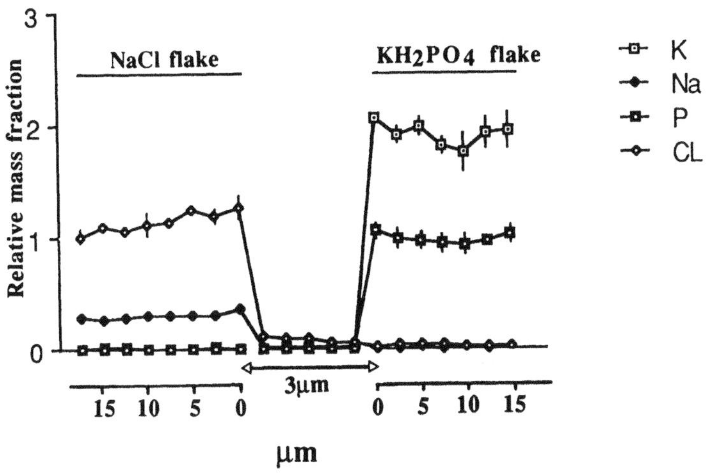

Fig. 4. Percent relative mass fractions of Na, P, Cl and K measured across the interface between dextran salt flakes as described in the text. Position of the flakes is indicated by the bars. Note that the scale changes for the ~3 μm gap between the flakes. The values are mean ±S.E. of 3 adjacent probes at each site

solution (20% wt./vol.) were rapidly frozen by allowing single drops to fall from a height of about 1 m onto the surface of a highly polished, pure copper mirror, cooled to −196°C in liquid nitrogen. The surface of the mirror was kept free of condensed water vapour by maintaining it about 10 mm below the "surface" of the cold nitrogen gas layer overflowing a dewar rim. Successfully frozen samples remained as flakes on top of the mirror. Flakes, one of NaCl and one of KH_2PO_4 were laid in pairs, freezing front to freezing front, in the wells of the low temperature freeze-drier, then dried and vacuum embedded in Araldite. Spectra were acquired from static probes (100 s live time and 0.4 nA) along a transect line across the adjacent flakes and the narrow intervening gap (3 μm) between them. Quantities, determined by the continuum normalisation method [30], of the elements Na, P, Cl and K found in the transect line are shown in Fig. 4. Although the probe was always several times larger than the size of ice crystal artefacts, local variation in the elemental quantities within each flake may be caused by the phase separation. There was no detectable diffusion or translocation of Na, P or K from their flake of origin and zero quantities of these elements were detected by probes placed at ~0.5 μm from the edge of the flakes (Fig. 4). Cl measurement is more problematical as Araldite itself contains a measurable quantity of the element. To correct for this, a block with only a KH_2PO_4 flake was analysed and the ratio of the Cl characteristic count in the flake to that in the pure peripheral resin gave a figure for the resin content in the flake of 77%. The measured Cl quantities in the Cl-containing flake were therefore corrected by this factor and the resin between the flakes was corrected for the Cl content of pure resin. The original and corrected Cl values are shown in Fig. 5 and these indicate that either there is a small residual error after correction for the Cl content of the Araldite or some minor translocation of Cl from the NaCl flake has occurred. From these data, which are in accord with the evidence of Coleman [9] and Ingram and Ingram [44], it has been concluded that there is no significant translocation of elements caused by the preparation procedures.

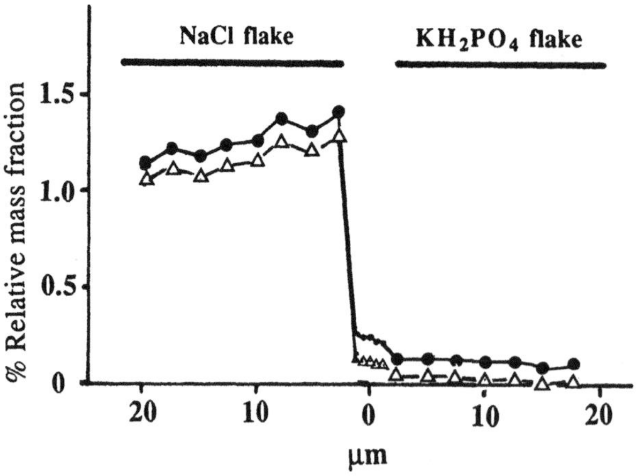

Fig. 5. Percent relative mass fractions of total Cl (circles) replotted from data in Fig. 4 and corrected by subtraction of the calculated Cl content of the resin (triangles), measured across the interface between dextran salt flakes as described in the text. Position of the flakes is indicated by the bars

Heavy Metal Staining

Use of any heavy metals in the preparative procedure to improve image contrast, e.g. osmium vapour fixation, is highly undesirable from the microanalytical point of view if continuum normalization is used in quantification. Roos and Barnard [80] report markedly lower values of most measured elements in freeze-dried osmium vapour-fixed resin embedded tissues in comparison with freeze-dried unfixed resin specimens. They attribute these to loss or displacement of elements because of interference of the osmium with resin polymerisation. These authors do not say whether the matrix factor in the Hall equation was altered to account for the heavy metal and this is likely to vary amongst probed sites since Os reacts differentially with cell organelles, e.g. reacting very little with zymogen granules [80]. Because of the Z^2 term in the matrix factor of the Hall equation, failure to correct for this in specimens where heavy metals have been introduced would result in significantly underestimated concentrations of measured elements and could at least partly explain the lower values obtained by Roos and Barnard [80].

Choice of Resin

The choice of resin may be more critical than has previously been suspected [10]. Wroblewski et al. [98] noted that Maraglas resin was unsuitable for resin-embedded microanalysis. The methacrylates, including the Lowicryl resins, are unstable under electron irradiation [24] and exhibit greater beam-induced mass loss than the epoxy resin Araldite [98], although image contrast for target location is superior in Lowicryl sections [97]. In a comparative study of Araldite and Lowicryl embedded tissue, Wroblewski et al. [98] found similar elemental quantities of all the diffusible elements retained in the tissues by both resins except for Na, the peak for which was significantly smaller in the Araldite sections. They suggested that this indicated that elemental translocation was a greater problem in Araldite during embedding and polymerisation. A more likely explanation, however, which these

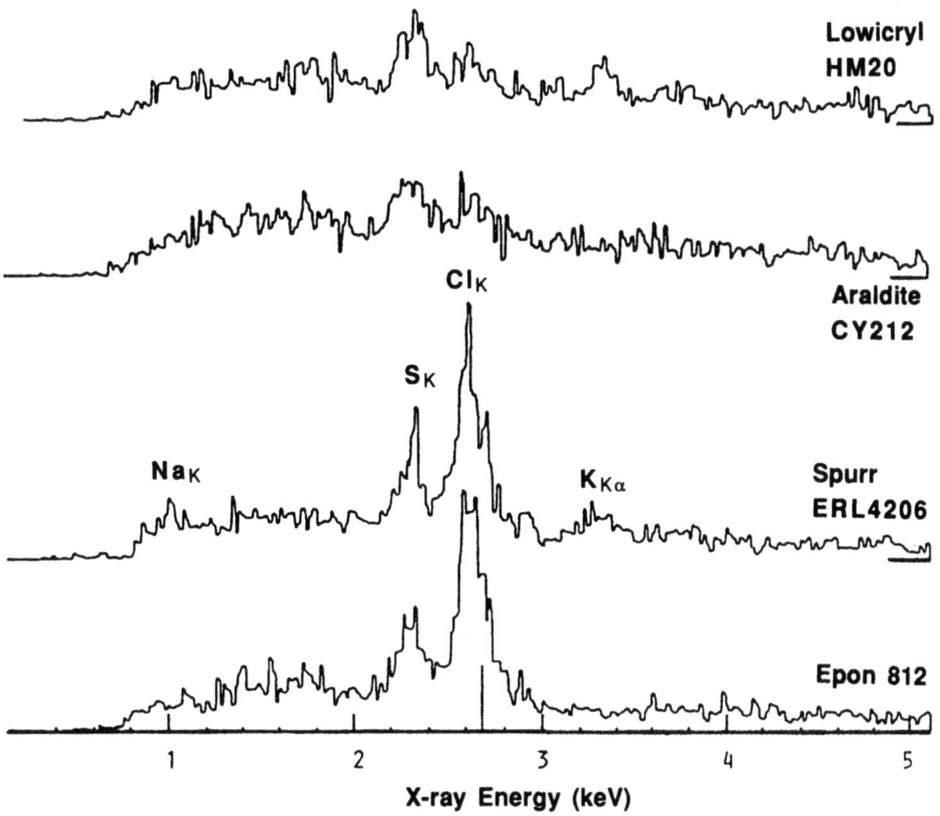

Fig. 6. Representative energy dispersive X-ray spectra taken from sections of four of the most commonly used resins for embedded specimen microanalysis. See text for further detail

authors also consider for the sections which were in the 0.2 to 1.5 μm range, is that these thick Araldite sections, suffering much less mass loss than the Lowicryl, are subject to greater self absorption of low energy photons (see peak-background subtraction section and Fig. 9 below).

Before quantitative analysis can be performed with resin embedded material, it is necessary to determine if the resin contains any of the elements which are to be measured. In our studies of transporting epithelia we wished especially to compare the quantities of sodium, phosphorus, chlorine, potassium and calcium in specimens at different states of activation and under normal and pathological conditions. Spectra from four of the non-polar resins most commonly used for these studies are shown in Fig. 6. The Lowicryl resin contained small but quantifiable amounts of Na, S, Cl and K. Spurr's resin contained larger amounts of these elements. Though relatively large peaks of S and Cl are present in Epon, quantities of Na and K are not distinguishable above noise. Araldite reveals the smallest peaks amongst the resins shown, though the S and Cl contents are still measurable. Lowicryl was not used in our studies because it suffers greater beam-induced mass loss than Araldite and produces uneven polymerisation through different tissue compartments [98]. Spurr's resin was not used since it contains the largest amounts of some of the elements of interest to us. Nor could we obtain evenly polymerised specimens using the low chloride version of Spurr's resin [68]. Of the epoxy resins, we therefore selected Araldite.

The only measurable peak in the Araldite amongst our elements of interest was chlorine. We corrected for the resin contribution to the total quantities of this element, measured in each spectrum, using the method of Ingram and Ingram, [45, 46], who doped their Epon 826 resin with known quantities of dibromoacetophenone. Since Br is not present in the tissues, the size of the bromine peak gave a measure of the resin content for comparison with the Br peak from pure resin outside the embedded specimen. However, it proved difficult to completely dissolve and evenly disperse dibromoacetophenone in unpolymerised Araldite CY212 and it may interfere with even polymerisation. In successfully brominated, resin embedded specimens, we obtained a mean resin content of 78% for gland cells. A quantity, 78% of that measured from the chlorine peak of pure peripheral resin of the same section was therefore subtracted from the measured specimen Cl peak of each spectrum.

Intracellular Elemental Gradients

Although intracellular elemental gradients can be retained in freeze-dried resin-embedded tissues [81], it would seem prudent to conclude that plastic-embedded materials should not be used for extracellular localization or measurement and, while there may be acceptable retention at the cellular level, faithful subcellular localization is doubtful [78]. Wroblewski et al. [99], however, have suggested not only that the polar resin Lowicryl K11M, as well as the non-polar Lowicryls HM20 and HM23, may be suitable for microanalytical studies but also that it is possible to retain subcellular localization with Lowicryl-embedded material. In our own studies [49] and Fig. 7, we have shown that physiologically relevant changes in intracellular K levels are detectable in Araldite-embedded tissues and that subcellular inhomogeneities, probably reflecting different potassium pools, are preserved in these sections. Fifteen spectra were taken from the cytoplasm of each of nine cells

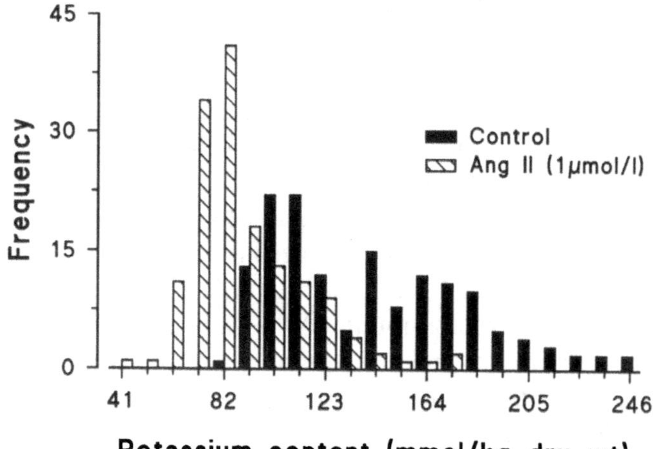

Fig. 7. Frequency histograms of the quantities of K measured in intracellular cytoplasmic sites of bovine adrenocortical cells unstimulated (black columns) and after stimulation with angiotensin II (cross hatched columns). Apart from the apparent fall in mean K concentration upon activation (significant at $P < 0.001$), there is evidence of a bimodal distribution in the K-frequencies found in the unstimulated cells. Replotted from Kenyon et al. [49]

of both control and experimental batches [49] and each of the control cells displayed a wide range of potassium concentrations amongst the sites probed, thus arguing against the alternative explanation that two cell populations were present. It is not known whether such gradients would be better preserved in freeze-dried cryosections, as found by Roos and Barnard [81].

Quantification

For thin and ultrathin biological specimens two quantification methods predominate, the "continuum normalisation" method, [30] and the "peripheral standards" method [68, 73, 36, 38]. The one most commonly employed is the continuum normalisation method. The rationale and formalisms of these and other techniques have been very fully described in the literature, e.g. [30, 31, 33, 34, 36, 37, 35, 76, 77]. The present discussion will be restricted to aspects of special relevance to plastic embedded specimens.

The continuum normalisation method was developed for thin biological sections, with two basic precepts, (i) that the mass of element x in a probed region of the section is proportional to the characteristic X-ray signal and (ii) the continuum intensity is proportional to the total mass of the analysed volume of section. It follows that the ratio of the characteristic to continuum intensity is a measure of the quantity of element x in the probed mass. Further, if a factor accounting for the mean atomic number of the matrix in the probed volume is included, then the measured quantities of element x are independent of local variations in section thickness and absolute values are obtained by reference to similarly measured standards.

The basic Hall equation is therefore:

$$C_x = C_{x\mathrm{st}} \cdot \frac{(P_x/W)}{(P_x/W)_{\mathrm{st}}} \cdot \frac{\overline{(Z^2/A)}}{\overline{(Z^2/A)}_{\mathrm{st}}} \quad (1),$$

where C_x is the quantity of element x per kg of section as represented by the analysed microvolume; P_x is the integral of the characteristic peak of element x; W is the continuum signal in a defined region of the *section* Bremsstrahlung spectrum, free of characteristic peaks; the subscript st denotes equivalent quantities of a comparably thin standard; the mean (Z^2/A) term, often called the "g-factor", accounts for Bremsstrahlung production in all of the elements in the specimen see [37, 76, 77] for discussion of its derivation.

Although in theory the continuum normalization method is elegantly simple, several practical difficulties have to be overcome if accurate results are to be obtained. For example, the quantity W from recorded spectra contains Bremsstrahlung X-rays not only from the specimen but also from the support film and from extraneous instrumental X-ray generation. Estimation of the latter remains one of the most problematical aspects of application of the Hall method [77]. Probably the most logical approach to the problem is first to systematically minimise all sources of extraneous instrumental X-ray generations [67]. This was done in the studies reported here from our laboratory. Modern TEMs are usually configured to minimise the extraneous contribution to the continuum but it should be checked that the contribution is less than 5–10% by use of appropriate test specimens (see

Nicholson et al. [66, 67]). Rather than solely ensuring that the remaining extraneous signal is devoid of characteristic peaks by constructing the supports closest to the specimen from materials of low atomic numbers such as C or Be which, however, contribute a small and unknown continuum, the remaining extraneous contribution should be from a specimen environment of a single defined element such as Al, Ti or Cu. Although the remaining extraneous contribution is larger than with the low atomic number elements, it can all be removed by computation from the measured spectra by scaling the continuum to its characteristic peak [65]. The continuum contribution of the support film can also be calculated and subtracted by measurement of spectra recorded from the support films alone, after correction for extraneous X-ray contribution [65].

A further difficulty lies in calculating the value of the g-factor since the precise proportions of the various atoms, particularly in biological specimen matrices, are rarely known. The most successful practical solution to the problem is an iterative approach [37, 76] in which best estimate values for constituent elements of unknown concentration in the matrix are first entered in the Hall equation (1) and the values obtained, used to calculate a better estimate of the (Z^2/A) term; successive values for the g-factor converge. Further discussion of this approach can be found in Roomans [76, 77].

From equation (1) we have the quantity of element x in mmol/kg of specimen. For frozen hydrated sections this comprises organic material plus the aqueous phase. In freeze-dried cryosections the specimen is composed of dry specimen only, while in plastic-embedded sections there is the specimen plus resin which has replaced the water. Measurements from resin-embedded sections might therefore be expected to yield values similar to those of frozen-hydrated sections. However, freeze-drying prior to embedding always involves shrinkage, which does not affect all tissue compartments and cell components equally. Further differences arise from the fact that resin does not penetrate evenly into all tissue compartments and that beam-induced mass loss in embedded sections differs from that in frozen-hydrated sections. While measured quantities in resin-embedded sections therefore more closely resemble those from frozen hydrated than from freeze-dried sections, there remains considerable uncertainty about the precise significance of the calculated values.

In a closely reasoned consideration of the problem, Hall [34] points out that if the resin could be tagged with an element which is absent from the tissue, e.g. Br as used by Ingram and Ingram [45], the resin content of each probed region could be measured and removed by computation, leaving unambiguously measured elements as quantities per kg of dry tissue, independent of shrinkage. The major practical problem in routine application of this method is the difficulty in achieving complete and even permeation of the tag compound through the resin and even then, as Hall [34] points out, it can only be assumed that the tag infiltrates uniformly. Our experience of uneven cutting properties with specimens embedded in doped resin suggest that the tagging compound may interfere with the infiltration or polymerisation processes. Hall's [34] suggestion, however, is a valuable one and should stimulate efforts to tag resins more easily and uniformly. Probably this should be done at the manufacturing stage. If the technique can be made routine, Hall [34] further points out that the tag element in the resin could serve as a standard and indeed, given knowledge of the K-factors (Cliff-Lorimer factors), no other standards need be used.

Specimens embedded in tagged resins could also be used for measurements by the peripheral standards method of Rick and colleagues [72, 73, 74, 12]. As originally applied to freeze-dried cryosections, the method simply scales the characteristic integral of element x in the specimen to that from a region of the same section, to which had been applied a solution of protein plus known concentration of x-containing electrolyte immediately prior to cryofixation. It gives the equation:

$$C'_x = C'_{x\,\mathrm{st}} \cdot \frac{P_x}{P_{x\,\mathrm{st}}} \qquad (2),$$

where C'_x is the amount of element x in mmoles per litre, P_x is the characteristic peak integral and st denotes the peripheral standard. Given knowledge of K-factors for elements of interest once more, the resin, tagged with an appropriate element, could equally well act as the "peripheral" standard, using a modified form of equation (2), see Hall [34] for details. Indeed, there are significant advantages to be obtained by parallel application of both formalisms. The continuum normalization method yields quantities in mmol/kg dry mass, independent of shrinkage but with problematical beam-induced mass loss and continuum measurement, whereas the peripheral standards method gives quantities in mmoles/l with shrinkage as a significant source of error; beam-induced mass loss should not matter and continuum measurement is unnecessary [34].

Improvements in Peak Background Subtraction

By whichever method calculations are made, an accurate measure of the area under the characteristic peaks is necessary and this involves subtraction of the background from the total signal in the appropriate energy band. The method employed in the programs used to compute the data by the continuum normalization method, in most of our studies, involves placement of background "windows", local to each peak [65]. More recent commercial software packages usually employ numerical methods, such as multiple least squares fitting, to remove the background for peak integral determination. The objective of these methods is to derive an accurate measure of the background under a given peak. An alternative and quite different approach is based upon theoretical prediction of the whole Bremsstrahlung spectrum shape for a given set of instrumental operating conditions. Although Kramers' [51] approximate theory of continuum generation has been shown to be quite adequate and simple to apply in the continuum normalization method [37], the modified Bethe-Heitler theory (MBH) has been shown to provide more accurate modelling of the Bremsstrahlung spectrum over the energy range 0–20 keV [8, 25]. A summed spectrum from nine cytoplasmic probes in bovine adrenal zona glomerulosa cells is shown in Fig. 8. Superimposed upon it is the Bremsstrahlung envelope modelled by the MBH method and scaled to a best amplitude match with the experimental spectrum.

For biological spectra, with numerous elemental peaks in the 1–4 keV range, there are difficulties in the location of clear background channels for peak determination by the windows method. The problem is most acute in the low energy range

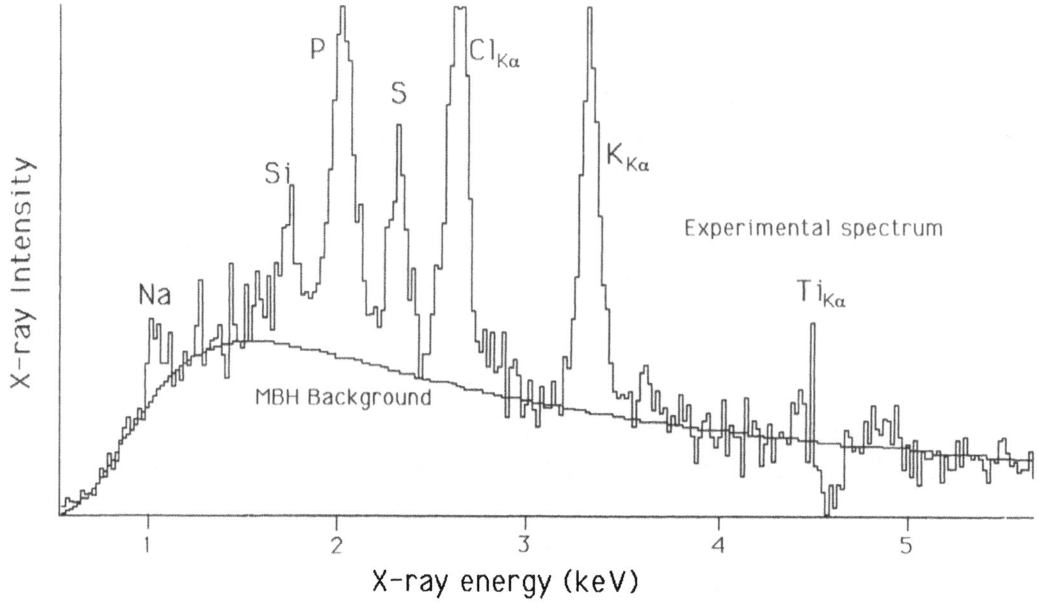

Fig. 8. A spectrum summed from nine cytoplasmic probes in bovine adrenal zona glomerulosa cells. A theoretical Bremsstrahlung spectrum modelled by the MBH method and scaled to a best amplitude match in the mid energy region with the experimental spectrum is superimposed. See text for further details. Replotted from Khan et al. [50]

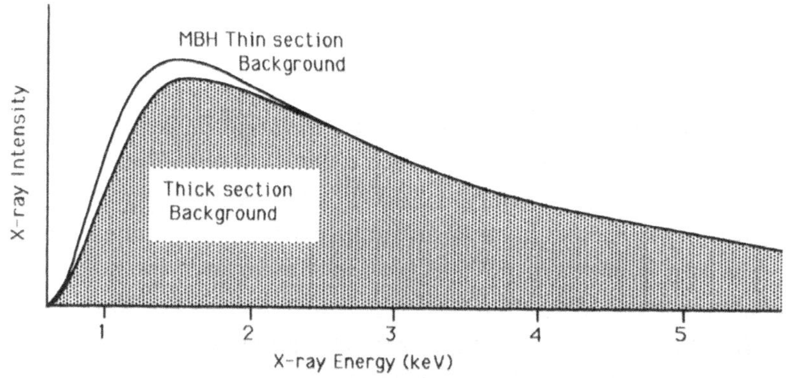

Fig. 9. The dotted area shows the same theoretical Bremsstrahlung spectrum as in Figure 8, for ultrathin sections, modelled by the MBH method and scaled, as in Figure 8, to the experimental data (not shown). The experimental data, in this instance, were from a section, dry-cut deliberately thick at ~250 nm and the filled spectrum is the MBH Bremsstrahlung spectrum best fitted, in the low energy region, to the experimental data. The apparent mismatch at the low energy end is caused by self absorption in the spectrum modelling the thick resin embedded section. See text for further details. Replotted from Khan et al. [50]

where the background is steep or inflected [65]. For the continuum normalization method we have therefore compared data computed using the windows and the MBH methods of background subtraction for peak integral calculation [50]. While significant changes in intracellular levels of Na, Cl and K between unstimulated and activated endocrine cells [69, 49] were demonstrated by both methods, the standard errors were smaller in each instance using the MBH method and only by this method was a small shift in intracellular P values shown to be statistically significant [50]. Work is currently in progress to compare the relative merits of numerical with modelling methods of background subtraction. One clear advantage of MBH modelling methods over numerical methods is illustrated in Fig. 9, which shows a significant mismatch at low energy between a Bremsstrahlung spectrum obtained from a resin-embedded tissue section, dry-cut deliberately thick at 0.25 μm, and the modelled MBH spectrum scaled to fit. The degree of mismatch is a function of the self absorption of the soft X-ray photons and the method would therefore seem to offer a convenient way of checking for the presence of unacceptable self absorption in the light elements, principally Na and Mg. It also provides the means to calculate the absorption path length and accurately correct for self-absorption. It could also, of course, provide a method for estimating the thickness of the section at the analysed area, though the requirement for accurate knowledge of the tilt angle and local specimen density would make this an awkward procedure; preliminary calculations show agreement to within 25% between nominal thickness of a range of sections and the thickness calculated from their degree of low energy self absorption.

Conclusions

Diffusible elements can be retained in their *in vivo* locations with a high degree of fidelity by plastic embedding and measured by X-ray microanalysis within acceptable limits of accuracy, particularly if relative quantities are the objective, despite the disadvantage that several characteristics of biological tissues are inherently unfavourable. Our evidence supports the conclusion that freeze-drying and vacuum embedding in resin gives reliable comparative results and offers advantages for X-ray microanalysis in intracellular locations of biological tissues which are difficult to target, provided that the principles are observed and appropriately monitored.

Acknowledgements. It is a pleasure to acknowledge constructive discussion of freeze-drying and assistance with computation by Prof. J. N. Chapman and Dr. J. D. Steele.

References

[1] T. Appleton, *J. Microsc.* **1974**, *100*, 49.
[2] W. B. Bald, *Quantitative Cryofixation*, Hilger, Bristol, 1987, pp. 185.
[3] J. R. Bellare, H. T. Davis, L. E. Scriven, Y. Talmon, in: *Proc. 12th Int. Conf. Electron. Microsc., Kyoto, II.*, (T. Imura, S. Maruse, T. Suzuki, eds.), 1986, 367.
[4] J. R. Bellare, H. T. Davis, L. E. Scriven, Y. Talmon, *J. Electron Microsc. Techn.* **1988**, *10*, 87.
[5] A. Boyde, in: *Scanning Electron Microsc.* **1978**, *II*, 303.
[6] E. J. Brooks, A. J. Tousimis, J. S. Birks, *J. Ultrastruct. Res.* **1962**, 7, 56.
[7] I. V. Burovina, F. G. Gribakin, A. M. Petroysan, N. B. Pivovarova, A. G. Pogorelov, A. D. Polyanovski, *J. Comp. Physiol.* **1978**, *127*, 245.
[8] J. N. Chapman, C. C. Gray, B. W. Robertson, W. A. P. Nicholson, *X-ray Spectrom.* **1983**, *12*, 153.
[9] J. R. Coleman, *Proc. 11th Ann. Conf. Microbeam Anal. Soc.*, 1976, pp. 58A–58H.
[10] R. J. Condron, A. T. Marshall, *Scanning Microsc.* **1990**, *4*, 439.
[11] M. J. Costello, J. M. Corless, *J. Microsc.* **1978**, *112*, 17.
[12] A. Dörge, F. X. Beck, R. Rick, W. Nagel, K. Thurau, in: *Electron Probe Microanalysis. Applica-*

tions in Biology and Medicine (K. Zierold, H. K. Hagler, eds.), Springer, Berlin, Heidelberg, New York, Tokyo, 1989, pp. 225–236.
[13] P. Echlin, *J. Microsc.* **1991**, *161*, 159.
[14] L. Edelmann, *J. Microsc.* **1978**, *112*, 243.
[15] L. Edelmann, *Scanning Electron Microsc.* **1986**, *IV*, 1337.
[16] L. Edelmann, in: *Electron Probe X-ray Microanalysis. Applications in Biology and Medicine*, (K. Zierold, H. K. Hagler, eds.), Springer, Berlin, Heidelberg, New York, Tokyo 1989, pp. 33–46.
[17] L. Edelmann, *J. Microsc.* **1991**, *161*, 217.
[18] H. Y. Elder, in: *Techniques in Immunocytochemistry, Vol. 4*, (G. R. Bullock, P. Petrusz, eds.), Academic Press, London, 1989, pp. 1–27.
[19] H. Y. Elder, D. L. Bovell, J. D. Pediani, S. M. Wilson, S. A. McWilliams, D. McE. Jenkinson, *Inst. Phys. Conf. Ser. No. 93 Vol. 3 EUREM 88*, York, 1988, pp. 575–576.
[20] H. Y. Elder, C. C. Gray, A. G. Jardine, J. N. Chapman, W. H. Biddlecombe, *J. Microsc.* **1982**, *126*, 45.
[21] H. Y. Elder, W. H. Biddlecombe, L. Tetley, S. M. Wilson, D. McE. Jenkinson, *EMSA Bull.* **1986**, *16*, 111.
[22] H. Y. Elder, D. McE. Jenkinson, S. A. McWilliams, S. M. Wilson, *J. Physiol.* **1985**, *367*, 74P.
[23] J. C. Gilkey, L. A. Staehelin, *J. Electron Microsc. Techn.* **1986**, *3*, 177.
[24] A. M. Glauert, *Fixation, Dehydration and Embedding of Biological Specimens. Practical Methods in Electron Microscopy, Vol. 3 (1)* (A. M. Glauert, ed.), 1974, pp. 208.
[25] C. C. Gray, J. N. Chapman, W. A. P. Nicholson, B. W. Robertson, R. P. Ferrier, *X-Ray Spectrom.* **1983**, *12*, 163.
[26] B. L. Gupta, T. A. Hall, *Tissue & Cell* **1981**, *13*, 623.
[27] B. L. Gupta, T. A. Hall, R. B. Moreton, in: *Transport of Ions and Water in Animals* (B. L. Gupta, R. B. Moreton, J. L. Oschman, B. J. Wall, eds.), Academic Press, New York, 1977, pp. 83–143.
[28] H. K. Hagler, L. M. Buja, in: *Science of Biological Preparation. Scanning Electron Microsc.* (J.-P. Revel, T. Barnard, G. H. Haggis, eds.), AMF O'Hare, Il., 1984, pp. 161–166.
[29] H. K. Hagler, A. C. Morris, L. M. Buja, in: *Electron Probe Microanalysis. Applications in Biology and Medicine.* (K. Zierold, H. K. Hagler, eds.), Springer, Berlin, Heidelberg, New York, Tokyo, 1989, pp. 181–197.
[30] T. A. Hall, in: *Physical Techniques in Biological Research, 2nd Ed., Vol. 1A* (G. Oster, ed.), Academic Press, New York, 1971, pp. 157–275.
[31] T. A. Hall, *J. Microsc.* **1979**, *117*, 145.
[32] T. A. Hall, *J. Microsc.* **1986**, *141*, 319.
[33] T. A. Hall, *Scanning Microsc.* **1989**, *3*, 461.
[34] T. A. Hall, *J. Microsc.* **1991**, *164*, 67.
[35] T. A. Hall, H. C. Anderson, T. Appleton, *J. Microsc.* **1973**, *99*, 177.
[36] T. A. Hall, B. L. Gupta, *J. Microsc.* **1982**, *126*, 333.
[37] T. A. Hall, B. L. Gupta, *Quart. Rev. Biophys.* **1983**, *16*, 279.
[38] T. A. Hall, B. L. Gupta, *J. Microsc.* **1984**, *136*, 193.
[39] V. Hanzon, L. H. Hermodson, *J. Ultrastruct. Res.* **1960**, *4*, 332.
[40] D. M. R. Harvey, *J. Microsc.* **1982**, *127*, 209.
[41] B. Humbel, M. Müller, in: *Science of Biological Specimen Preparation. Scanning Electron Microsc.* (M. Müller, R. P. Becker, A. Boyde, J. J. Wolosewick, eds.), AMF O'Hare, Il., 1986, pp. 175–183.
[42] T. E. Hutchinson, D. E. Johnson, A. P. MacKenzie, *Ultramicroscopy* **1978**, *3*, 315.
[43] F. D. Ingram, M. J. Ingram, *J. Microsc. Biol. Cell.* **1975**, *22*, 193.
[44] F. D. Ingram, M. J. Ingram, in: *Scanning Electron Microsc.* **1980**, *IV*, 147.
[45] M. J. Ingram, F. D. Ingram, in: *Scanning Electron Microsc.* **1983**, *III*, 1249.
[46] F. D. Ingram, M. J. Ingram, in: *Science of Biological Preparation. Scanning Electron Microsc.* L.-P. Revel, T. Barnard, G. H. Haggis, AMF O'Hare, Il., 1984, pp. 167–174.

[47] F. D. Ingram, M. J. Ingram, C. A. M. Hogben, *J. Histochem. Cytochem.* **1972**, *20*, 716.
[48] F. D. Ingram, M. J. Ingram, C. A. M. Hogben, in: *Microprobe Analysis as Applied to Cells and Tissues* (T. Hall, P. Echlin, P. Kaufmann, eds.), Academic Press, New York, 1974, pp. 119–146.
[49] C. J. Kenyon, R. M. Shepherd, R. Fraser, J. D. Pediani, H. Y. Elder, *Endocrinol. Res.* **1991**, *117*, 19.
[50] K. M. Khan, W. A. P. Nicholson, J. P. Pediani, H. Y. Elder, *Inst. Phys. Conf. Ser., No. 98 (2)*, BRISTOL 1990, pp. 691–694.
[51] H. A. Kramers, *Phil. Mag. 46*, 1923, pp. 836–871.
[52] G. A. J. Kuijpers, I. G. P. Van Nooy, J. J. H. H. M. De Pont, A. L. H. Stols, *Ultramicroscopy* **1984**, *14*, 414.
[53] M. K. Lamvik, *J. Microsc.* **1991**, *161*, 171.
[54] R. D. Leapman, S. B. Andrews, *J. Microsc.* **1991**, *161*, 3.
[55] C. P. Lechene, R. R. Warner, *Ann. Rev. Biophys. Bioeng.* **1977**, *6*, 57.
[56] S. A. Livesey, A. A. del Campo, A. W. McDowall, J. T. Stasny, *J. Microsc.* **1991**, *161*, 205.
[57] A. P. MacKenzie, *Cryobiology* **1967**, *3*, 387.
[58] A. P. MacKenzie, in: *Microprobe Analysis of Biological Systems*, (T. E. Hutchinson, A. P. Somlyo, eds.), Academic Press, New York, 1981, pp. 397–421.
[59] S. A. McWilliams, I. Montgomery, H. Y. Elder, D. McE. Jenkinson, S. M. Wilson, *Tissue & Cell* **1988**, *20*, 109.
[60] S. A. McWilliams, I., Montgomery, D. McE. Jenkinson, H. Y. Elder, S. M. Wilson, A. Sutton, *Br. J. Dermatol.* **1987**, *117*, 617.
[61] H. Moor, in: *Cryotechniques in Biological Electron Microscopy* (R. A. Steinbrecht, K. Zierold, eds.), Springer, Berlin, Heidelberg, New York, Tokyo 1987, pp. 175–191.
[62] A. J. Morgan, T. W. Davies, D. A. Erasmus, *Micron* **1975**, *6*, 11.
[63] W. A. P. Nicholson, in: *Microprobe Analysis as Applied to Cells and Tissues*, (T. Hall, P. Echlin, R. Kaufmann, eds.), Academic Press, London, 1974, pp. 239–248.
[64] W. A. P. Nicholson, *J. Microsc.* **1981**, *121*, 141.
[65] W. A. P. Nicholson, D. W. Dempster, *Scanning Electron Microsc.* **1980**, *II*, 517.
[66] W. A. P. Nicholson, W. H. Biddlecombe, H. Y. Elder, *J. Microsc.* **1982**, *126*, 307.
[67] W. A. P. Nicholson, C. C. Gray, J. N. Chapman, B. W. Robertson, *J. Microsc.* **1982**, *125*, 25.
[68] C. K. Pallaghy, *Aust. J. Biol. Sci.* **1973**, *26*, 1015.
[69] J. D. Pediani, C. J. Kenyon, H. Y. Elder, *J. Physiol.* **1990**, *426*, 25P.
[70] H. Plattner, L. Bachmann, *Int. Rev. Cytol.* **1982**, *79*, 237.
[71] N. Reid, J. E. Beesley, *Sectioning and Cyrosectioning for Electron Microscopy. Practical Methods in Electron Microscopy, Vol. 13* (A. M. Glauert, ed.), Elsevier, Amsterdam, Oxford. 1991, pp. 322.
[72] R. Rick, A. Dörge, A. D. C. Macknight, A. Leaf, K. Thurau, *J. Membr. Biol.* **1978**, *29*, 257.
[73] R. Rick, A. Dörge, K. Thurau, *J. Microsc.* **1982**, *125*, 239.
[74] R. Rick, W. Schratt, in: *Electron Probe Microanalysis. Applications in Biology and Medicine.* (K. Zierold, H. K. Hagler, eds.), Springer, Berlin, Heidelberg, New York, Tokyo, 1989, pp. 213–224.
[75] A. W. Robards, U. Sleytr, *Low Temperature Methods in Biological Electron Microscopy. Practical Methods in Electron Microscopy, Vol. 10* (A. M. Glauert, ed.), Elsevier, Amsterdam, Oxford, 1985, 551.
[76] G. M. Roomans, *J. Electron Microsc. Techn.* **1988**, *9*, 19.
[77] G. M. Roomans, *Scanning Microsc.* **1990**, *4*, 1055.
[78] N. Roos, in: *Electron Probe X-ray Microanalysis. Applications in Biology and Medicine.* (K. Zierold, H. K. Hagler, eds.), Springer, Berlin, Heidelberg, New York, Tokyo, 1989, pp. 17–32.
[79] N. Roos, T. Barnard, *Ultramicroscopy* **1984**, *15*, 277.
[80] N. Roos, T. Barnard, *Ultramicroscopy* **1985**, *17*, 335.
[81] N. Roos, T. Barnard, in: *Scanning Electron Microsc.* **1986**, *II*, 703.
[82] L. Sevéus, *Scanning Electron Microsc.* **1980**, *IV*, 161.
[83] H. Sitte, K. Neumann, L. Edelmann, in: *Science of Biological Specimen Preparation. Scanning*

Electron Microsc. (M. Müller, R. P. Becker, A. Boyde, J. J. Wolosewick, eds.), AMF O'Hare, Il., 1986, pp. 103–118.

[84] H. Sitte, L. Edelmann, K. Neumann, in: *Cryotechniques in Biological Electron Microscopy*, (R. A. Steinbrecht, K. Zierold, eds.), Springer Berlin, Heidelberg, New York, Tokyo, 1987, pp. 87–113.

[85] A. P. Somlyo, A. V. Somlyo, M. Bond, H. Shuman, in: *Microbeam Analysis* (A. D. Romig, W. F. Chambers, eds.), San Francisco Press, San Francisco, 1986, pp. 199–204.

[86] A. V. Somlyo, M. Bond, J. C. Silcox, A. P. Somlyo, in: *Proc. 43rd Conference Electron Microsc. Soc. Amer.* (G. W. Bailey, ed.), San Francisco Pres, San Francisco, 1985, pp. 10–13.

[87] A. V. Somlyo, J. C. Silcox, in: *Microbeam Analysis in Biology* (C. P. Lechene, R. R. Warner, eds.), Academic Press, New York, 1979, pp. 535–555.

[88] R. A. Steinbrecht, *MEM (Zeiss Information)* **1984**, *3*, 9.

[89] R. A. Steinbrecht, M. Müller, in: *Cryotechniques in Biological Electron Microscopy* (R. A. Steinbrecht, K. Zierold, eds.), Springer, Berlin, Heidelberg, New York, Tokyo, 1987, pp. 149–172.

[90] Y. Talmon, *J. Microsc.* **1982**, *125*, 277.

[91] K. E. Tvedt, G. Kopstad, J. Halgunset, O. A. Haugen, *Am. J. Clin. Pathol.* **1989**, *92*, 51.

[92] W. Umrath, *Mikroskopie* **1983**, *40*, 9.

[93] T. von Zglinicki, M. Bimmler, H.-J. Purz, *J. Microsc.* **1986**, *141*, 79.

[94] S. M. Wilson, H. Y. Elder, D. McE. Jenkinson, S. A. McWilliams, *J. Exp. Biol.* **1988**, *136 (a)*, 489.

[95] S. M. Wilson, H. Y. Elder, S. M. Sutton, D. McE. Jenkinson, F. Cockburn, I. Montgomery, S. A. McWilliams, D. L. Bovell, *Tissue & Cell*, **1988**, *20*, *(b)* 691.

[96] S. M. Wilson, H. Y. Elder, D. McE. Jenkinson, A. M. Sutton, F. Cockburn, in: *Inst. Phys. Conf. Ser. No. 98*, EMAG-MICRO 89, London, 1990, pp. 735–738.

[97] J. Wroblewski, R. Wroblewski, *J. Microsc.* **1986**, *142*, 351.

[98] R. Wroblewski, J. Wroblewski, M. Anniko, L. Edström, *Scanning Electron Microsc.* **1985**, *I*, 447.

[99] R. Wroblewski, J. Wroblewski, S.-O. Wikström, M. Anniko, *Scanning Microsc.* **1990**, *4*, 787.

[100] K. Zierold, in: *Science of Biological Specimen Preparation. Scanning Electron Microsc.* (M. Müller, R. P. Becker, A. Boyde J. J. Wolosewick, eds.), AMF O'Hare, Il., 1986, pp. 119–127.

[101] K. Zierold, in: *Cryotechniques in Biological Electron Microscopy* (R. A. Steinbrecht, K. Zierold, eds.), Springer Berlin, Heidelberg, New York, Tokyo, 1987, pp. 132–148.

[102] K. Zierold, *J. Electron Microsc. Techn.* **1988**, *9*, 65.

[103] K. Zierold, *J. Microsc.* **1991**, *161*, 357.

[104] K. Zierold, D. Schäfer, *Scanning Microsc.* **1988**, *2*, 1775.

Mikrochim. Acta (1992) [Suppl.] 12: 75–92

Analytical and High-Resolution Electron Microscopy Studies at Metal/Ceramic Interfaces

Manfred Rühle

Max-Planck-Institut für Metallforschung, Seestrasse 92, D-W-7000 Stuttgart, Federal Republic of Germany

Abstract. Both analytical electron microscopy and high-resolution electron microscopy are widely used to investigate the structure of and defects in materials. The spatial resolution of the analytical electron microscope (AEM) is < 5 nm and the point-to-point resolution of the high-resolution electron microscope (HREM) is small enough to image lattice planes in metals and ceramics directly. AEM and HREM are described in this paper with special emphasis on the study of interfaces between single crystal Nb-films grown by molecular beam epitaxy (MBE) on (0001) planes of sapphire (Al_2O_3)-substrates. From these specimens, cross-sectional specimens with thickness < 20 nm were prepared, so that the Nb/Al_2O_3-interface could be investigated by AEM and HREM. AEM studies revealed that Al_2O_3 was dissolved by the Nb so that Al could be detected in the Nb adjacent to the interface. The atomistic structure of the interface was identified by HREM. Defects at or close to the interface were analyzed and a model of the atomistic arrangements of the interface in the relaxed, "perfect" regions is described with a discussion of the results.

Key words: Analytical electron microscopy, high-resolution electron microscopy, interfaces.

Special properties of materials are required for many applications which cannot be fulfilled for single phase materials. Therefore, the processing, characterization and understanding of properties of composite materials are important aims in materials science [1]. The composites can exist either of a metal matrix material strengthened by ceramic particles (MMC) or ceramic matrix composites (CMC) strengthened by metal particle laminates, etc. Recently it was established that mechanical properties, such as strength and toughness, can be improved up to a factor of ten compared to the corresponding pure materials [1, 2].

For any type of metal/ceramic composite, the properties of the metal/ceramic interface dictate the properties of the composite [3, 4]. It is required that the metal/ceramic interfaces be tailored, which means that decohesion is required for specific loading [1, 2]. Metal/ceramic interfaces play an important role also in the

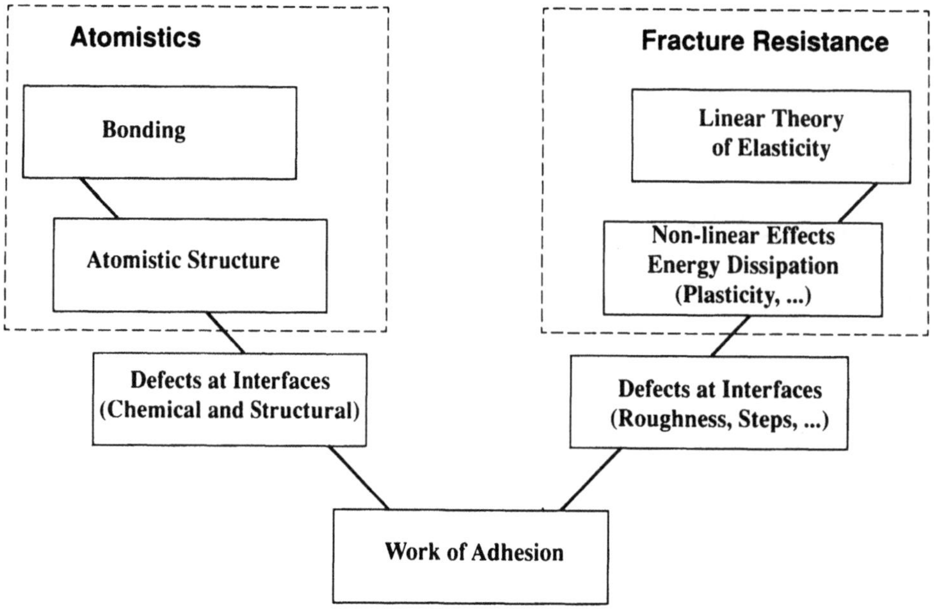

Fig. 1. The fundamental correlation of the work of adhesion, W_{ad}, with microscopical and macroscopical properties

bonding of metallic to ceramic structural components [5, 6], during the process of oxidation of metals and metallic alloys [7, 8] and in the technologically important area in electronic packaging [9, 10]. In this paper, model studies will be reported exclusively for the niobium/alumina interfaces (Nb/Al_2O_3) [11–16]. Investigations on other systems are published in recent conference proceedings [3, 4, 17]. Some general remarks on microstructural studies will precede the detailed results on Nb/Al_2O_3.

The research involves processing, modelling, characterization and measurement of properties of metal/ceramic interfaces. It also requires that these studies are carried out on different levels (Fig. 1). One important task is the correlation between atomistic studies and mechanical studies, such as fracture resistance. Both research activities are connected through the basic physical quantity: the work of adhesion (Fig. 1). In this paper, results on the atomistic structure of Nb/Al_2O_3 interfaces and some remarks on defects at interfaces will be included.

Methods to Produce Metal/Ceramic Interfaces

There are essentially three ways to produce metal/ceramic interfaces. Firstly, the most simplistic way is to prepare metal/ceramic interfaces by the diffusion bonding process [18]. During this process a flat metal object is bonded to a flat ceramic object by applying a small load at elevated temperatures. For this method the interface orientation is predetermined so that no relaxations with respect to preferable orientation relationships are possible. The diffusion bonding process usually requires a high temperature. If the thermal expansion coefficient of the metal and the ceramic components are different, thermal stresses develop and the interface is under a stress state after cooling. This often leads to a fracture at the interface. Fortunately, the thermal expansion coefficients of Nb and Al_2O_3 are about the

Table 1. Conditions of diffusion bonding (see text for explanation)

Bonding Pressure:	1 MPa
Bonding Temperature:	1600 °C ... 1950 °C
Bonding Time:	1 h ... 8 h
Cooling Rates:	"quenching" (1000 K/min) "fast" (230 K/min) "slow" (20 K/min)
Cross Sectioning for:	TEM HREM AEM

same resulting in very small residual stresses. Diffusion bonding of Nb and Al_2O_3 requires temperatures > 1650 K [5]. Typical diffusion bonding conditions selected for several studies are summarized in Table 1.

Secondly, metal/ceramic interfaces can also be formed by internal oxidation [12, 15]. The internal oxidation of metallic alloys requires that the oxygen partial pressure outside the alloy is selected so that diffusion of oxygen into the material is faster than the diffusion of the minor component of the metallic alloy to the surface. For example, a Nb–Al 3.5 at.% requires an oxidation temperature of 1773 K and a controlled atmosphere of oxygen pressure of 10^{-4} millibars which then leads to α-Al_2O_3 precipitates within the Nb matrix. Interfaces created by this process of internal oxidation have well defined geometric characteristics. Precipitates of α-Al_2O_3 in Nb possess a penny-shaped geometry with a diameter of $\sim$300 nm and a thickness of 50 nm. A fixed orientation relationship exists between the Nb matrix and the Al_2O_3 precipitate so that close-packed planes of Al_2O_3 are parallel to close-packed planes in Nb [21]. High resolution electron microscopy (HREM) studies of interfaces of internally oxidized Nb–Al_2O_3 alloys reveal the atomistic structure [12, 15, 21].

Thirdly, metal/ceramic interfaces can be formed by evaporation of a thin metallic film on a ceramic substrate [22, 23]. For this technique Nb-layers were fabricated in an MBE-growth chamber, equipped with electron beam sources for evaporating refractory metals. The substrates had different orientations. For these studies, the orientations parallel to [0001] sapphire were used. The substrates were preheated and cleaned by annealing at 1500 K for 1 h in a dynamic vacuum of $5 \cdot 10^{-9}$ Pa. The growth of the Nb-films was performed at a substrate temperature of 1123 K. Typical growth rates were one monolayer per second. The monolayers were grown to a thickness ranging from 10 to 500 nm.

Transmission Electron Microscopy

The different possibilities of transmission electron microscopy are summarized schematically in Fig. 2. Conventional TEM techniques (CTEM) [24] involve bright

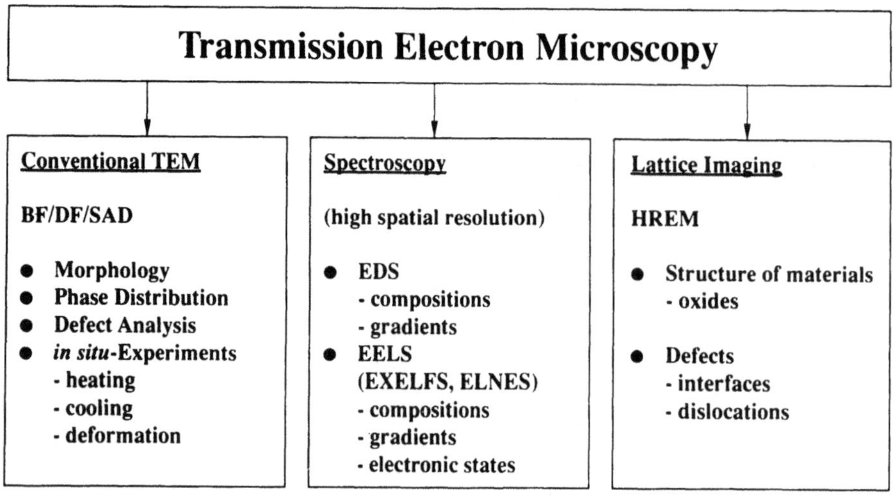

Fig. 2. Schematic representation of different TEM techniques (BF = bright field imaging; DF = dark field imaging; EDS = energy dispersive spectroscopy; EELS = electron energy-loss spectroscopy; EXELFS = extended energy-loss fine structure; ELNES = electron energy loss near-edge structure; HREM = high-resolution electron microscopy; SAD = selected Area Diffraction

field and dark field images and selected area diffraction. These are used for morphological analysis, phase and defect identification studies. Spectroscopy can be performed with high spatial resolution (AEM) [25]. The probe size in scanning electron microscopy (STEM) ranges from ~1 to 15 nm. Energy dispersive spectrosopy (EDS) uses X-rays which are emitted from the specimen, for chemical characterization. Characteristic energy losses are used for identifying qualitatively and quantitatively light elements by energy loss spectroscopy [26]. The surrounding of atoms can be probed by extended energy loss fine structure (EXELFS), whereas energy loss near edge structure (ELNES) investigations result in information on the excitation states of atoms or ions. HREM studies result in structure images [27, 28].

This paper focuses on studies of Nb–Al_2O_3 interfaces by AEM and HREM.

Cross Section Specimen Preparation for TEM Studies

A special technique was used to obtain transmission electron microscopy (TEM)-cross sections of a Nb–Al_2O_3 interface [29]. Firstly, a region of the specimen containing the interface was cut out of the bulk material. Two pieces were glued together face to face. The sandwich was cut into strips and glued into a slit within a cylindrical aluminium holder (2.0 mm diameter), which was stiffened by a thin alumina tube with an outer diameter of 3.0 mm. Disks, cut from these holders, were carefully polished to a thickness of 0.3 mm, and dimpled. During ion milling, shields were used to protect the interface and to reduce the effect of the different milling rates of Al_2O_3 and Nb, respectively. This preparation technique resulted in specimens suitable for AEM and HREM studies. The foil thicknesses were less than ~20 nm in the region which included the interface. Foil thicknesses down to 5 nm could be obtained.

Analytical Studies

Concentration and concentration profiles of different species were measured in Al_2O_3 and Nb. In Al_2O_3 no Nb could be detected. The concentration of Al in Nb, $c_{Al}[Nb]$, however, could be measured by EDS [30]. It should be emphasized that prior to bonding no Al could be detected in Nb: $c_{Al}[Nb]$ lies below the limit of detectability. Electron energy loss spectroscopy (EELS) should allow the detection of oxygen. The concentration of oxygen in Nb ($c_O[Nb]$) was, however, below the limit of detectability for both bonded and unbonded Nb. Figure 3a shows $c_{Al}[Nb]$ as a function of the distance from the interface for fast cooled specimens (230 K/min). Special attention was drawn to $c_{Al}[Nb]$ close to the interface (Fig. 3b). The distribution of Al in Nb can be readily determined from Fig. 3 a, b. At the interface $c_{Al}[Nb]$ is smaller than the level of detectability which is ~0.13 at.%

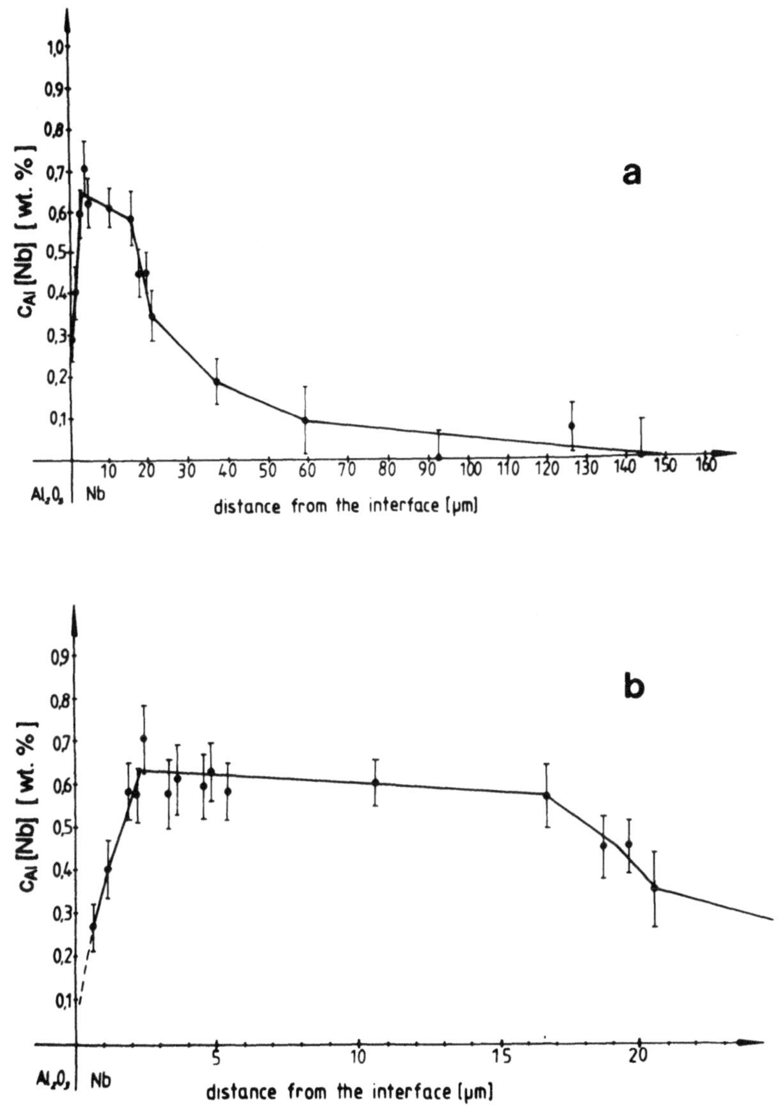

Fig. 3. **a** Aluminum concentration profile $c_{Al}[Nb]$ in Nb of a "fast" cooled bond as a function of the distance from the interface. **b** Detail ($o < x < 22$ μm) of the concentration profile $c_{Al}[Nb]$ for small distances from the interface

Al. With increasing distance from the interface, $c_{Al}[Nb]$ increases to a maximum value of ~0.65 wt% at a distance of 2.5 μm. The concentration $c_{Al}[Nb]$ is then almost constant up to a distance of $x = 16$ μm from the interface. For larger distances, exceeding 16 μm, $c_{Al}[Nb]$ decreased approximately exponentially with increasing distance from the interface. The concentration of interstitially dissolved O, $c_O[Nb]$ is smaller than the limit of detectability of a serial EELS (~1 at.% O).

For the evaluation of the data one has to assume that at the bonding temperature, $c_{Al}[Nb]$ possesses a value which can be obtained by extrapolation of the plateau in Fig. 3 to the position of the interface. This assumption can be justified theoretically since $c_{Al}[Nb]$ should have its largest value at the interface during bonding. Al_2O_3 is dissolved at the interface and Al and O diffuse in the Nb lattice. The concentration of Al($c_{Al}[Nb]$) is highest at the interface, however, $c_{Al}[Nb]$ cannot exceed the solution limit of $c_{Al}[Nb]$ which is a function of temperature and oxygen concentration. The amount of dissolved Al_2O_3 does not depend on the thermodynamics at the interface only, but also on the diffusivity of Al in Nb and O in Nb, respectively. The width where $c_{Al}[Nb]$ reaches its maximum value depends on bonding temperature and bonding time. These dependencies are complex functions. For the bonding described in Fig. 3 the total layer thickness of dissolved Al_2O_3 can be evaluated by integrating the total amount of Al dissolved in Nb and converting it to Al_2O_3. The evaluation leads to a layer thickness of about 900 nm dissolved during bonding.

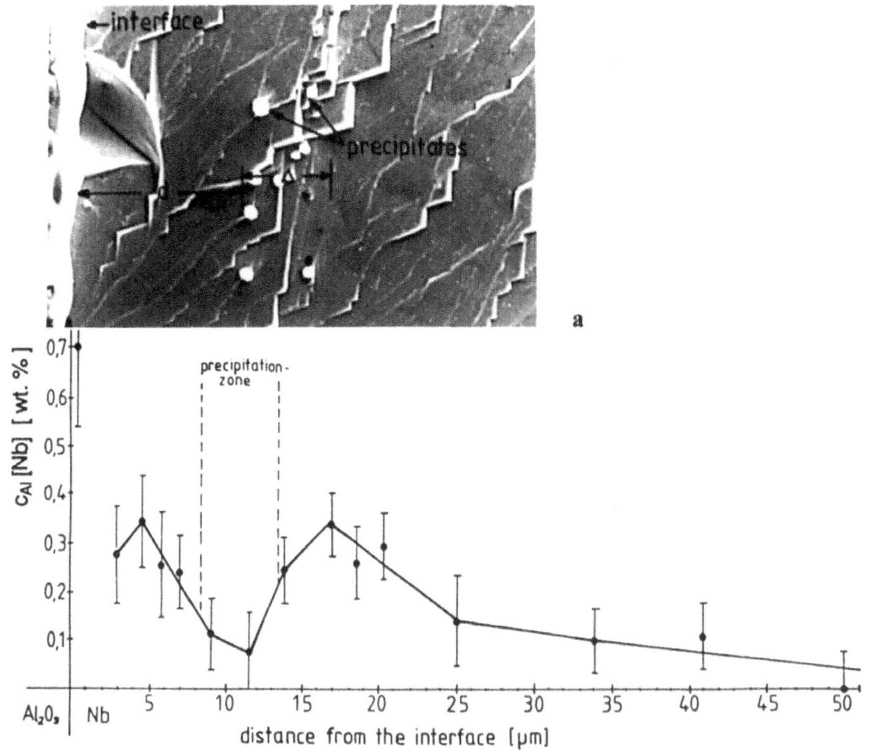

Fig. 4. **a** SEM picture of the near interface region of niobium of a slowly cooled bond. Precipitates are formed which can easily be identified. **b** Concentration profile $c_{Al}[Nb]$ of a slowly cooled Nb/Al_2O_3-bond

The decrease of $c_{Al}[Nb]$ for very small distances from the interface suggests that during "fast" cooling a back diffusion of Al (and also of O) occurs which leads to a recondensation of Al_2O_3 at the interface. Experimental studies show [11, 14, 16] that only Al_2O_3 is present at the interface, and no other phase could be identified. This finding supports the validity of assumptions mentioned above. The amount of condensed Al_2O_3 also can be evaluated from the concentration profile, which results in a layer of 13 nm thickness.

"Slow" cooling (20 K/min) leads to completely different observations. Figure 4 shows a SEM micrograph of a cross section of regions near the interface perpendicular to the image plane (cross section). The interface is at the left side of the micrograph. The near-interface part of Nb is shown which was chemically etched and small precipitates became clearly visible at distances between 8 and 14 μm from the interface. An evaluation of $c_{Al}[Nb]$ profiles (Fig. 4b) does not show a constant value for certain distance intervals as it does for fast cooling (c.f. Fig 3). Close to the precipitation zone $c_{Al}[Nb]$ is small since most of the dissolved Al precipitated in the particles. The maximum value of Al concentration, $c_{Al}[Nb]$, is about a factor of 2 smaller than in the fast cooled material. This indicates that during diffusion bonding and cooling different processes occurred. During "slow" cooling a nucleation and growth process of precipitates was possible. The chemical composition of the precipitates was obtained by EDS and EELS. Diffraction studies showed that the particles precipitated in the structure of metastable θ-Al_2O_3 [30–32].

High-Resolution Electron Microscopy Studies

Comments on Direct Lattice Imaging of Distorted Materials Using HREM

The geometric beam path through the objective lens of a TEM is shown in Fig. 5a. Beams from the lower side of the object travel both in the direction of the incident and diffracted beams. All beams are focused by the objective lens in the back focal plane to form the diffraction pattern. In the image plane, the image of the object is produced by interference of the incident and diffracted beams. Fig. 5b uses wave optics to describe physical processes which contribute to the image formation. From the lower side of the foil, a wave field emerges that can be described by a transmission function $q(x, y)$. For an undistorted lattice, $q(x, y)$ represents a simple, sine-shaped amplitude and intensity distribution. The transmission function of a complex lattice with a large periodicity length is very complicated and $q(x, y)$ is a non-periodic (or "quasi-periodic") function for the distorted region of a crystal.

The intensity distribution in the diffraction pattern is given by the Fourier transformation $Q(u, v)$ of the transmission function $q(x, y)$, where u, v are the coordinates in the diffraction plane (reciprocal space). Since spherical aberration cannot be avoided with rotationally symmetrical electromagnetic lenses [27, 28], the beams emerging from an object at a certain angle (Fig. 5a) are subjected to a phase shift relative to the incident beam. Imaging under a slightly defocussing mode, Δf, leads also to a phase shift which depends on the sign and magnitude of the defocus value, Δf. The influence of the lens errors and the defocus on the amplitude of the diffraction patterns can be described by the contrast transfer function $\chi(u, v)$. The

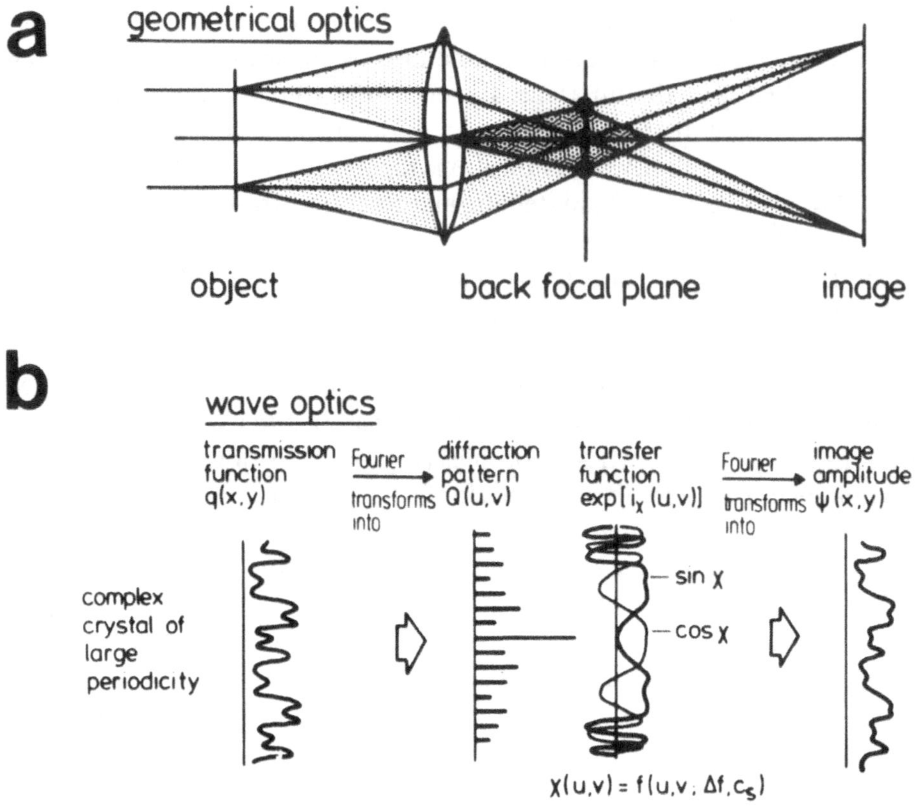

Fig. 5. Image formation by the objective lens of a transmission electron microscope. **a** Geometrical optical path diagram. **b** Wave optical description (see text for explanation)

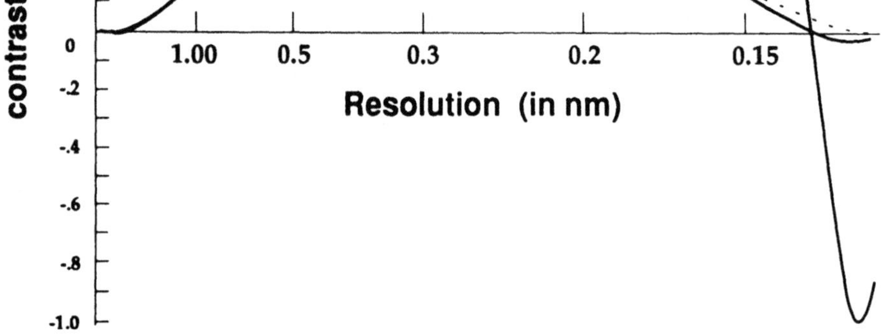

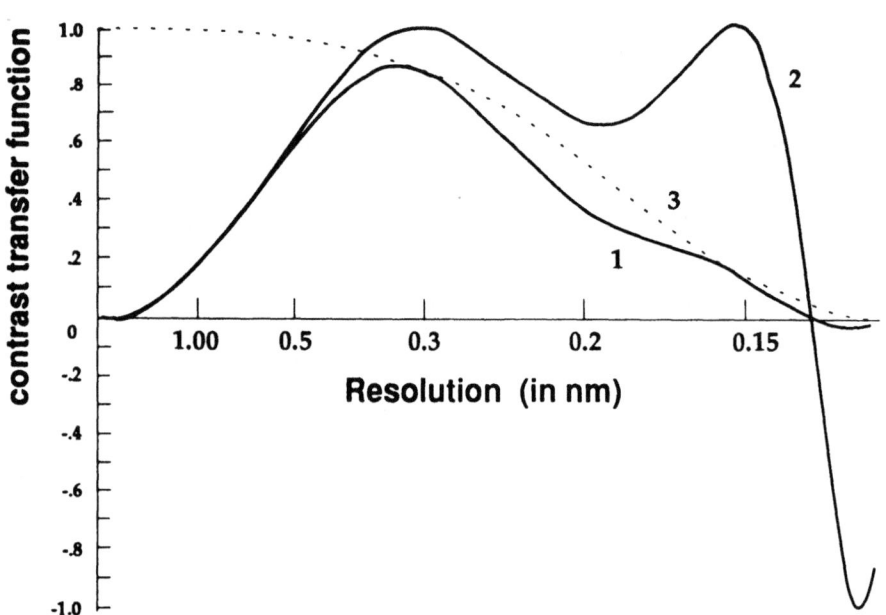

Fig. 6. Actual calculated contrast transfer function (CTF). (1) undamped CTF, (2) and damping envelope, (3) of the atomic resolution microscope (ARM) at the Center for Electron Microscopy, Berkeley, CA. Operating conditions: voltage 800 kV, defocus −55 nm

contrast transfer function as the function of reciprocal distance ("space frequency") for the instrument used, is shown in Fig. 6.

The image is formed by a second Fourier transformation of the amplitude distribution in the diffraction pattern multiplied by the contrast transfer function. The amplitude in the image plane, $\psi(x, y)$, is therefore not identical to the wave field in the object plane (transmission function $q(x, y)$). The image is severely modified if scattering to large angle occurs, since the influence of the spherical aberration increases strongly with increasing scattering angle. The modification is most severe if the components in the diffraction pattern coincide with the oscillating part of the contrast transfer function (large values of u and v). If, however, the wave vectors lie within the first wide maximum of the contrast transfer function $\chi(u, v)$, it can be assumed that characteristic features and properties of the object can be directly recognized in the image [27, 28].

From these considerations it can be concluded that the objective lens of a high-resolution instrument acts as a band pass filter and tuning in on a specific range of spatial frequencies to form a sensitive interference pattern. By using a suitable low-index zone axis and choosing an optimum range of specimen thicknesses, the patterns can be extremely sensitive to the strength and periodicity of the projected potentials. However, the results also depend critically on other experimental parameters including specimen tilts and strains. It is customary to establish the optimum experimental conditions from a sequence of image simulations based on the anticipated structure and chemistry of the systems under investigation. This type of pretuning may, however, reduce the sensitivity of the interference pattern to unforeseen structural effects which might easily be overlooked in the image.

HREM represents a coherent imaging process in which the image is reconstructed from a set of diffracted beams. The phases of these beams are sensitive to objective lens defocus and specimen thickness so that the image can take many different forms. This is demonstrated in Fig. 7 where simulated phase contrast images for Nb are shown for different thicknesses and defocus values.

A flow chart for quantitative HREM is shown in Fig. 8. It should be emphasized that hitherto the comparison between simulated and experimentally obtained images is done by simple visualization.

The analysis of interfaces requires that the lattice to be investigated is periodic in the direction in the incident beam on both sides of the interface so that the projected potential on the lower surface is formed by the superposition of identical atoms that lie one above the other in the direction of the incident beam. Furthermore, the interface has to be parallel to the incident beam.

Structural Studies at Nb–Al_2O_3 Interfaces

The theoretical and experimental study of metal/ceramic interfaces is rather complex since usually metal/ceramic interface structures are incommensurate. Several features can be evaluated by HREM for a "perfect" interface (Fig. 9). They represent quasi-periodic structures and in cases where no atomistic relaxations occur (formation of misfit dislocation), no periodic structure can be identified at the interface. This would indeed complicate the evaluation of HREM images and restrict it to a rather low accuracy since the image simulation of those high-resolution images

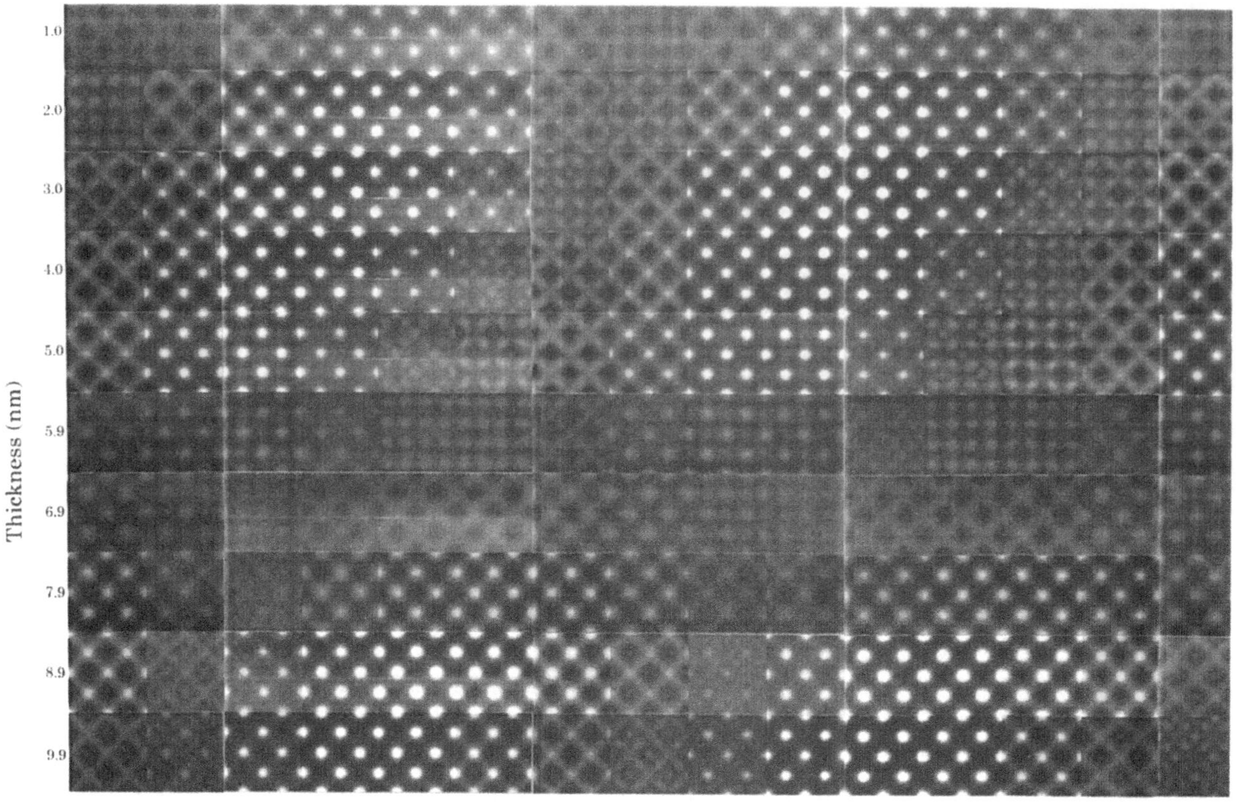

Fig. 7. HREM phase contrast simulated images for Nb [100]. The acceleration voltage of the HREM is 400 kV and the objective lens spherical aberration coefficient, Cs, is 1 mm (Scherzer resolution limit 0.17 nm). The dependency of the contrast on thickness and defocus is clearly demonstrated

would be extremely difficult. Fortunately, metal/ceramic systems often possess a small misfit and relaxations occur, hence regions with good matching alternate with regions of poor matching.

Another important question for an understanding of the properties of a material is the terminating plane of the ceramic (oxide). For example, the basal plane of Al_2O_3, consists of alternating planes between oxygen and aluminum ions. Bonding at the interface strongly depends on the nature of the terminating plane which, in turn, depends on the activities of the components of the ceramic material in the metal [13, 17, 37]. For Nb–Al_2O_3 interfaces, the nature of the terminating plane depends on the ratio of the activities, a_{Al}/a_O, in Nb. This observation was also verified for Pd/Al_2O_3 interfaces [33, 34]. Finally, atomistic relaxation may occur which leads to a certain translation stage between the metal and the ceramic where different planes adjacent to the ceramic may have distances different from the lattice plane distances in the bulk material. Single atom columns may also relax resulting in a very complex "perfect" interface structure.

In addition to this perfect interface, defects may also occur at the interfaces as summarized in Fig. 10. These defects can be split into chemical and structural defects. The chemical defects exist of localized defects such as impurities, segregated at the interface and chemical gradients resulting from chemical reactions (c.f. Fig. 3, 4). At elevated temperatures reactions may occur between metals and ceramics

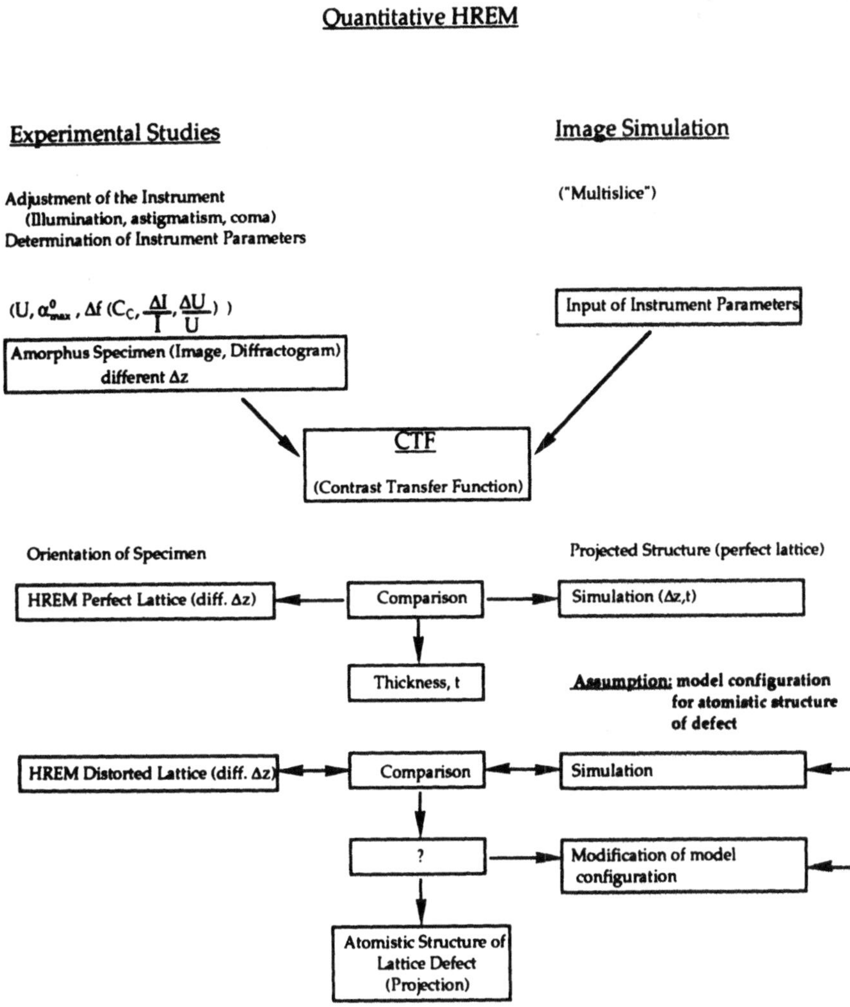

Fig. 8. Flow chart demonstrating schematically quantitative HREM (see text for explanation)

Crystallography
- incommensurate structures
- quasiperiodic structures

Terminating plane of ceramic
- anions or cations?
- dependancy on activities of ceramic
- components in metal

Atomic Relaxations (HREM, X-ray diffraction)
- planes
- atoms

Fig. 9. Features which can be evaluated by HREM on "perfect" metal/ceramic interfaces

resulting in new phases which may lead to more than one interface [35, 36]. Under certain conditions, morphological instability of interfaces may appear [37], which means that small fluctuations which exist at an interface, increase in amplitude during annealing at high temperatures with increasing periods of time. In addition, structural defects often form at metal/ceramic interface. They may exist of steps,

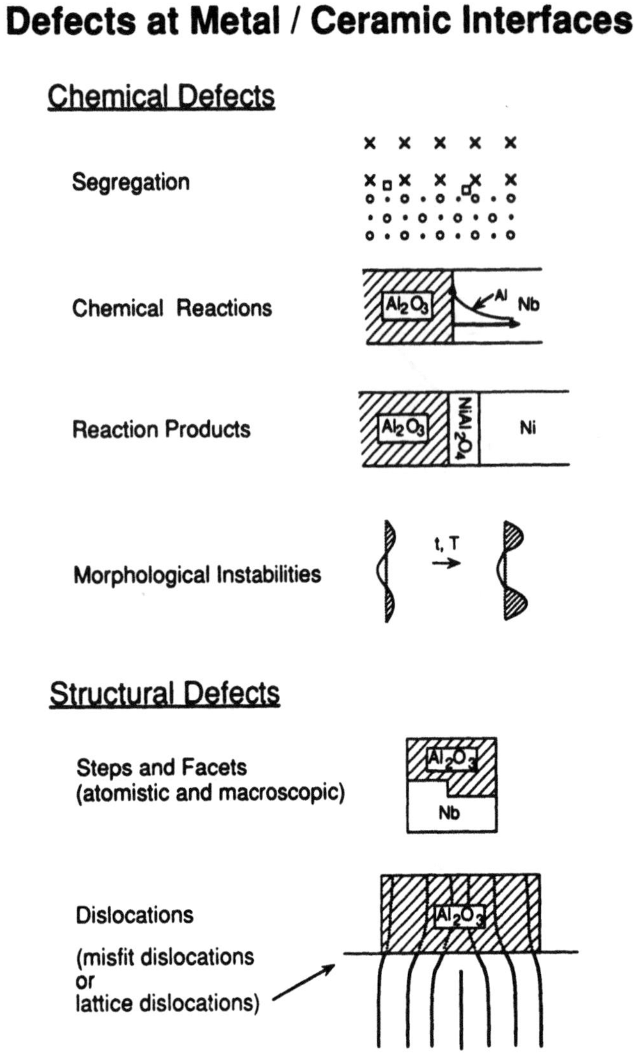

Fig. 10. Schematic presentation of possible defects at metal/ceramic interfaces

which can be atomistic or microscopic, and of dislocations, which can be misfit dislocations as well as lattice dislocations.

High-Resolution Studies

HREM studies were performed at Nb–Al_2O_3 interfaces grown by various techniques. In these studies results are reported for Nb-films (thickness: ~500 nm) and a sputter-cleaned (0001)-plane of Al_2O_3. The HREM studies were performed on the atomic resolution microscope (ARM) at the National Center for Electron Microscopy, Berkeley, CA, USA. The microscope was operated at 800 kV. The point-to-point resolution of the instrument is in the range of 0.17 nm. Image simulations were performed using Stadelmann's program [38]. The contrast transfer function of the ARM is shown in Fig. 6. The comparison between experimentally obtained and simulated images was performed by optical inspection.

Diffraction studies reveal that for this condition the (111)-plane of the Nb is parallel to the basal plane of the Al with $[1\bar{1}0]_{Nb} \parallel [2\bar{1}\bar{1}0]_S$ (direction A) or along $[1\bar{2}1]_{Nb} \parallel [10\bar{1}0]_S$ (direction B) [39]. HREM micrographs were taken from the same

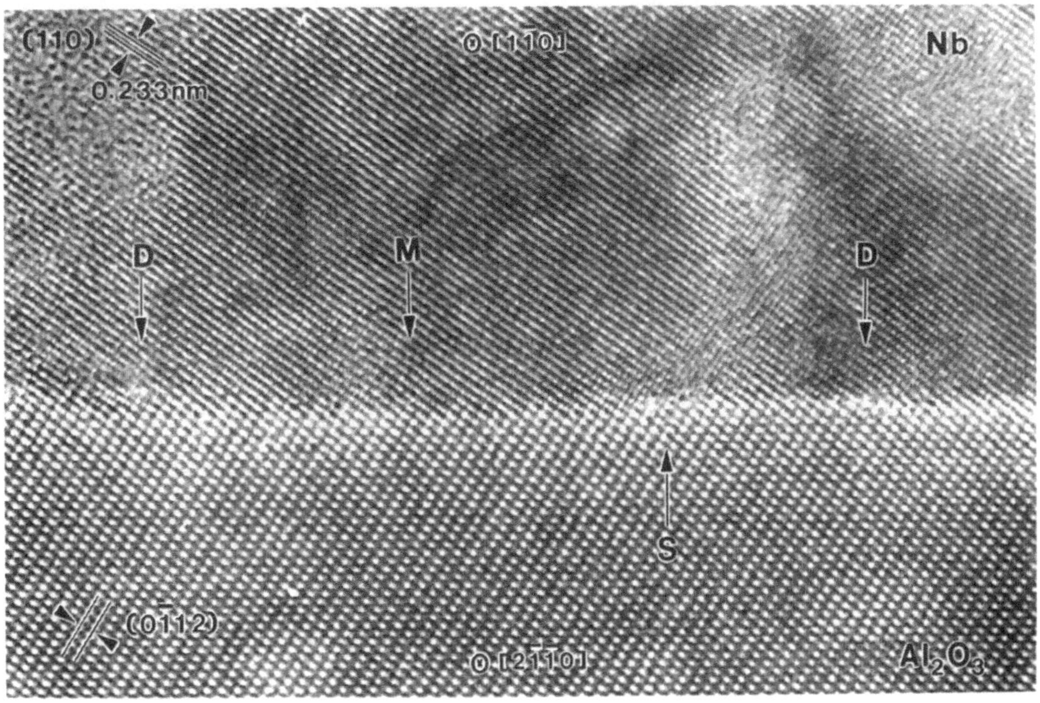

Fig. 11. High resolution image of a Nb/Al_2O_3 interface. Lattice planes can be clearly identified in both sapphire and Nb. Foil thickness is 7 nm. At the interface, regions of good matching (M) and poor matching (D) alternate. S: Step in the substrate

interface area in both directions A and B by simply tilting the specimen inside the ARM.

Figure 11 shows an overview of a large area of a near interface region. The defocus of the objective lens is slightly more positive ($\Delta f = -40$ nm) than the Scherzer defocus ($\Delta f = -55$ nm). Lattice planes can be readily identified in Nb and Al_2O_3. The foil thicknesses of Nb and Al_2O_3 are the same. In Nb, regions of good matching (M) and poor matching (D) alternate at the interface. Steps can also be identified (S). The regions of poor matching (D) cover less than 10% of the interface plane and can be described as slightly delocalized misfit dislocations [39]. This paper deals only with the regions of good matching, which cover more than 90% of the interfacial area.

The region of good matching (Fig. 11, M) is imaged at a higher magnification in Fig. 12. Figure 12a shows the interface with the electron beam parallel to direction A. Lattice planes transfer continuously from Nb to Al_2O_3. Fig. 12b is a micrograph of the same interface viewed along orientation B. Only $(10\bar{1})_{Nb}$ lattice planes with a spacing of 0.233 nm are visible in the Nb crystal. The (222) lattice planes (perpendicular to $(10\bar{1})$) possess a spacing of 0.095 nm which is beyond the information limit of the ARM. In both orientations (Fig. 12a,b) a perfect match of the Nb- and Al_2O_3-lattice at the interface is visible. The lattice mismatch of the $(0110)_S$ and $(11\bar{2})_{Nb}$ planes, which are perpendicular to the interface, is only $\sim 1.9\%$ and this misfit is accommodated by localized defects (misfit dislocations) in the regions of poor matching. This allows the Nb-lattice in between these defects to expand slightly along the interface (the Nb-lattice possesses the smaller lattice plane spacing),

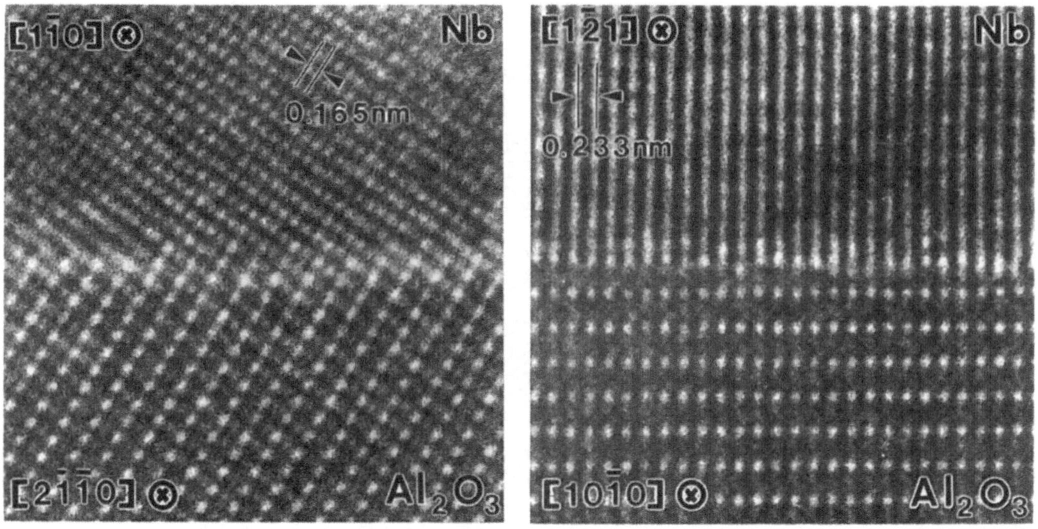

a b

Fig. 12. High resolution image of Nb/Al_2O_3 interface. Region of good matching (M). **a** Direction A with $[1\bar{1}0]_{Nb} \| [2\bar{1}\bar{1}0]_S$, **b** Direction B with $[1\bar{2}1]_{Nb} \| [10\bar{1}0]_S$. Note that contrast fringes cross the interfaces without steps or interphases

resulting in extended regions of perfect matching. The expansion of the lattice plane spacing parallel to the interface is limited to regions close to the interface. The lattice planes of the Nb, especially near the misfit dislocations, are bent resulting in a continuous transition to the undistorted Nb-lattice further away from the interface (this can be seen by viewing Fig. 11 under grazing incidence) [39]. No dislocations or lattice distortions could be seen in the Al_2O_3-lattice.

Quantitative HREM requires computer simulation. The atomistic configuration is obtained if experimental images are identical to images simulated for specific atomistic model configurations. The analysis requires the knowledge of the exact focusing value and the foil thickness [27]. The determination of the foil thickness is most accurate if observed images of the Nb and Al_2O_3 are compared to presentations of calculated lattice images for varying thicknesses and focus values for both Nb and Al_2O_3. Fig. 13 shows that good agreement between the simulated images and the experimental image is obtained for a defocus of $\Delta f = -40$ nm and a foil thickness of $t = 7$ nm in Nb and Al_2O_3. From the simulated images it is also possible to determine the positions of the atoms in both crystals with respect to the intensity distribution in the experimental image (Fig. 13).

The next step is to identify the translational state $\underline{T}$ of the two crystals with respect to each other. Such a translational state only exists in the areas of good matching because the two crystals become commensurate at the interface by expanding the lattice plane spacing of the Nb. The vector $\underline{T}(T_1, T_2, T_3)$ can be constructed so that the components T_1 and T_2 lie within the interface plane and T_3 is perpendicular to it.

An inspection of the HREM images (Figs. 11–13) reveals that certain lattice fringes in Nb and Al_2O_3 transfer continuously into each other across the interface. Furthermore, the positions of the atoms with respect to these lattice fringes in the experimental images are known (Fig. 13). If we assume that the lattice continues

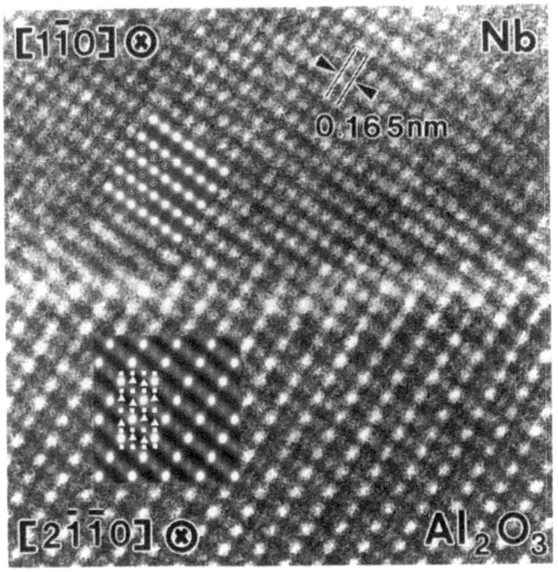

Fig. 13. HREM micrograph of Fig. 12 at higher magnification. The inserts represent simulated images for a specimen thickness of $t = 7$ nm and a defocus value of $\Delta f = -40$ nm. The atom positions are marked. Nb: stars, Al: white triangles, O: black on white dots

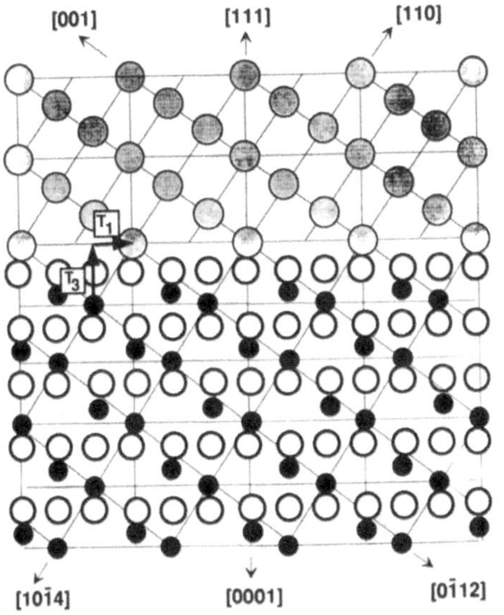

Fig. 14. Schematic drawing of the atomic positions at the Nb/Al_2O_3 interface. The result is obtained by quantitative HREM. O^{2-} ions: light large circles, Al^{3+}: black circles, Nb atoms: grey circles. Note the approximate parallelity of $[110]_{Nb}$ to $[10\bar{1}4]_S$ planes and $[001]_{Nb}$ to $[0\bar{1}\bar{1}2]_S$ plane

undisturbed up to the interface, then the atomistic structure represented in Fig. 14 and 15 is obtained: the Al_2O_3 is terminated by an oxygen layer and the Nb atoms of the first Nb layer fit accurately in the A sites (Fig. 15) under which no Al ions are positioned. An obvious choice for the origin of the translation vector $\underline{T}$ is given by one of the Al-sublattices as indicated in Figs. 14 and 15. The first Nb layer has exactly the same threefold symmetry and the same atomic distances as the individual layers, parallel to the interface. From this T_1, T_2 and T_3 result in:

$$T_1 = a_0\sqrt{3}/6 = 0.137 \text{ nm}, \qquad T_2 = a_0/2 = 0.238 \text{ nm}, \qquad T_3 = c_0/6 = 0.216 \text{ nm}$$

where a_0 and c_0 are the lattice constants of sapphire. The values of T_1, T_2, T_3 were calculated according to the model shown in Fig. 14–15, not taking into account an experimental error of ± 0.02 nm in determining the location of the lattice planes from the high resolution images.

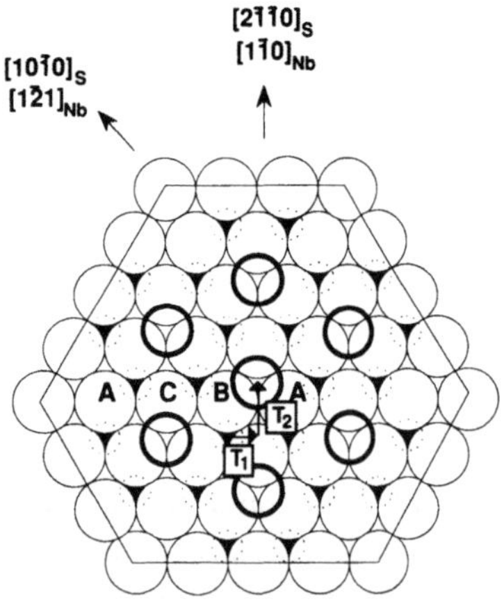

Fig. 15. Schematic drawing of a non-reconstructed $(0001)_S$ basal surface. Symbol for atom (ion) columns: O^{2-} ions: light small squares, Al^{3+}: light triangles, Nb atoms: circles. It is assumed that the O^{2-} ions form the outermost layer. The Nb atoms of the first Nb layer are positioned above the empty sites of the first Al^{3+} layer

Qualitatively, the Nb film formation on $(0001)_S$ substrates was described in [38]. Steps exist at the sapphire surface which may act as nucleation centers for the Nb film. The growth conditions are selected such that two-dimensional crystal growth (ledge growth) occurs. The initially reconstructed, Al-rich surface of the sapphire [40], still possesses 3-fold symmetry with characteristic distances equivalent to atomistic distances of $(111)_{Nb}$ planes. For symmetry reasons it is expected that the first deposited monolayer of Nb atoms has an orientation relationship which is dictated by the symmetry of the sapphire surface leading to $(111)_{Nb} \| (0001)_S$. Our experiments show no evidence that the surface reconstruction is still present at our Nb/Al_2O_3 interface. From this we conclude that a transformation back to the bulk structure occurs during growth of the Nb film, by this presumably lowering the energy of the interface.

No twins could be identified in the (nearly) perfect Nb film. It was demonstrated that a unique atomistic relationship exists between the sapphire surface and the monoatomic Nb layer. There exists only one set of Nb atom positions on the unreconstructed sapphire surface which leads to a twin free film (Fig. 14, 15). Nb atoms must be located on the terminating O^{2-} layer of sapphire on top of "empty sites." The quantitative evaluation of the HREM images verified this hypothesis. In a concluding step, experimental HREM images of regions which included the interface have to be compared to corresponding simulated images.

Summary

AEM and HREM reveal important quantities which are necessary in understanding the structure and chemistry of interfaces. In this paper, results were reported with special emphasis on Nb/Al_2O_3 interfaces. Chemical gradients were identified as well as the atomistic arrangement at or near the interface. The spatial resolution of AEM studies is mainly dictated by the probe geometry, while HREM results are limited by the resolution of the instrument. Valuable information can be obtained by the application of various techniques at the same interface. This information is important for understanding the properties of the material.

Acknowledgements. This work was supported in part by the Bundesministerium für Forschung und Technologie, BMFT contract NTS-02300. The author gratefully acknowledges helpful discussions with Dr. Joachim Mayer who performed the HREM studies at the interfaces. Furthermore, the use of the ARM at the National Center for Electron Microscopy in Berkeley, CA, is acknowledged.

References

[1] A. G. Evans, *J. Am. Ceram. Soc.*, **1990**, *73*, 187.
[2] M. Rühle, A. G. Evans, *Prog. Mat. Sci.*, **1989**, *33*, 85.
[3] S. Suresh, A. Needleman (eds.), *Interfacial Phenomena in Composites: Processing, Characterization and Mechanical Properties*, Elsevier, London, New York, 1989.
[4] R. G. Brandt, S. G. Fishman, J. S. Murday, Y. Rajapakse, A. K. Vasudevan, K. J. Wynne (eds.), *Science of Composite Interfaces*, Elsevier, London, New York, 1990.
[5] G. Elssner, G. Petzow, *ISIJ International*, **1990**, *30*, p. 1011.
[6] M. Doyama, S. Somiya, R. P. H. Chang (eds.), *Proc. Mat. Res. Int. Meetings*, 1988, 8.
[7] N. Birks, G. H. Meier, *Introduction to High Temperature Oxidation in Metals*, Edward Arnold, London, 1983.
[8] W. E. King (ed.), The Reactive Element Effect on High Temperature Oxidation-After Fifty Years, *Materials Science Forum*, 1989, 43.
[9] E. A. Giess, K.-N. Tu, D. R. Uhlmann (eds.), Electronic Packaging Materials Science, *Mat. Res. Soc. Symp. Proc.*, 1985, 40.
[10] K. A. Jackson, R. C. Pohanka, D. R. Uhlmann, D. R. Ulrich (eds.), Electronic Packaging Materials Science II, *Mat. Res. Soc. Symp. Proc.*, 1986, 72.
[11] M. Florjancic, W. Mader, M. Rühle, M. Turwitt, *J. de Physique* **1985**, *46*, C4.
[12] W. Mader, *Z. Metallkunde* **1989**, *80*, 139.
[13] M. Rühle, A. G. Evans, *Mat. Res. Soc. Symp. Proc.* **1988**, *120*, 293.
[14] W. Mader, M. Rühle, *Acta Metall.* **1988**, *37*, 253.
[15] M. Kuwabara, J. C. H. Spence, M. Rühle, *J. Mat. Res.* **1989**, *4*, 972.
[16] D. Knauss, W. Mader, *Ultramicroscopy* (in press.)
[17] M. Rühle, A. G. Evans, M. F. Ashby, J. P. Hirth (eds.), Metal-Ceramic Interfaces, Proc. Intern. Workshop, Santa Barbara, CA, *Acta-Scripta Metall. Proc. Series, Vol. 4*, Pergamon Press, Oxford, 1990.
[18] Y. Ishida (ed.), *Fundamentals of Diffusion Bonding*, Elsevier, Amsterdam, Oxford, New York, Tokyo, 1987.
[19] C. Wagner, *Z. Elektrochem.* **1959**, *63*, 772.
[20] J. L. Meijering, Internal Oxidation in Alloys, in: *Advances in Materials Research, Vol. 3*, (H. Hermann, ed.) Wiley, New York, 1971.
[21] W. Mader, *Mat. Res. Symp. Proc.* **1987**, *82*, 403.
[22] S. M. Durbin, J. E. Cunningham, M. E. Mochel, C. P. Flynn, *J. Phys. F* **1981**, *11*, L223.
[23] S. M. Durbin, J. E. Cunningham, C. P. Flynn, *J. Phys. F* **1982**, *12*, L75.
[24] P. Hirsch, A. Howie, R. B. Nicholson, D. W. Pashley, M. J. Whelan, *Electron Microscopy of Thin Crystals*, Robert E. Krieger, Huntington, New York, 1965.
[25] D. C. Joy, A. D. Romig Jr., J. I. Goldstein, *Principles of Analytical Electron Microscopy*, Plenum, New York, London, 1986.
[26] R. F. Egerton, *Electron Energy Loss Spectroscopy in the Electron Microscope*, Plenum Press, New York, London, 1986.

[27] J. C. H. Spence, *Experimental High-Resolution Electron Microscopy*, Oxford University Press, Oxford 1987.
[28] P. Buseck, J. Cowley, L. Eyring, *High-Resolution Transmission Electron Microscopy and Associated Techniques*, Oxford University Press, Oxford, 1988.
[29] A. Strecker, U. Salzberger, J. Mayer, *TEM Specimen Preparation: A New Method to Produce Cross-Sectioned Specimens of Brittle Materials*, (to be published).
[30] K. Burger, M. Rühle, *Ceram. Eng. Sci. Proc.* **1989**, *10*, 1549.
[31] K. Burger, *Doctoral Thesis*, Universität Stuttgart, 1987.
[32] K. Burger, W. Mader, M. Rühle, *Ultramicroscopy* **1987**, *12*, 1
[33] X. Y. Huang, W. Mader, J. A. Eastman, R. Kirchheim, *Scripta Metall.* **1988**, *22*, 1109.
[34] T. Muschik, M. Rühle, *Phil. Mag.* (in press.)
[35] K. P. Trumble, M. Rühle, *Acta Metall.* **1991**, *39*, 1924.
[36] K. P. Trumble, M. Rühle, *Z. Metallkunde* **1990**, *81*, 749.
[37] M. Backhaus-Ricoult, *Ber. Bunsenges. Phys. Chem.* **1980**, *80*, 684.
[38] P. A. Stadelmann, *Ultramicroscopy* **1987**, *21*, 131.
[39] J. Mayer, C. P. Flynn, M. Rühle, *Ultramicroscopy* **1990**, *33*, 51.
[40] T. M. French, S. A. Somorjai, *J. Phys. Chem.* **1970**, *74*, 2409.

Mikrochim. Acta (1992) [Suppl.] 12: 93–97

Quantitative Electron Probe Microanalysis of Multi-layer Structures

Guillaume F. Bastin, Johannes M. Dijkstra,* Hans J. M. Heijligers, and Dick Klepper

Laboratory of Solid State Chemistry and Materials Science, University of Technology Eindhoven, P.O. Box 513, NL-5600 MB Eindhoven, The Netherlands

Abstract. A method is discussed to use the basics of quantitative electron probe microanalysis (EPMA) programs for the analysis of thin films and multi-layer structures. The program described yields good results for both simple cases like a single film on a substrate, as well as for multi-layer systems and for the determination of concentration profiles in depth.

Key words: EPMA, multi-layers, thin-film-analysis.

The idea to use an electron probe microanalyzer for the analysis of thin films is nearly as old as the instrument itself. For many years, however, the computer programs, necessary to convert the measured intensity ratios into film thicknesses and compositions were lacking.

For the determination of the thickness and the composition of thin films an accurate knowledge of X-ray ionisation (Φ) versus mass-depth (ρz) must be available. Unfortunately, these so-called $\Phi(\rho z)$ models were first developed in the early eighties, because then enough experimental information had been gathered to base these models on.

Compared to conventional ZAF models, which are invariably based on separate calculations of the atomic number effect (Z) and the absorption effect (A), and which use an unrealistic description of the $\Phi(\rho z)$-curve for absorption purposes only, these new models have already shown highly improved performance under extreme conditions like very low and very high overvoltages, and for heavy absorption. This makes these programs especially capable of analyzing ultra-light elements (boron to oxygen) [1].

The Gaussian $\Phi(\rho z)$ Model

One of the more successful $\Phi(\rho z)$ models for bulk specimen analysis is the surface-centred Gaussian $\Phi(\rho z)$ model, originally developed by Packwood and Brown [2].

---

* To whom correspondence should be addressed

They discovered that most of the measured $\Phi(\rho z)$-curves can be fitted with an equation of the type:

$$\Phi(\rho z) = \{\gamma - (\gamma - \Phi_0)\cdot\exp[-\beta\rho z]\}\cdot\exp[-\alpha^2(\rho z)^2],$$

in which Φ stands for the number of X-ray ionisations, ρz is the mass thickness, and α, β, γ and Φ_0 are parameters, dependent on the elements involved like atomic number, atomic weight, etc and on the accelerating voltage of the impinging electrons [3]. Integration of this formula from zero to infinity yields the intensity of the X-rays generated in the specimen. Multiplication of the $\Phi(\rho z)$-curve by the absorption factor $\exp(-\mu/\rho\cdot\mathrm{cosec}(\varphi)\cdot\rho z)$, in which μ/ρ is the mass-absorption-coefficient and φ is the X-ray take-off angle, followed by integration from zero to infinity, yields the intensity emitted by the specimen. We have extended this model to make thin film analysis possible.

Thin Film Analysis

In the case of a single film with mass-thickness T on a substrate the basic idea of thin-film analysis is to simply integrate the $\Phi(\rho z)$-curve of the film elements from zero to T, and to integrate the $\Phi(\rho z)$-curve of the substrate elements from T to infinity. For the emitted intensities, corrections have to be made for both self-absorption in an element's own layer and for absorption of the substrate element's X-rays on the path to the surface. This works very well for elements with nearly the same atomic weight, like for an aluminium film on a silicon substrate, assuming again that the $\Phi(\rho z)$-curve is very close to reality.

However, with increasing differences in atomic number, the changing backscattering characteristics of the substrate become more and more prominent and will have to be taken into account. The following assumptions have been made to deal with this problem:

1) The ρz-scale in each $\Phi(\rho z)$-curve is continuous across the interface between film and substrate.

2) The $\Phi(\rho z)$-curves of the elements in the film are affected by the substrate, those of the substrate elements are however not influenced by the nature of the film elements.

3) The Gaussian parameters α, β, γ and Φ_0 for any film element have a value in between two extremes. For an infinitely thin film the parameters can be approximated by assuming that the elements are solved in the substrate, whereas for an infinitely thick film the bulk parameters will have to be approached. In all intermediate cases the parameters can be calculated as follows:

$$\alpha = \alpha_b + (\alpha_s - \alpha_b)\cdot\exp(-1.5\cdot(\alpha_b\cdot T\cdot 10^{-6})^2)$$

$$\beta = \beta_b + (\beta_s - \beta_b)\cdot\exp(-1.5\cdot\alpha_b\cdot T\cdot 10^{-6})$$

$$\gamma = \gamma_b + (\gamma_s - \gamma_b)\cdot\exp(-2.0\cdot(\alpha_b\cdot T\cdot 10^{-6})^2)$$

$$\varphi_o = \varphi_{ob} + (\varphi_{os} - \varphi_{ob})\cdot\exp(-2.5\cdot(\alpha_b\cdot T\cdot 10^{-6})^2)$$

where T is the layer thickness in $\mu g/cm^2$ and the indices s and b indicate the solved and bulk parameters respectively.

This empirical weighting procedure ensures a smooth variation of the parameters and takes into account the different effects of the substrate on each of those parameters.

In the case of a multi-layered specimen a more complex method has to be followed, by taking into account the distance from each film to the film under observation.

Once the four parameters have been found for each element in each layer the generated and emitted intensities can again be calculated by partial integration of the $\Phi(\rho z)$-curves.

Structure of the Programs

The most basic option in the program is the calculation of emitted intensities from a given layered sample. In addition there is a possibility to perform calculations at different accelerating voltages or with slowly varying layer thicknesses to show the sensitivity of the k-ratios to these parameters, thus allowing the operator to choose the optimum measuring conditions. If each element is only present in one single layer, an iterative procedure is also possible, allowing the computer to calculate the composition and the thickness of each layer from a measured set of k-ratios at a fixed accelerating voltage.

In more complex cases, measurements at a few different accelerating voltages have to the made. One can start measuring at a very low voltage to excite only the first layer(s), and by increasing the voltage, information from the buried layers can be gathered. From all this information one can thus determine all layer thicknesses and compositions of the specimen in a trial and error procedure. In this way concentration profiles in depth can also be determined.

Some Applications

Two examples are discussed which illustrate the capabilities of the analysis method described. In example 1 the measurements of two platinum films (thicknesses of 84 and 519 $\mu g/cm^2$, or 39.2 and 242.0 nm) on substrates of silicon and gold [4] were analyzed. Figure 1 shows that the predictions of our program agree very well with the measurements, indicating that the influence of the atomic number of the substrate is properly taken into account.

In example 2 we have a monocrystalline silicon substrate, covered by a titanium layer, with an amorphous silicon film on top. Results of the measurements are shown in Fig. 2. The program predicted mass-thicknesses for the silicon and the titanium films of 19 and 17 $\mu g/cm^2$. Analysis with a transmission electron microscope revealed linear thicknesses of 74 and 41 nm respectively (Figure 3). Assuming densities of 2.5 g/cm^3 and 4.3 g/cm^3 for silicon and titanium respectively this would result in mass-thicknesses of 18.5 and 17.6 $\mu g/cm^2$ respectively. This brings us to one of the problems of this analysis method: it is very hard to verify your conclusions with other techniques like direct imaging in TEM, because these techniques usually yield the linear thickness of a film, whereas the method described yields the mass-thickness of the film. The density of the film is the link between them, however quite often densities are not known within ten percent, especially when dealing with CVD-thin

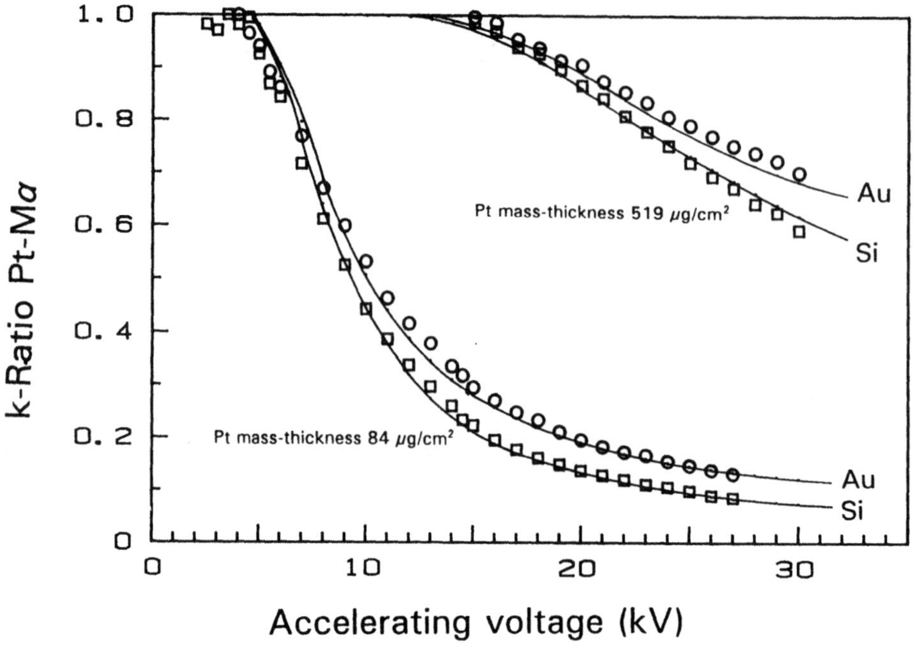

Fig. 1. Agreement between the predictions of our program and the experimental data of Reuter et al. [4]. (Example 1.) X-ray take-off angle is 52.5 degrees. (○) Pt on Au-substrate; (□) Pt on Si-substrate

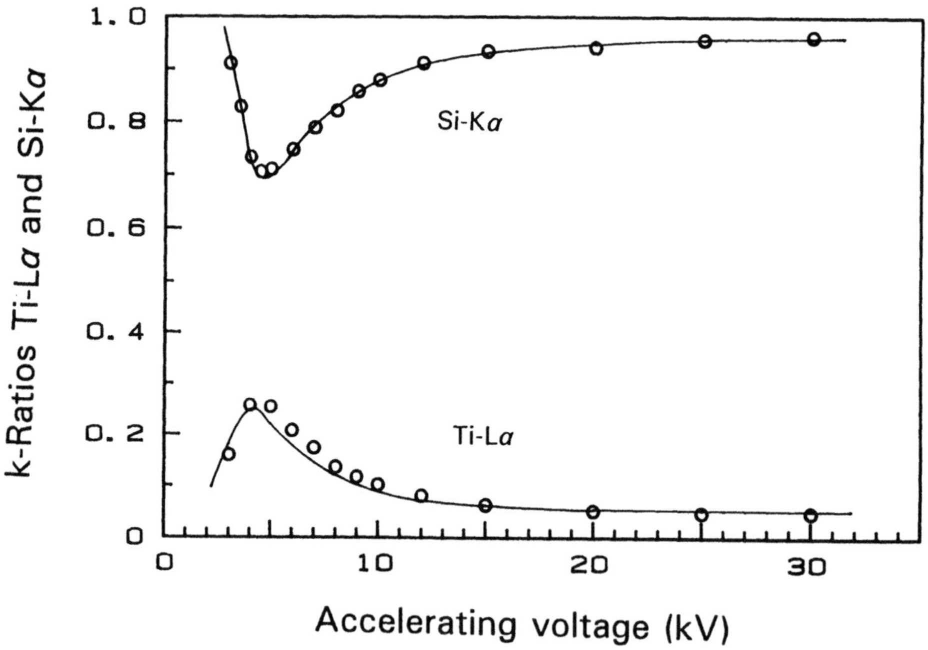

Fig. 2. Agreement between measured and calculated data for silicon-titanium-silicon sample. (Example 2.) X-ray take-off angle is 40.0 degrees

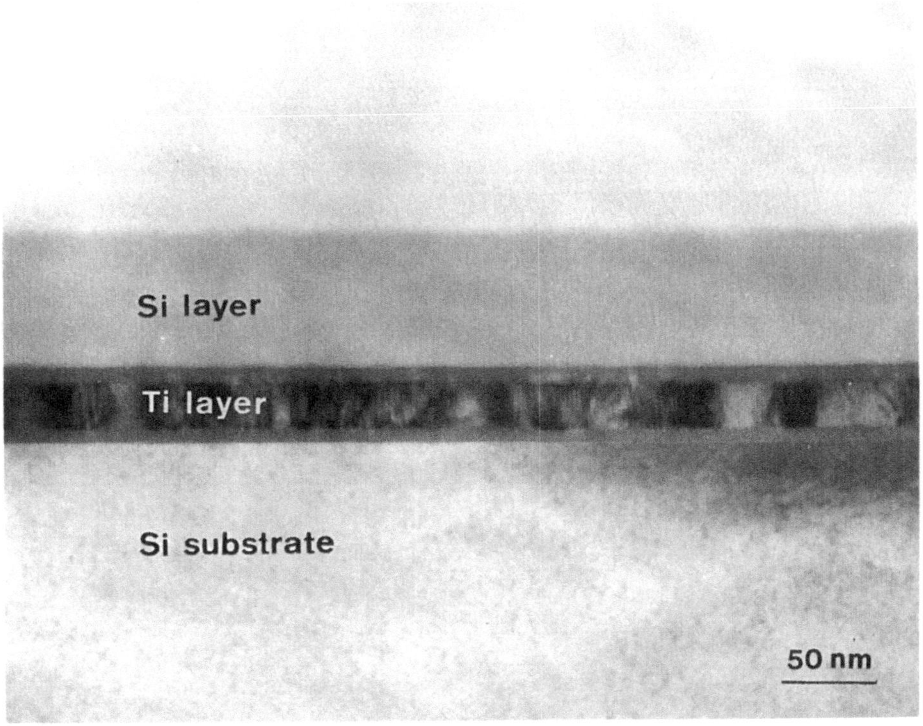

Fig. 3. TEM micrograph of silicon-titanium-silicon sample. (Example 2.) Note the different phases of titanium

films. In more complex cases, like the analysis of concentration profiles, the situation can be even more difficult because the density might have a gradient.

Conclusion

The analysis of thin films and multi-layers with an electron probe microanalyzer or a scanning electron microscope is an elegant and non-destructive technique yielding accurate results as is shown in examples.

References

[1] G. F. Bastin, H. J. M. Heijligers, *Scanning* **1990**, *12*, 225.
[2] R. H. Packwood, J. D. Brown, *X-Ray Spectrom.* **1981**, *10*, 138.
[3] G. F. Bastin, H. J. M. Heijligers, in: Electron Probe Quantitation (K. F. J. Heinrick, D. E. Newbury, eds.), Plenum, New York, 1991, p. 145.
[4] W. Reuter, J. D. Kuptsis, A. Lurio, D. F. Kyser, *J. Phys. D: Appl. Phys.* **1978**, *11*, 2633.

Mikrochim. Acta (1992) [Suppl.] 12: 99–105

Comparison of $\Phi(\rho z)$ Curve Models in EPMA

José A. Riveros*, Gustavo E. Castellano, and Jorge C. Trincavelli**

Facultad de Matemática, Astronomíay Física, Universidad Nacional de Córdoba, Argentina

Abstract. Several authors have proposed different models for the ionization distribution function $\Phi(\rho z)$. This paper presents a comparison of three of the most successful models for $\Phi(\rho z)$: Packwood and Brown's Gaussian model including Bastin et al.'s and Riveros et al.'s versions, the quadrilateral model proposed by Sewell et al. and Pouchou and Pichoir's model. In general, all the tested models showed similar performances. Finally the advantages of the models related to basic principles over the mathematically optimized ones are considered.

Key words: EPMA, ZAF correction models, ionization distribution function.

In electron probe microanalysis (EPMA), concentration determinations may be carried out through different correction models relating the concentration C to the measured intensity ratio k:

$$k = C \cdot \mathrm{ZAF}, \tag{1}$$

where $k = I_{sp}/I_{st}$, I_{sp} and I_{st} are the intensities emerging from sample and standard respectively, and ZAF is the combined correction for atomic number Z, absorption A and fluorescence F. The combined ZA correction is usually expressed through $\Phi(\rho z)$, the distribution of ionizations with mass depth ρz, ρ being the target density. After a description of three of the most successful models for $\Phi(\rho z)$, a mutual comparison of them will be performed using a compilation of 680 microanalyses published by Bastin et al. [1]. This set was chosen because of its advantages with respect to other data sets, because it includes: absorption effects which are somewhat stronger (11% of data with ZA > 50% and 30% with ZA > 20%), greater range of experimental conditions and average experimental errors between 2% and 3%. On the other hand, mass absorption coefficients (MACs) will be assessed, combining the algorithms by McMaster et al. [2] or MAC30 [3] with the experimental data for low energies given by Henke et al. [4] or Bastin et al. [1].

---

* To whom correspondence should be addressed

** Present address: University of Antwerp (UIA), Universiteitsplein 1, B-2610 Antwerp-Wilrijk, Belgium

Ultralight elements are not included in the present test because their analysis involves serious problems, as pointed out by Bastin and Heijligers [5], the most important being: shift and shape alteration of characteristic lines due to chemical effects, experimental inconveniences, and inaccurate knowledge of MACs.

Gaussian Model (Packwood and Brown)

Packwood and Brown [6] proposed a Gaussian model for $\Phi(\rho z)$ assuming a random walk for the incident electrons within the sample; the resulting normal distribution is modified close to the surface by a transient function taking into account the variation of X-ray production with depth. The corresponding equation is:

$$\Phi(\rho z) = \exp[-\alpha^2(\rho z^2)]\cdot[\gamma_0 - (\gamma_0 - \Phi_0)\exp(-\beta\rho z)], \tag{2}$$

in which the assessment of the parameters α, β, γ_0 and Φ_0 has been described by Packwood and Brown in their original paper. As these original parameters did not produce successful results, modifications were proposed by Bastin et al. [1, 5, 7] and by Riveros and co-workers [8–10], providing better performances.

Modifications by Bastin et al.

Bastin et al. [7] optimized the expressions for the original parameters using a set of 430 microanalyses compiled by Love et al. [11]; they found limitations to their first version of the parameters and produced new expressions [1] (Table 1: BASTIN 86) for β and γ_0 based on experimental data and Monte Carlo simulations for $\Phi(\rho z)$.

Recently, Bastin and Heijligers [5] proposed a drastic change in the $\Phi(\rho z)$ parameterizations (Table 1: BASTIN 89). In their previous model, independent equations had been developed on physical bases for the parameters. Now, a new mathematical optimization has forced the parameters α, β, γ_0 and Φ_0 to 'cooperate' in a consistent way in order to provide a specified value for the total generated intensity in the specimen, by means of the atomic number correction of Pouchou and Pichoir [12]. The equation for Φ_0 was also taken from these authors.

Modifications by Riveros and co-workers.

Bearing in mind the physical meaning of each parameter of the Gaussian distribution, Riveros and co-workers [8–10] derived new expressions for them, without mathematical optimizations for a particular set of microanalysis data. More careful calculations were performed for the parameters Φ_0 and γ_0 [8], especially for the mean free path length of electrons and for the contribution of backscattered electrons to the surface ionizations. The obtained expressions depend on the spectral energy distribution of backscattered electrons $d\eta/dU$, U being the overvoltage. However, good results are produced when approximating $d\eta/dU$ by a constant function proportional to the total fraction of backscattered electrons η (Table 1: GAUSS 1). Expressions involving $d\eta/dU$ have also been evaluated using the expression proposed by del Giorgio et al. [13] (Table 1: GAUSS 2) [10]. The parameter β was assessed by relating the mean depth of diffusion to the value for which the transient function approaches unity [9].

Table 1. Performance of the tested correction models. "Sym 1" and "Sym $\langle k'/k \rangle$" indicate the symmetry of the histogram around unity and mean value, respectively. The bold lines show the best results obtained with each model

Model	Mac references	Rejected data	U_0 range	Average	RMS	Sym 1	Sym $\langle k'/k \rangle$
	[5]	23	$U_0 > 1$	0.094	2.61	0.685	1.040
Gaussian	[5]	23	$U_0 > 1.5$	0.994	2.57	0.691	1.030
BASTIN 89	[2, 5]	24	$U_0 > 1$	0.990	2.25	0.451	0.964
	[3, 5]	**27**	$\boldsymbol{U_0 > 1}$	**0.993**	**2.19**	**0.536**	**1.028**
	[5]	**39**	$\boldsymbol{U_0 > 1}$	**1.003**	**2.79**	**1.462**	**1.180**
Gaussian	[5]	25	$U_0 > 1.5$	1.003	2.72	1.469	1.184
BASTIN 86	[3, 5]	20	$U_0 > 1$	0.999	3.63	1.129	1.143
	[2, 5]	22	$U_0 > 1$	0.997	3.39	1.092	1.317
	[2, 5]	**31**	$\boldsymbol{U_0 > 1}$	**0.999**	**2.51**	**0.870**	**0.949**
Gaussian	[2, 5]	20	$U_0 > 1.5$	0.998	2.39	0.826	0.937
GAUSS 2	[2, 4]	66	$U_0 > 1$	0.995	2.78	0.689	1.040
	[3, 5]	29	$U_0 > 1$	1.002	2.62	1.093	0.943
	[3, 4]	66	$U_0 > 1$	0.998	2.89	0.875	0.974
	[2, 5]	**64**	$\boldsymbol{U_0 > 1}$	**1.001**	**2.53**	**0.875**	**0.839**
Gaussian	[2, 4]	80	$U_0 > 1$	0.995	2.61	0.656	0.987
GAUSS 1	[3, 5]	54	$U_0 > 1$	1.005	2.66	1.133	0.830
	[3, 4]	66	$U_0 > 1$	0.999	2.60	0.858	0.949
	[2, 4]	21	$U_0 > 1$	0.994	2.49	0.721	1.053
	[2, 5]	58	$U_0 > 1$	0.994	2.51	0.737	1.152
LOS II	[2, 5]	58	$U_0 > 1.5$	0.994	2.52	0.788	1.126
	[5]	61	$U_0 > 1$	0.998	2.70	1.056	1.142
	[3, 5]	60	$U_0 > 1$	0.997	2.49	1.006	1.168
	[3, 4]	**20**	$\boldsymbol{U_0 > 1}$	**0.997**	**2.51**	**0.970**	**1.095**
	[2, 5]	29	$U_0 > 1.5$	0.994	2.24	0.573	1.017
	[2, 5]	**45**	$\boldsymbol{U_0 > 1}$	**0.994**	**2.28**	**0.586**	**0.991**
	[2, 4]	53	$U_0 > 1$	0.991	2.67	0.524	1.090
PAP	[3, 5]	46	$U_0 > 1$	0.996	2.37	0.668	0.933
	[3, 4]	51	$U_0 > 1$	0.993	2.88	0.605	1.042
	[5]	43	$U_0 > 1$	0.999	2.37	0.846	0.996

The parameter α has not been modified from that given by Packwood and Brown, but the model for the mean ionization potential J, proposed by Brizuela and Riveros [14], is considered, taking into account shell effects. Parameters are atomically averaged except for $d\eta/dU$ which is mass averaged, according to del Giorgio et al. [13].

Quadrilateral Model (Sewell, Love, Scott)

In this model, $\Phi(\rho z)$ was approximated by means of two straight lines, determined by the surface ionization Φ_0, the position and height of the peak of the $\Phi(\rho z)$-curve, ρz_m and Φ_m; and ρz_r, a value related to the electron range in the sample. The involved parameters were derived from tracer and Monte Carlo determinations of $\Phi(\rho z)$, as well as from an optimization over a set of microanalyses [15]; using the Z-correction proposed by Love et al. [16], Bloch's model for J [17] and Reed's fluorescence correction [18]. The resulting function is:

$$\Phi(\rho z) = [\Phi_m - \Phi_0]\cdot(\rho z/\rho z_m) + \Phi_0 \qquad \text{for} \qquad 0 < \rho z < \rho z_m, \tag{3}$$

$$\Phi(\rho z) = \Phi_m\cdot(\rho z_r - \rho z)/(\rho z_r - \rho z_m) \qquad \text{for} \qquad \rho z_m < \rho z < \rho z_r, \tag{4}$$

(See Table 1 section LOS II for data).

PAP Model (Pouchou and Pichoir)

With the purpose of obtaining an expression for $\Phi(\rho z)$ which simplifies EPMA calculations, Pouchou and Pichoir [12] searched for a function consisting in two parabolae, that should accomplish the following requirements: predict the total generated radiation, begin with the $\Phi(0)$ value at the surface and vanish with a zero slope at a certain depth related to the electron range. The resulting curve is given by:

$$\Phi(\rho z) = A_1\cdot(\rho z - R_m)^2 + B_1 \qquad \text{for} \qquad 0 < \rho z < R_c, \tag{5}$$

$$\Phi(\rho z) = A_2\cdot(\rho z - R_x)^2 \qquad \text{for} \qquad R_c < \rho z < R_x, \tag{6}$$

where A_1, A_2 and B_1 are expressed in terms of Φ_0; R_m, the maximum of the function $\Phi(\rho z)$; R_x, the electron range and R_c, the crossover point of the parabolae. These parameters were derived from physical considerations, from experimental and simulated data for $\Phi(\rho z)$, from experimental analyses of specimens of known composition and from measurements of the electron range. The authors gave alternative procedures for low overvoltages; in such cases an additional degree of freedom can be introduced in the distribution by suppressing the parametric relation between R_c and R_x. This feature suggests that two parabolae may not be enough to fit completely the function $\Phi(\rho z)$.

Results and Discussion

The procedure usually followed in order to evaluate the performance of the different models for $\Phi(\rho z)$ in microanalysis consists in studying the distribution of quotients between calculated intensity ratios k' and experimental k-ratios for a large set of specimens of known composition. Samples whose k'/k-ratio deviate from the mean value more than three times the relative root mean square (rms)-error are rejected. By means of this criterion, it has been found that most rejected data are discarded by all models, without showing any systematic trend, that is, they are not indicated by a high ZAF correction, overvoltage, etc. In the present paper, Reed's fluorescence correction factor [18] is used and for each model, average values for Z, A and η have been taken following the original papers.

Gaussian Model

Modifications by Bastin et al.

Bastin et al. claimed an rms error of 2.99% around a mean value of 1.001 when testing the BASTIN 86-model in their compilation of 680 analyses [1], in which data with incident overvoltage is lower than 1.5 were rejected. They later tested the BASTIN 89-model upon a data file of 877 measurements [5] (the previous set of 680 data supplemented with 197 metal analyses in borides), obtaining an rms error of 2.44% around a mean value of 0.9955. It should be noted that for the subset with 197 analyses, an rms error around 1.3% is obtained, and most correction factors for these data are relatively unimportant: only in 7% of them the ZA correction is larger than 10% (none of them is above 15%).

According to Bastin et al., better performances in the BASTIN 86-model are obtained when selecting $U_0 > 1.5$ but no significant variation for the whole set of microanalyses was found. It can be seen that in this case, 39 data are rejected, whilst the test discards 25 for the subset with $U_0 > 1.5$. It must be emphasized that from the 50 samples with $U_0 < 1.5$, only in 14 the k'/k values lie three times the rms error beyond the mean value and the 36 left remain close to the mean value. As it can be seen from Table 1, the BASTIN 89 model does not modify substantially the performance of the BASTIN 86 model.

It can also be seen that the rms error deteriorates in BASTIN 86, when the set of MACs chosen is different from that suggested by Bastin et al. in their compilation of microanalyses. Although they suggest the use of their own set of MACs, in BASTIN 89 the algorithms by McMaster et al. [2] and MAC30 [3] produce better results.

Modifications by Riveros and co-workers

Performances are not strongly improved when avoiding the approximation of uniform backscattered distribution, except in the number of rejected data (see Table 1). Significant improvements could only be evidenced with a more accurate data set. The use of different expressions for J based on either experimental data of the stopping power or theoretical assessments does not modify significantly the performance of the model.

Quadrilateral Model (Sewell, Love, Scott)

Sewell et al. [15] quoted an rms error of 2.94% around a mean value of 0.994 when testing this model on their compilation of 554 data. Bastin et al. [1] tested the original quadrilateral model on the data set used in this paper obtaining an rms error of 4.33% around 0.990, considering the MACs proposed by them.

According to the present evaluation (see Table 1) results worsen when the complete set of MACs proposed by Bastin are used, as well as when the model chosen for J is different from that given by Bloch [17], except when Wilson's model [19] for J is used (both models have a similar origin). Bishop's model [20] for η was chosen in this test, since it slightly improves the performances produced when the model given by Love et al. [21] is used. No improvement is introduced if the range of overvoltages is limited.

PAP Model (Pouchou and Pichoir)

Pouchou and Pichoir [22] have recently evaluated their model in a set of 826 analyses, quoting an rms error of 1.91% around 0.998. They have used the MAC30 algorithm if the emitter is not a very light element, correcting some values of MACs corresponding to situations in which the line is close to an absorption edge or to particular resonance situations which are ignored by the MAC30 algorithm. For very low energies they have used the absorption coefficients of Henke and replaced some values by other measurements made by Bastin and Heijligers. It should be pointed out that this set of 826 analyses does not present matrix effects as important as those appearing in Bastin's 680 data base, since only in a 15% of them the ZA correction is larger than 20%.

When this model is tested upon Bastin's data base, the best results are obtained replacing the MAC values by those given by McMaster et al. [2] for energies larger than 1.6 keV (see Table 1). No substantial variation is observed for different sets of MACs, J or η formulae or when limiting the range of overvoltages to $U_0 > 1.5$.

Final Comments

– For low energies, in most correction models, MACs proposed by Bastin et al. [1] produce better performances than those given by Henke et al. [4]. On the other hand, for high energies the best results are obtained with the algorithms of McMaster et al. [2] and MAC30 [3], with slight differences between them. Only for the Gaussian model BASTIN 86 important improvements are achieved when using MACs proposed by these authors. This behavior may be due to the fact that these MACs were obtained by means of the function $\Phi(\rho z)$ proposed by them.
– The correction models LOSII and Gaussian-BASTIN 86 strongly depend on the expression for J taken, while PAP and Gaussian models BASTIN 89 and GAUSS are quite insensitive to the model for J used.
– Except for MACs, the parameters should be atomically averaged, since the incident electrons interact with atoms; however, the only models in which most of the parameters are atomically averaged are GAUSS 1 and 2. A remarkable case is the PAP model, in which the parameter Z is averaged in several different ways.
– The limitation $U_0 > 1.5$ does not improve the performance of the different models over the whole data set.
– The better performances of each model are set bold in Table 1. There is no significant variation in the rms error, while in those models in which a better performance in the rms error is observed, the values for the average worsen.
– Finally, it should be emphasized that provided the models are closely related to basic principles (which means no optimizations in the shape of $\Phi(\rho z)$ or in the parameters), any advance in the description of the process of interaction of electrons with matter will be reflected in advances in the performance of the models. On the other hand, if $\Phi(\rho z)$ or the parameters are developed through optimizations, every advance will need new optimizations in the models.

Acknowledgements. The authors acknowledge support of the Consejo Nacional de Investigaciones Científicas y Técnicas de la República Argentina and the Consejo de Investigaciones Científicas y Técnicas de la Prov. de Córdoba and the University of Antwerp (UIA).

References

[1] G. Bastin, H. Heijligers, J. van Loo, *Scanning* **1986**, *8*, 45.

[2] W. H. McMaster, N. Kerr del Grande, J. H. Mallet, J. H. Hubell, *Compilation of X-Ray Cross Sections*, Doc. UCRL 50174, California, 1969.

[3] K. Heinrich, in: *X-Ray Optics and Microanalysis* (J. Brown, R. Packwood, eds.), University of Western Ontario Press, London, Canada 1987, p. 67.

[4] B. Henke, P. Lee, T. Tanaka, R. Shimabukuro, B. Fujikawa, *At. Data Nucl. Data Tables* **1982**, *27*, 1.

[5] G. Bastin, H. Heijligers, *Report Eindhoven Univ. of Techn., Netherlands*, Nr. 73, ISBN 90-6819-013-X, 1990.

[6] R. Packwood, J. Brown, *X-Ray Spectrom.* **1981**, *10*, 138.

[7] G. Bastin, J. van Loo, H. Heijligers, *X-Ray Spectrom.* **1984**, *13*, 91.

[8] J. Tirira, J. Riveros, *X-Ray Spectrom.* **1987**, *16*, 27.

[9] J. Tirira, M. del Giorgio, J. Riveros, *X-Ray Spectrom.* **1987**, *16*, 243.

[10] M. del Giorgio, J. Trincavelli, J. Riveros, *X-Ray Spectrom.* **1990**, *19*, 261.

[11] G. Love, M. Cox, V. Scott, *J. Phys. D.* **1975**, *8*, 1686.

[12] J. Pouchou, F. Pichoir, in: *X-Ray Optics and Microanalysis*, (J. Brown, R. Packwood, eds.), University of Western Ontario Press, London, 1987, p. 249.

[13] M. del Giorgio, J. Trincavelli, J. Riveros, *X-Ray Spectrom.* **1989**, *18*, 229.

[14] H. Brizuela, J. Riveros, *X-Ray Spectrom.* **1990**, *19*, 173.

[15] D. Sewell, G. Love, V. Scott, *J. Phys. D.* **1985**, *18*, 1245.

[16] G. Love, M. Cox, V. Scott, *J. Phys. D.* **1978**, *11*, 7.

[17] F. Bloch, *Z. Phys.* **1933**, *81*, 363.

[18] S. J. Reed, *Br. J. Appl. Phys.* **1965**, *16*, 913.

[19] R. Wilson, *Phys. Rev.* **1941**, *60*, 749.

[20] H. Bishop, in: *Optique des Rayons X et Microanalyse* (R. Castaing, P. Descamps, J. Philibert, eds.), Hermann Paris, 1966, p. 153.

[21] G. Love, V. Scott, *J. Phys. D.* **1978**, *11*, 106.

[22] J. Pouchou, F. Pichoir, *Proc. NBS Workshop* 1989 (in press).

Mikrochim. Acta (1992) [Suppl.] 12: 107–115

Quantitative Electron Probe Microanalysis: New Accurate $\Phi(\rho z)$ Description

Claude Merlet

Centre Géologique et Géophysique CNRS, Case 060, Université de Montpellier II, Sciences et Techniques du Languedoc, Pl. E. Bataillon, F-34095 Montpellier Cedex 5, France

Abstract. This study presents a new method of matrix correction, combining atomic number and absorption correction [ZA]. The procedure requires an accurate knowledge of the $\Phi(\rho z)$-distribution with mass depth (ρz) of primary X-rays generated in the target. The $\Phi(\rho z)$-expression is evaluated using experimental measurements based on tracer layers and a theoretical approach involving the Monte Carlo method.

Key words: X-ray depth distribution, Monte Carlo, EPMA, matrix correction.

Classically, quantitative electron-probe microanalysis is carried out by comparing intensities of characteristic X-rays of element A emitted from a specimen (I_A^e) and a standard (I_A^s) under identical instrumental conditions. The ratio of these intensities (K-ratio) can be expressed as a function of the specimen concentration C_A:

$$K_A = \frac{I_A^e}{I_A^s} = C_A.f(C_A),$$

where $f(C_A)$ is a correction function including various factors, such as distribution of ionisation in depth, electron backscattering, absorption of X-rays, and the influence of fluorescence. Usually, this correction function is divided in three factors that are calculated separately—atomic number effect [Z], absorption effect [A] and fluorescence effect [F]. However, as shown by Castaing [1], the atomic number and absorption effects can be expressed by a single equation describing the intensity I_A^e emitted from the element A in an homogeneous sample:

$$I_A^e = cst.C_A \int_0^\infty \Phi_e(\rho z).\exp(-\chi_e \rho z)\, d\rho z, \tag{1}$$

where C_A is the weight fraction of element A, ρz is the mass depth in the sample, $\chi - \mu/\rho$ cosec θ, with μ/ρ being the mass absorption coefficient and θ the X-ray take-off angle, $\Phi(\rho z)$ is the depth distribution of X-rays. The total intensity generated in the specimen is given by:

$$\int_0^\infty \Phi_e(\rho z).d\rho z, \tag{2}$$

which corresponds to the atomic number effect. If there is no significant fluorescence effect, the K-ratio is simply the ratio of equation (1) and a similar equation for a pure element standard:

$$K_A = C_A \frac{\int_0^\infty \Phi_e(\rho z).\exp(-\chi_e \rho z)\, d\rho z}{\int_0^\infty \Phi_s(\rho z).\exp(-\chi_s \rho z)\, d\rho z} = C_A[ZA]. \tag{3}$$

If the $\Phi(\rho z)$-curves are known, the concentration C_A can be exactly calculated from the intensity ratio.

Castaing and Descamps [2] described a tracer method for measuring $\Phi(\rho z)$ curves. In their procedure they measured a sandwich sample, in which a tracer element is deposited at various depths within a matrix element. Several authors [3–10] have measured $\Phi(\rho z)$-curves using this technique. Thus, $\Phi(\rho z)$-curves are known nowadays for several matrices and X-ray energies. An alternate method for obtaining $\Phi(\rho z)$ is the Monte Carlo calculation, which is a theoretical approach for determination of electron trajectories. In this the X-ray distribution may be determined by introducing an appropriate ionisation cross-section. The Monte Carlo approach produces data that are in good agreement with experimentally determined $\Phi(\rho z)$ curves.

On the other hand, some formulae have been used to approximate the $\Phi(\rho z)$-curves. The simplest of these assumes a constant intensity down to some depth, which is considered to represent the total intensity generated in the specimen. Philibert [11] has proposed an exponential expression, while Tanuma and Nagashima [12, 13] proposed to use a Gaussian profile with the centre displaced along the (ρz)-axis and to subtract a second Gaussian profile in order to fit the profile shape to Φ_0. Sewell et al. [14] applied a simple quadrilateral shape. These $\Phi(\rho z)$-curves are valid for absorption correction only. Some authors have proposed more accurate $\Phi(\rho z)$-curves for atomic number and absorption. Pouchou and Pichoir [15, 16] used a double parabolic shape or a combination of exponential shapes [17]. Packwood and Brown [18] suggested to use a Gaussian, centred at the surface, and to modify its shape close to the $\Phi(\rho z)$ axis. The parameters, used in this method, have been improved by Bastin et al. [19, 20, 21], Tirira et al. [22] and Rehbach and Karduck [23].

This paper presents a new expression of $\Phi(\rho z)$ which closely approximates the true X-ray depth distribution for a wide range of analytical conditions.

The Proposed Correction Method

In this model the X-ray distribution is approximated by a double partial Gaussian profile ($\Phi 1$, $\Phi 2$; Fig. 1), described by three coordinates:

- at the surface, $[0, \Phi(0)]$,
- at the maximum of the $\Phi(\rho z)$-curve, $[\rho z_m, \Phi(m)]$,
- at the X-ray range, $[\rho z_x, 0.01]$.

Thus $\Phi(\rho z)$ may be expressed as:

$$\Phi(m) \exp\left\{-\left(\frac{\rho z - \rho z_m}{\beta}\right)^2\right\} \quad \text{for} \quad \rho z \in [0, \rho z_m] \quad \text{with} \quad \beta = \frac{\rho z_m}{\sqrt{\ln(\Phi(m)/\Phi(0))}}$$

and (4)

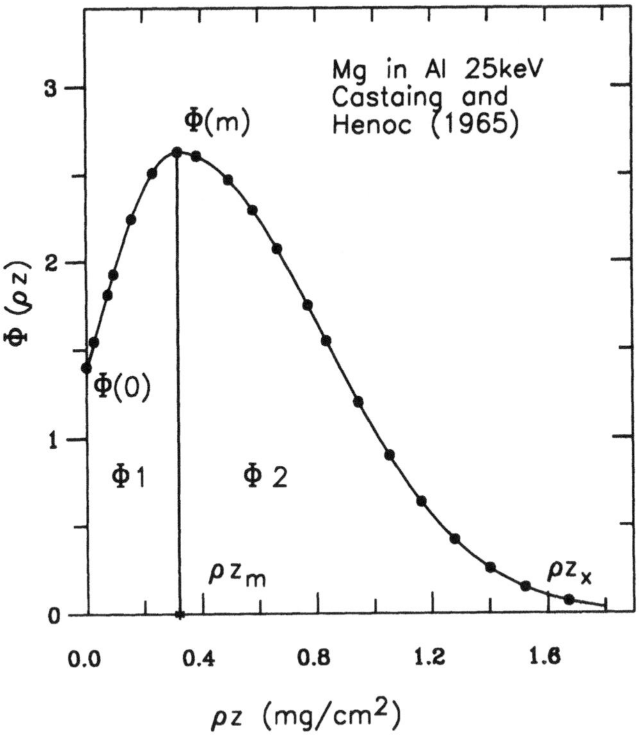

Fig. 1. $\Phi(\rho z)$-curve for Mg $K\alpha$ in an aluminium matrix for 25 keV electron energy. The data points are from Castaing and Henoc [3]. The solid line is the predicted curve using the double Gaussian expression ($\Phi 1$ and $\Phi 2$) and the parameters: $\Phi(0)$, ρz_m, $\Phi(m)$, and ρz_x (see text for further explanation)

$$\Phi(m)\exp\left\{-\left(\frac{\rho z - \rho z_m}{\alpha}\right)^2\right\} \quad \text{for} \quad \rho z \in [\rho z_m, \rho z_x] \quad \text{with} \quad \alpha = \frac{\rho z_x - \rho z_m}{\sqrt{\ln(\Phi(m)/0.01)}}.$$

The double Gaussian profile was chosen because: **(1)** it may ensure a good fit of the experimental data, **(2)** it can easily be integrated, **(3)** it is physically acceptable. The maximum of the $\Phi(\rho z)$-curve corresponds to the complete diffusion of the electrons in the sample. At the maximum the electrons are randomized in energy and direction. Therefore the random behavior of the electrons on this point can be explained by a Gaussian centred at the maximum. **(4)** it varies in a realistic way from light to heavy targets, as well as from high to low overvoltage. In comparison with the Gaussian approach of Packwood and Brown [18], this double Gaussian does not present boundary problems. Rehbach and Karduck [23] showed the limit of the Packwood and Brown [18] model, which needs unrealistic parameters in the low matrix. Using the same Monte Carlo simulation [10], the flexibility of our model is shown in Fig. 2 for a low atomic number element in various matrices. For the very low overvoltages ($E_0/E_c = U_0 \to 1$) we have $\Phi(m) \to 1$ and $\Phi 1 \to 0$. Thus, there are no boundary problems if the parameters are well defined (see Fig. 3). **(5)** at the connecting point $\Phi(m)$, the conditions $\Phi 1 = \Phi 2$ and $d\Phi 1/d\rho z = d\Phi 2/d\rho z$ are respected.

The proposed $\Phi(\rho z)$-expression was tested with experimental $\Phi(\rho z)$ data from the literature and Monte Carlo calculations. The ability of the expression in accurately describing the $\Phi(\rho z)$ data can be judged from Fig. 4 which shows calculated curves and experimental measurements for various matrices. A good fit was obtained for two extreme matrices: Carbon at 15 keV with aluminium tracer $K\alpha$ [9] and gold at 25 keV with a cadmium $L\alpha$ tracer [7].

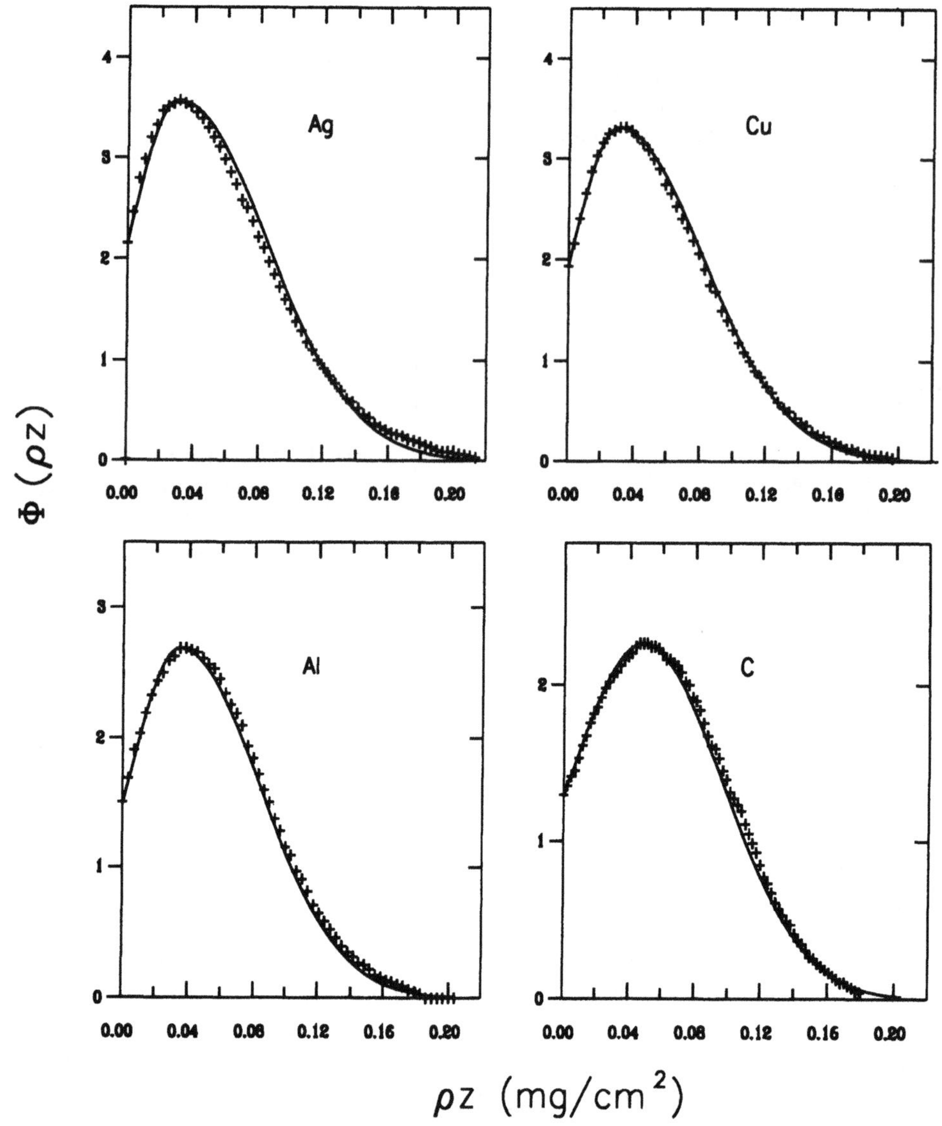

Fig. 2. Comparison between Monte Carlo curves (+) calculated by Karduck and Rehbach [10], and the corresponding double Gaussian expression (solid lines) for C $K\alpha$ at $E_0 = 7$ keV in Ag, Cu, Al, and C. The parameters $\Phi(0)$, ρz_m, $\Phi(m)$, ρz_x are taken from the Monte Carlo curves

Using expressions 4, equation 1 becomes:

$$I_A^e = cst.C_A\Phi(m)\int_0^{\rho z_m} \exp\left\{-\left(\frac{\rho z - \rho z_m}{\beta}\right)^2\right\}.\exp(-\chi_e \rho z)\, d\rho z$$
$$+ \int_{\rho z_m}^{\infty} \exp\left\{-\left(\frac{\rho z - \rho z_m}{\alpha}\right)^2\right\}.\exp(-\chi_e \rho z)\, d\rho z\Bigg]$$

This equation can be solved by using the error function (erf):

$$\Phi(m)\int_0^{\rho z_m} \exp\left\{-\left(\frac{\rho z - \rho z_m}{\beta}\right)^2\right\}.\exp(-\chi_e \rho z)\, d\rho z$$
$$= \frac{\Phi(m)\beta\sqrt{\pi}\exp(c)}{2}\left[erf\left(\frac{\rho z\,1}{\beta}\right) + erf\left(\frac{\chi_e \beta}{2}\right)\right],$$

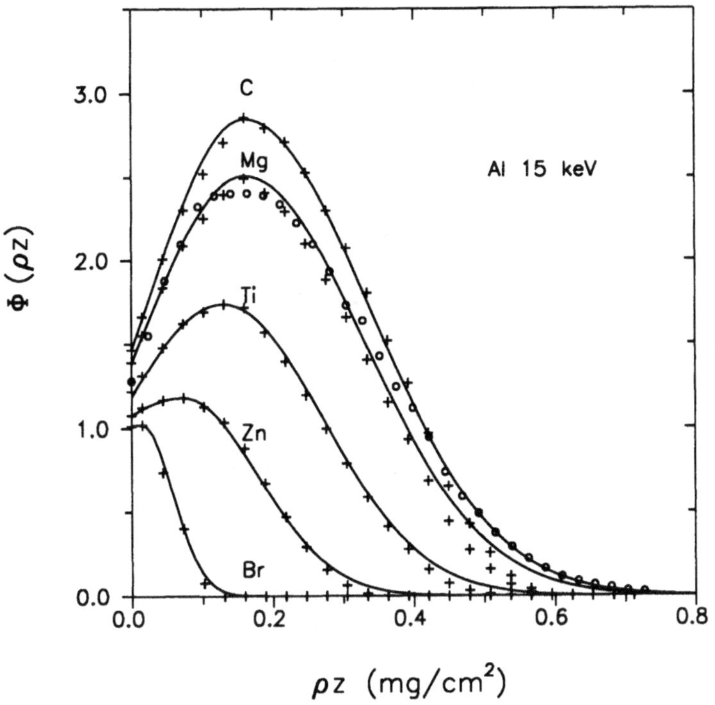

Fig. 3. $\Phi(\rho z)$ curves for various overvoltages in an aluminium matrix for 15 keV electron energy. The data points (o) are from Castaing and Henoc [3], (+) are Monte Carlo calculations obtained in this study and the solid line is the predicted curve using the double Gaussian expression. C $K\alpha$, $U_o = 52.8$; Mg $K\alpha$, $U_o = 11.5$; Ti $K\alpha$, $U_o = 3.02$; Zn $K\alpha$, $U_o = 1.55$; Br $K\alpha$, $U_o = 1.12$

with

$$c = \left(\frac{\chi\beta}{2}\right)^2 - \rho z_m \chi \qquad \text{and} \qquad \rho z_1 = \rho z_m - \frac{\chi\beta^2}{2},$$

followed by:

$$\Phi(m)\int_{\rho z_m}^{\infty} \exp\left\{-\left(\frac{\rho z - \rho z_m}{\alpha}\right)^2\right\}.\exp(-\chi_e \rho z)\, d\rho z$$
$$= \frac{\Phi(m)\alpha\sqrt{\pi}\exp(d)}{2}\left[1 - erf\left(\frac{\chi_e \alpha}{2}\right)\right],$$

with

$$d = \left(\frac{\chi_e \alpha}{2}\right)^2 - \rho z_m \chi_e.$$

The combined atomic number and absorption correction $[ZA]$ is then given by:

$$[ZA] = \frac{\Phi(m)_e\left[\beta\exp(c)\left\{erf\left(\frac{\rho z\,1}{\beta}\right) + erf\left(\frac{\chi_e\beta}{2}\right)\right\} + \alpha\exp(d)\left\{1 - erf\left(\frac{\chi_e\alpha}{2}\right)\right\}\right]_e}{\Phi(m)_s\left[\beta\exp(c)\left\{erf\left(\frac{\rho z\,1}{\beta}\right) + erf\left(\frac{\chi_s\beta}{2}\right)\right\} + \alpha\exp(d)\left(1 - erf\left(\frac{\chi_s\alpha}{2}\right)\right\}\right]_s}. \tag{5}$$

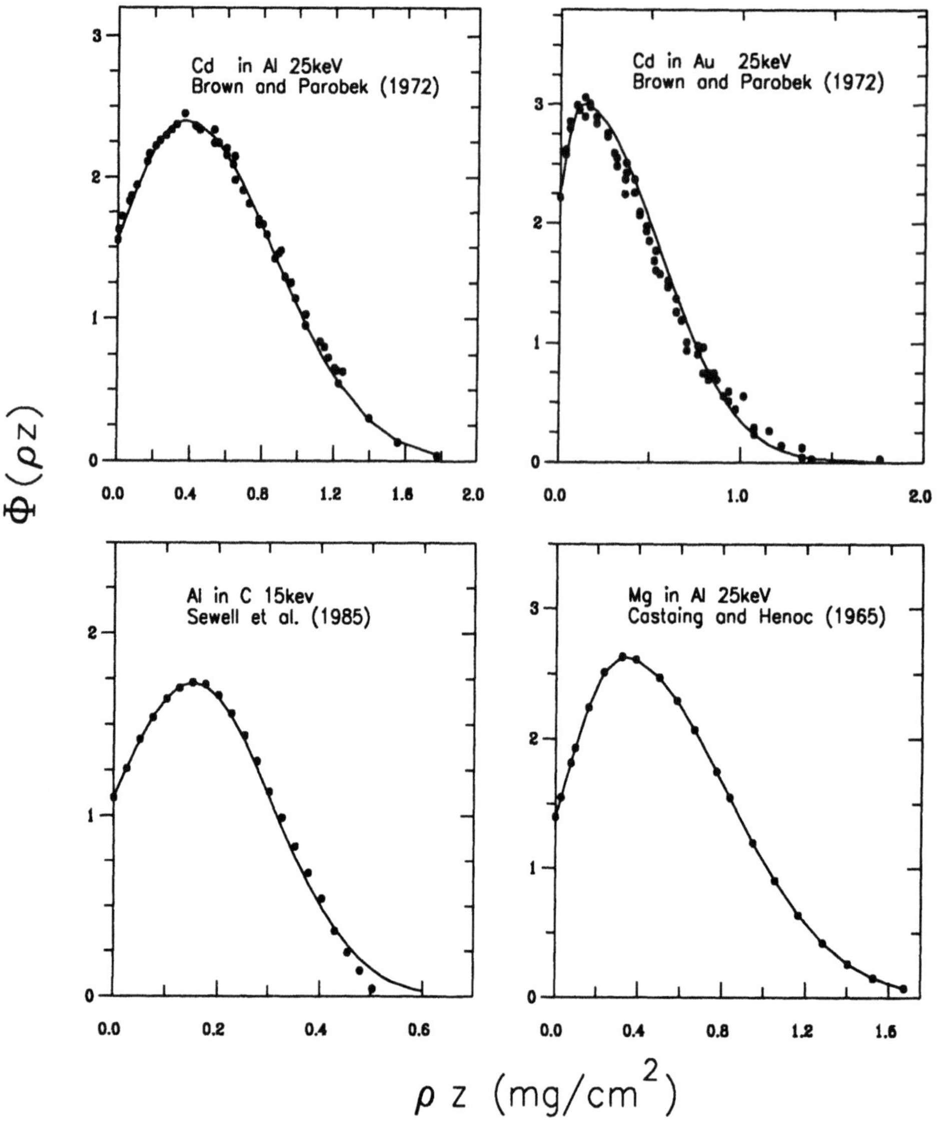

Fig. 4. Comparison between calculated curves using the double Gaussian expression (solid lines) and experimental measurements (•) obtained by tracer measurements from various matrices [3, 7, 9]

The error function can be approximated with a good accuracy ($<5.10^{-7}$) by analytical expressions.

Similarly the atomic number correction factor [Z] can be obtained by integrating $\Phi(\rho z)$ for a sample and a standard in the absence of absorption.

$$[Z] = \frac{\int_0^\infty \Phi_e(\rho z).\, d\rho z}{\int_0^\infty \Phi_s(\rho z).\, d\rho z} = \frac{\Phi(m)_e \cdot \left[\beta \cdot erf\left(\frac{\rho z_m}{\beta}\right) + \alpha\right]_e}{\Phi(m)_s \cdot \left[\beta \cdot erf\left(\frac{\rho z_m}{\beta}\right) + \alpha\right]_s}. \tag{6}$$

Computing the $[ZA]$-factor requires accurate knowledge of the following parameters: $\Phi(0)$ (the surface value of $\Phi(\rho z)$), $\Phi(m)$ and ρz_m (the maximum value of $\Phi(\rho z)$ and its depth position), and the X-ray range ρz_x.

Theoretical Derivation of the $\Phi(\rho z)$ Parameters

In order to develop exact analytical expressions of the $\Phi(\rho z)$ parameters, experimental data ($\Phi(\rho z)$ and $\Phi(0)$ measurements, transmitted and backscattered electrons intensities) and Monte Carlo simulation data, have been used.

The parameter $\Phi(0)$ represents the amplitude of the depth distribution of X-ray production at the surface of the sample and can be considered to be composed of two terms: (i) ionization by the incident electron in the surface layer, and (ii) contribution to ionization by the backscattered electrons that leave the specimen. Thus $\Phi(0)$ may be expressed as:

$$\Phi(0) = 1 + \frac{2}{(\ln(U_0)/U_0^m)} \cdot \int_1^{U_0} (\ln(U)/U^m) \cdot \frac{d\eta}{dU}\, dU\,, \qquad (7)$$

with $d\eta/dU$ being the energetic distribution of the backscattered electrons ($U = E/E_c$) (m is a factor in the ionisation cross section expression).

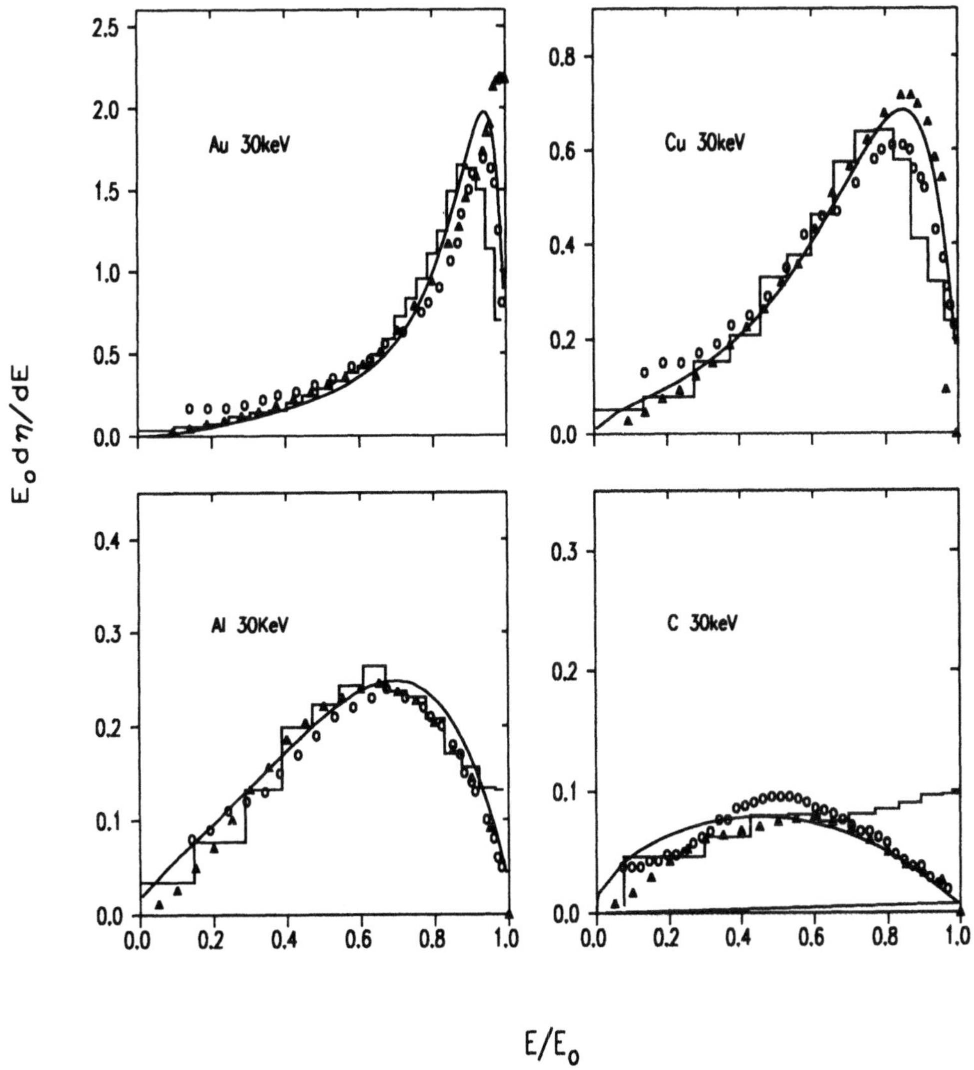

Fig. 5. Experimental and calculated energy spectra $E_0 d\eta/dE$ for Au, Cu, Al and C at $E_0 = 30$ keV. (▵) Darlington [24], (o) Matsukawa et al. [25], (histograms) Monte Carlo, (solid lines) calculated $E_0 d\eta/dE$

An analytical expression of $d\eta/dU$ has been developed which satisfies the following conditions:

- the expression is only dependent on the atomic number Z, U and the acceleration voltage E_0
- it is straightforward and represents with good accuracy the experimental and Monte Carlo data,
- the integral of $d\eta/dU$ has an analytical solution and accurately represents η,
- the integral of $(\ln(U)/U^m)\, d\eta/dU$ has an analytical solution and accurately represents $\Phi(0)$,
- the boundary conditions of $\Phi(0)$ must be respected for $U = 1$ and $U > 100$.

Figure 5 shows the energy spectra calculated with this equation and compared to the experimental data obtained by Darlington [24] and Matsukawa et al. [25] as well as to Monte Carlo data.

The analytical expression of $\Phi(0)$ obtained from $d\eta/dU$ may be written as:

$$\Phi(0) = 1 + 1.87a.[0.27b_1 + (1.1 + 5/Z)(b_2 - 1.1b_3)], \tag{8}$$

with

$$b_i = \frac{1}{d_i}\left(1 - \frac{(1 - 1/U_0^{d_i})}{d_i \ln(U_0)}\right)$$

$$d_1 = 0.02\,Z - m + 1 \quad d_2 = 0.1\,Z - m + 1 \quad d_3 = 0.4\,Z - m + 1$$

$$a = 0.001\,Z[-3.91\ln(z+1) + 7.21(\ln(z+1))^2 - 1.067(\ln(z+1))^3](E_0/30)^{p(Z)}$$

with $m = 1$ for the K and L lines and $m = 0.8$ for the M line.

A complete description of the expression for $\Phi(0)$ is proposed in a separate paper [26]. Considerations similar to those used in deriving $\Phi(0)$ can be applied to give the expressions of $\Phi(m)$, ρz_m and ρz_x. These parameters, and the performance of this model are presented in a forthcoming paper [Merlet, in preparation].

References

[1] R. Castaing, *Thesis*, University of Paris, Publication ONERA No. 55, 1951.

[2] R. Castaing, and J. Descamps, *J. Phys. Rad.* **1955**, *16*, 304.

[3] R. Castaing, and J. Henoc, in: *Optique des Rayons X et Microanalyse, 4th Congrès Int.* (R. Castaing, P. Deschamps, J. Philibert, eds.), Hermann, Paris, 1965, p. 120.

[4] R. Shimizu, K. Murata, G. Shinoda, in: *Optique des Rayons X et Microanalyse, 4th Congrès Int.* (R. Castaing, P. Deschamps, J. Philibert, eds.), Hermann, Paris, 1965, p. 127.

[5] J. D. Brown, *Ph.D. Thesis*, University of Maryland, U.S.A., 1966.

[6] A. Vigne, G. Dez, *J. Phys. D.* **1968**, *1*, 1309.

[7] J. D. Brown, L. Parobek, in: *Proc. 6th Int. Conference on X-Ray Optics Microanal* (G. Shinoda, K. Kohra, T. Ichinokawa, eds.), University of Tokyo Press, Tokyo, 1972, p. 163.

[8] J. D. Brown, L. Parobek, *X-Ray Spectrom.* **1976**, *5*, 36.

[9] D. A. Sewell, G. Love, V. D. Scott, *J. Phys. D.* **1985**, *18*, 1233.

[10] P. Karduck, W. Rehbach, in: *Microbeam Analysis* (D. E. Newbury, ed.), San Francisco Press, San Francisco, 1988, p. 277.

[11] J. Philibert, in: *Proc. 3rd Int. Conference X-Ray Optics Microanal.*, (H. H. Patee, V. E. Cosslett, A. Engström, eds.), Academic, New York, 1963, p. 379.

[12] S. Tanuma, K. Nagashima, *Mikrochim. Acta* [*Wien*] **1983**, *I*, 299.

[13] S. Tanuma, K. Nagashima, *Mikrochim. Acta [Wien]* **1984**, *III*, 265.
[14] D. A. Sewell, G. Love, V. D. Scott, *J. Phys. D.* **1985**, *18*, 1245.
[15] J. L. Pouchou, F. Pichoir, *La Recherche Aérospatiale*, **1984**, *3*, 167.
[16] J. L Pouchou, F. Pichoir, in: *11th Int. Congress X-Ray Optics Microanal.* (J. D. Brown, R. H. Packwood 1986, p. 249.
[17] J. L. Pouchou, F. Pichoir; in: *Microbeam Analysis* (D. E. Newbury, eds.), San Francisco Press, San Francisco, 1988, p. 315.
[18] R. H. Packwood, J. D. Brown, *X-Ray Spectrom.* **1981**, *10*, 138.
[19] G. F. Bastin, F. J. J. Van Loo, H. J. M. Heijligers, *X-Ray Spectrom.* **1984**, *13*, 91.
[20] G. F. Bastin, H. J. M. Heijligers, F. J. J. Van Loo, *Scanning* **1986**, *8*, 45.
[21] G. F. Bastin, H. J. M. Heijligers, *Report Eindhoven University of Technology, The Netherlands*, 73, ISBN 90-6819-013-X 1990.
[22] J. Tirira, M. del Giorgio, J. Riveros, *X-Ray Spectrom.* **1987**, *16*, 243.
[23] W. Rehbach, P. Karduck, in: *Microbeam Analysis* (D. E. Newbury, ed.), San Francisco Press, San Francisco, 1988, p. 285.
[24] E. H. Darlington, *J. Phys. D.* **1975**, *8*, 85.
[25] T. Matsukawa, R. Shimizu, H. Hashimoto, *J. Phys. D.* **1974**, 7, 695.
[26] C. Merlet, *X-Ray Spectrom.* **1991** (accepted).

Mikrochim. Acta (1992) [Suppl.] 12: 117–124

A Modular Universal Correction Procedure for Quantitative EPMA

Ian Farthing[1,2,*], Glyn Love[2], Victor D. Scott[2], and Clive T. Walker[1]

[1] Commission of the European Communities, Joint Research Centre, European Institute for Transuranium Elements, D-W-7500 Karlsruhe, Federal Republic of Germany
[2] University of Bath, School of Materials Science, BA2 7AY Bath, U.K

Abstract. An EPMA correction method is outlined which offers significant improvements in performance compared to earlier models. The method is versatile being capable of dealing with the analysis of elements ranging from atomic number 4 to 96 while using a wide range of experimental conditions including non-normal electron beam incidence. The correction factors-atomic number (Z), X-ray absorption (A) and X-ray fluorescence (F) (characteristic and continuum)—are calculated separately using equations which accurately reflect the physical processes occurring in EPMA. As a result it is shown that most experimental data can be corrected with errors less than 3% relative and even for difficult analyses involving ultra-light elements, errors are typically less than 5%. The computer program incorporating this method consists of modules allowing microanalysts to tailor input and output to suit their own requirements. The correction program has been written to run on standard IBM-PC compatible machines in a compiled version of BASIC.

Key words: Correction procedure, EPMA, quadrilateral model, computer program.

Continuing developments in instrumentation and in the automation of EPMA are leading to improvements in the precision of X-ray measurements. If however, commensurate improvements are to be made in the chemical characterization of the specimen itself, attention must be focussed on updating the conventional methods of correcting the experimental data. In this paper we outline a rigourous correction procedure capable of dealing with data obtained using an extremely wide range of experimental conditions and from specimens of nearly all types. It is being developed using the BASIC language so that microanalysts can easily modify the program to suit their own particular need if necessary. Within this correction procedure we have used the atomic number and absorption corrections of Love-Scott [1], based upon

* To whom correspondence should be addressed

their $\Phi(\rho z)$ quadrilateral model, extended to deal with tilted specimens as proposed by Sewell [2], together with the characteristic fluorescence correction of Castaing [3] and the continuum fluorescence correction of Springer [4].

Physical Basis

In the computer program the three correction factors, atomic number (Z), X-ray absorption (A) and X-ray fluorescence (F), are treated as separate entities. Thus the weight concentration c_A of element A in the specimen may be determined from:

$$\frac{I_A^{sp}}{I_A^{st}} = c_A ZAF, \tag{1}$$

where I_A^{sp} and I_A^{st} are the characteristic X-ray intensities from element A emitted from the specimen and pure element standard respectively under identical analysis conditions. Although the basis of the correction procedure described here is a traditional one, calculation of the magnitude of individual factors has been carried out using revised equations which reflect the physical processes occurring in electron-beam induced X-ray emission more accurately.

The atomic number correction factor is evaluated by considering separately (a) the loss in X-ray generation arising from high energy electrons being backscattered from the specimen and standard and (b) the efficiency with which electrons remaining in the specimen and standard generate characteristic X-rays. These two components are known as the backscatter factor (R) and the stopping power factor (S) respectively. From data accumulated from Monte Carlo calculations new empirical equations for R have been derived which describe its variation with the electron backscatter coefficient η, incident electron energy, E_0, and the critical excitation energy, E_c. The final form of the equation is;

$$\left(\frac{1-R}{\eta}\right)^{0.6} = I(U_0) + \eta G(U_0), \tag{2}$$

where $U_0 = E_0/E_c$ and both $I(U_0)$ and $G(U_0)$ are functions of the overvoltage ratio, U_0. The value of the stopping power factor (S) is evaluated from:

$$\frac{1}{S} = \int_{E_0}^{E_c} Q \cdot \left[\frac{dE}{d\rho s}\right]^{-1} \cdot dE, \tag{3}$$

where Q is the ionisation cross-section and $dE/d\rho s$ is the rate of electron energy loss. The Bethe expression [5], usually adopted to represent $dE/d\rho s$ has also been modified [6], in order to render it more physically realistic at low electron energies. The modification has the added advantage of allowing the above integral to be evaluated in closed form using the Green and Cosslett [7] expression for Q.

To determine the X-ray absorption correction the distribution of generated X-rays with depth in the specimen and standard must first be modelled. These distributions are depicted in the form of X-ray intensity versus mass depth (ρz) and are usually referred to as $\Phi(\rho z)$ curves. The absorption correction factor, $f(\chi)$ may be computed from the curve as follows:

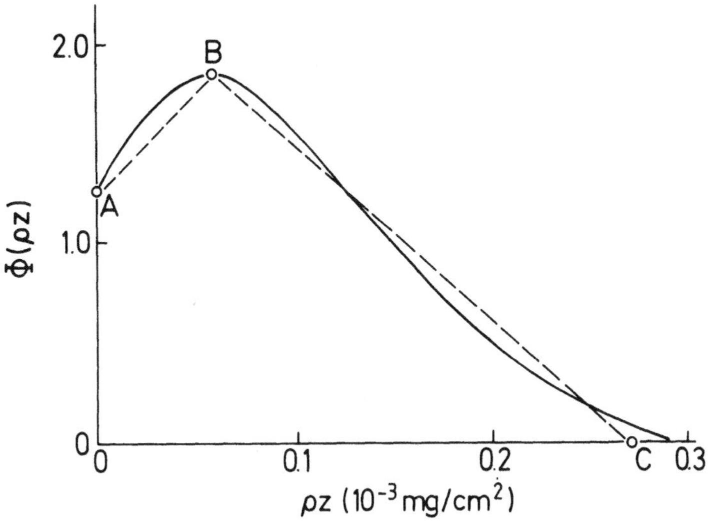

Fig. 1. The variation of generated X-ray intensity $\Phi(\rho z)$ with mass depth ρz. The broken line represents the quadrilateral profile and is defined by the coordinates of points A, B, C and their origin

$$f(\chi) = \frac{\int_0^\infty \Phi(\rho z)\exp(-\chi\rho z)\,d\rho z}{\int_0^\infty \Phi(\rho z)\,d\rho z}. \tag{4}$$

Where χ is the product of the X-ray mass absorption coefficient (μ/ρ) and the cosecant of the X-ray take-off angle, ψ. In our model [1] the distribution is described using a quadrilateral profile (Fig. 1). This profile is defined by three parameters, the ratio (h) of the maximum height of the $\Phi(\rho z)$ curve to its value at the surface, the mean mass depth at which the peak in the distribution occurs (ρz_m) and the maximum depth of X-ray generation, (ρz_r). For the quadrilateral profile, $f(\chi)$ becomes:

$$f(\chi) = \frac{\int_0^{\rho z_m}\left(\frac{(h-1)\rho z}{\rho z_m}+1\right)\exp(-\chi\rho z)\,d\rho z + 2\int_{\rho z_m}^{\rho z_r}\left(\frac{(\rho z_r-\rho z)h}{(\rho z_r-\rho z_m)}\right)\exp(-\chi\rho z)\,d\rho z}{(h\rho z_r+\rho z_m)}. \tag{5}$$

In practice ρz_m and ρz_r are expressed in terms of the mean depth of X-ray generation, $\overline{\rho z}$, which is the single largest factor controlling the magnitude of $f(\chi)$ when X-ray absorption is moderate (>0.5). Equations for each of these have been derived from data extracted from a combination of Monte Carlo simulations and measurements using a tracer technique [8, 9], the final equations being expressed in terms of E_0, E_c, h and atomic number, Z.

At first sight the quadrilateral profile may seem to be a somewhat crude approximation to the more realistic X-ray depth profile as calculated from the Monte Carlo program (Fig. 1). However the simplified approach may be readily justified by calculating the absorption correction factor by numerical integration of the Monte Carlo profile and also using the formula for the quadrilateral model to establish

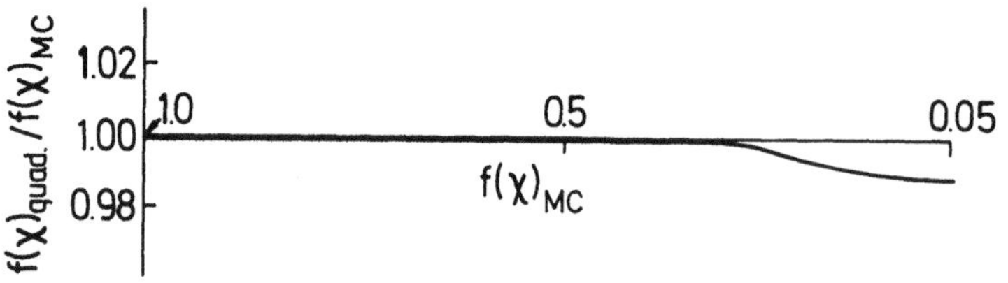

Fig. 2. The ratio $f(\chi)_{\text{quad}}/f(\chi)_{\text{Mc}}$ plotted as a function of $f(\chi)_{\text{Mc}}$. Where the ratio is unity the quadrilateral model may be judged to be working perfectly

$f(\chi)$. When the two sets of results are compared (Fig. 2), they may be seen to be identical for all practical values of $f(\chi)$.

In computing the absorption correction, accurate mass absorption coefficients (MACs) are essential. In this paper MACs for elements 5 to 92 are calculated using the algorithms developed recently by Heinrich [10] together with X-ray line energies taken from Bearden [11]. The MACs for elements 92–96 are determined by extrapolation using line energies taken from Kleykamp [12]. When calculating values for some of the ultra-light elements, where the line energy falls below the highest N-edge of the absorber, the multiplying factor of 1.02 used in equation 4 of Heinrich's paper is changed to 0.727 in order to show better agreement with the MACs of Henke et al. [13]. It should be noted that facilities exist within the program for introducing alternative MAC data such as those of Bastin and Heijligers [14, 15] when carrying out ultra-light element analysis; these new values can significantly improve the quality of the corrected data.

The characteristic fluorescence correction is based upon the work of Castaing [3] and the form used in the program is similar to that described fully in Love and Scott [16]. The fractional fluorescence contribution from an element B to the electron-excited X-ray intensity of emission from element A is given as:

$$\frac{I_f}{I_a} = 0.5 \cdot C_B \frac{r_k(A) - 1}{r_k(A)} \cdot \omega_k(B) \cdot \frac{A_A}{A_B} \left[\frac{U_{0B} \ln U_{0B} - U_{0B} + 1}{U_{0A} \ln U_{0A} - U_{0A} + 1} \right] \frac{(\mu/\rho)_B^A}{(\mu/\rho)_B} \cdot \left(\frac{\ln(1+x)}{x} + \frac{(\ln(1+y)}{y} \right), \tag{6}$$

with $\quad x = \dfrac{(\mu/\rho)_A}{(\mu/\rho)_B} \operatorname{cosec}\theta; \qquad y = \dfrac{4.05 \cdot 10^5}{(E_0^{1.65} - E_c^{1.65})(\mu/\rho)_B};$

where $(\mu/\rho)_j^i$ is the MAC of element i for X-rays from element j, $(\mu/\rho)_j$ is the MAC for the compound as a whole for X-ray from element j, θ is the take-off angle, A_j is the atomic weight of element j, $\omega_k(j)$ is the fluorescence yield of the k line of element j and C_j is the weight concentration of the jth element.

In equation (6) the term $[(U_{0B} \ln U_{0B} - U_{0B} + 1)/(U_{0A} \ln U_{0A} - U_{0A} + 1)]$ replaces $[(U_{0B} - 1)/(U_{0A} + 1)]^{1.67}$ given in [18]. As pointed out by Reed [17] this change has a significant effect for high atomic number elements in the analysis with low accelerating voltages and vice-versa, producing a stronger dependence on beam energy. The factor $(r_k(A) - 1)/r_k(A))$ has been tabulated using Heinrich's MACs [10] with the absorption jump ratio being calculated from points 0.001 eV on either side of the absorption edge. These values are plotted as a function of atomic number,

Table 1. Values of the coefficients used in the $(r_k(A) - 1)/r_k(A)$ Polynomial

Coefficient	Value	Standard Error
C_1	$6.88 \cdot 10^{-1}$	$2.68 \cdot 10^{-2}$
C_2	$4.15 \cdot 10^{-3}$	$2.00 \cdot 10^{-5}$
C_3	$-4.22 \cdot 10^{-4}$	$5.44 \cdot 10^{-5}$
C_4	$6.13 \cdot 10^{-6}$	$6.40 \cdot 10^{-7}$
C_5	$-2.61 \cdot 10^{-8}$	$2.75 \cdot 10^{-9}$

and follow a smooth curve which is described by the polynomial $C_1 + C_2Z + C_3Z^2 + C_4Z^3 + C_5Z^4$. The coefficients of this equation are given in Table 1. For the Lα line the absorption edge jump ratio which is based upon the excitation of the L3 level is given by $(r_{L3} - 1)/(r_{L1} \cdot r_{L2} \cdot r_{L3})$ and for the L_β line the ratio is based upon excitation of the L2 level and is given by $(r_{L2} - 1)/(r_{L1} \cdot r_{L2})$ [18]. The fluorescence yield ω_k for the k shell is determined using the polynomial given by Reed [17] and for the L shell by $\omega_L = Z^4/(10^8 + Z^4)$. When dealing with L line excitation there is some uncertainty regarding the ionisation cross-section to be used and therefore the formula for K radiation is applied and a multiplication factor (P_{ij}) introduced into equation (6), with $P_{kk} = P_{LL} = 1$, $P_{kL} = 0.25$ and $P_{Lk} = 4$.

Continuum fluorescence is often ignored in correction procedures and this can result in large errors ($\sim 10\%$), if there is a large difference between the elements absorption coefficients. In the present program the continuum fluorescence correction derived by Springer [4] has been adopted with the constant terms (4.34×10^{-6} for K and 3.13×10^{-6} for L radiation) modified in line with the suggestion of Reed [17, 18] to 3.85×10^{-6} and 1.08×10^{-6} for K and L radiation, respectively. The correction has been constructed to include all cases where absorption edges (K, L_1, L_2, L_3, M_1, M_2, M_3, M_4, M_5, N_1) lie between the incident electron beam energy and the critical excitation energy of the X-ray line of interest.

Testing of the Method

The method has been rigorously tested on a wide range of microanalysis data [1]. For assessing the performance on elements $11 < Z < 93$, data on well characterized binary alloys collated by Sewell et al. [8] have been used. The RMS error is $\sim 3\%$, which is not substantially larger than the estimated experimental error of $\sim 2\%$. Ultra-light element performance has been assessed using oxide and fluoride data of Sewell et al. [8]. and the carbon and boron measurements of Bastin and Heijligers [14, 15]. Again, the results of the tests show the corrected data to agree with the stochiometric composition within 5% relative in most cases provided Bastin and Heijligers' mass absorption coefficients are used for carbon and boron K_α radiation.

Having proved the accuracy of the method for the correction factors in the next stage the principles were incorporated in a stand-alone computer program which was designed to be user friendly, and versatile.

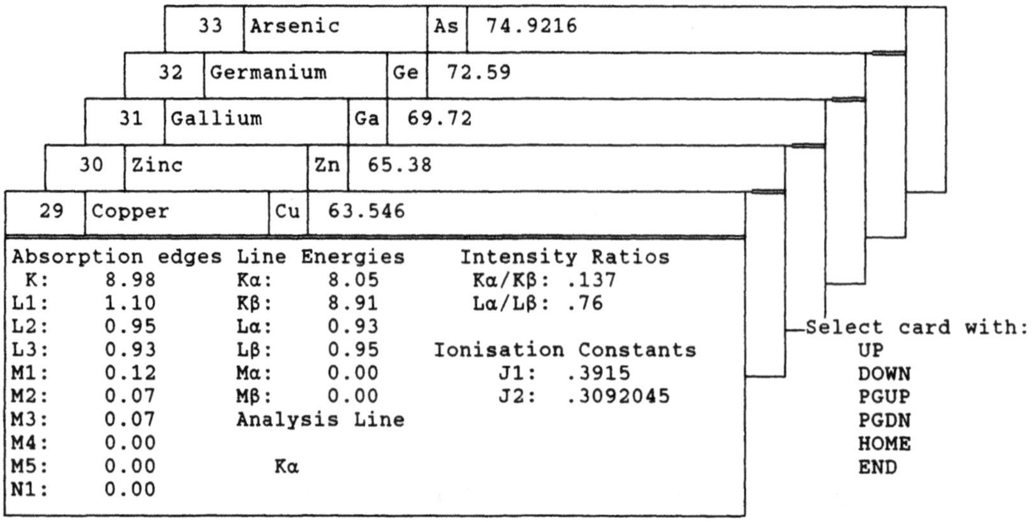

Fig. 3. Example of the computer card file Z-dependent data base. Intensity ratios should read K_β/K_α, L_β/L_α

Program Description

The program is a menu-based system allowing the operator to select the input/output parameters and includes a comprehensive data base of atomic number (Z)-dependent variables, which can be updated using a simple computer card filing system (Fig. 3). All variables not dependent on concentration are calculated outside the iteration process (thus enhancing processing speed). It is structured in such a way that any particular part can be updated easily without any extensive knowledge of computing or a detailed understanding of the operation of the program itself. To do this the program is divided into modules, each of which has a specific function. The correction procedure flowchart (Fig. 4) illustrates the main modules comprising the procedure. Each module has specific inputs and outputs; for example, the module containing the MAC algorithms has input comprising of line or critical energy and absorbent atomic number with the output being the MAC.

The program is initially being written in an extended version of BASIC (Microsoft QuickBASIC), which will run on standard IBM-PC compatible machines. The BASIC language is used because it is widely understood and the version selected, Microsoft QuickBASIC, has powerful editing and debugging facilities which make program modification straightforward. QuickBASIC also has the following advantages:

(1) Variables may be of single and double precision.
(2) DO loops are available as in FORTRAN.
(3) CASE statements and procedures (modules) may be used as in PASCAL.
(4) Data files may be randomly accessed thereby greatly enhancing data flow rate.
(5) Programs are compilable which substantially reduces the processing time.

Currently, the program is being tested on a series of well characterized specimens showing promising initial results. The final version of the program will be available soon.

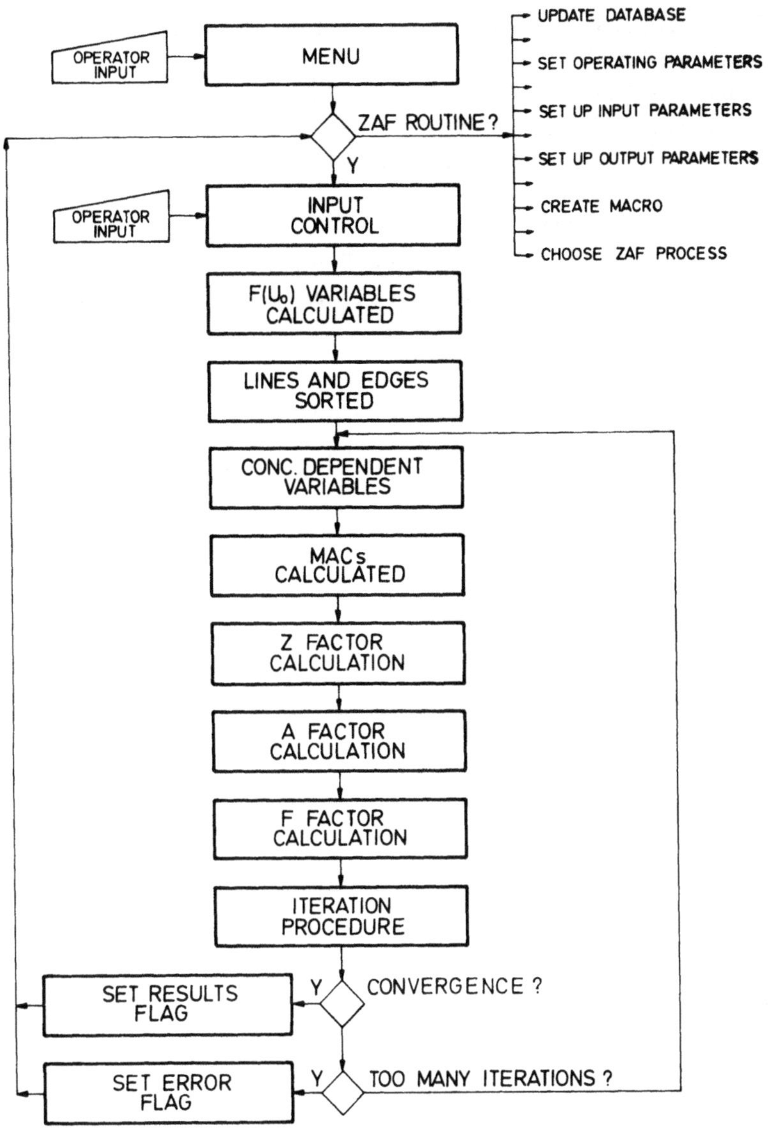

Fig. 4. Flowchart showing the structure of the correction procedure

References

[1] G. Love, V. D. Scott, *X-ray Spectrom.* (in press).
[2] D. A. Sewell, G. Love, V. D. Scott, *J. Phys. D.* **1987**, *20*, 1567.
[3] R. Castaing *Ph.D. Thesis*, University Paris, 1951.
[4] G. Springer, *N Jahrb. Miner. Abh.* **1967**, *106*, 241.
[5] H. A. Bethe, J. Ashkin, *Experimental Nuclear Physics*, Wiley, New York, 1953, p. 252.
[6] G. Love, M. G. C. Cox, V. D. Scott, *J. Phys. D.* **1978**, *9*, 7.
[7] M. Green, V. E Cosslett, *Proc. Phys. Soc.* **1961**, *78*, 1206.
[8] D. A. Sewell, G. Love, V. D. Scott, *J. Phys. D.* **1985**, *18*, 1245.
[9] D. A. Sewell, G. Love, V. D. Scott, *J. Phys. D.* **1985**, *18*, 1233.
[10] K. F. J. Heinrich, *Proc. of 11th Int. Cong. X-ray Optics and Microanalysis* (J. D. Brown, A. H. Packwood, eds.) University of Western Ontario, 1986, p. 67.
[11] J. A. Bearden, *X-ray Wavelengths*, U.S. Atomic Energy Commission, Report NYO 10586, 1964.
[12] H. Kleykamp, *Z. Naturforsch.* **1981**, *36A*, 1388.

[13] B. L. Henke, P. Lee, T. J. Tanaka, R. L. Shimabukuro, B. K. Fujikawa, *Atomic Data and Nuclear Data Tables, Vol. 27*, Academic, New York, 1982, p. 1.
[14] G. F. Bastin, H. J. M. Heijligers, *Quantative Electron Probe Microanalysis of Carbon in Binary Carbides*, University of Technology, Eindhoven, 1984.
[15] G. F. Bastin, H. J. M. Heijligers, *Quantative Electron Probe Microanalysis of Boron in Binary Borides*, University of Technology, Eindhoven, 1986.
[16] G. Love, V. D. Scott, *Scanning* **1981**, *4*, 111.
[17] Ṣ. J. B. Reed, *Microbeam Analysis*, San Francisco Press, San Francisco, 1990, p. 109.
[18] S. J. B. Reed, *Electron Microprobe Analysis*, Cambridge University Press, Cambridge, 1975, p. 280.

Mikrochim. Acta (1992) [Suppl.] 12: 125–129

Monte Carlo Simulation of Backscattered and Secondary Electron Profiles

Christian Eisenschmidt* and Ulf Werner

Max-Planck-Institut für Mikrostrukturphysik, Weinberg 2, D-O-4020 Halle, Federal Republic of Germany

Abstract. A model of the Monte Carlo simulation is used for calculating profiles of the backscattered electron intensity and the secondary electron yield caused by the scanning electron beam. In order to get a fast yield calculation a simple procedure has been developed to describe the escape of secondary electrons. In the present paper first simulations are discussed with the aim of comparing backscattered electron and secondary electron profiles of wall and trench structures. Special attention is given to the influence of the shape of the target surface topography on the electron backscattering and secondary emission.

Key words: Monte Carlo simulation, secondary electrons, backscattered electrons, topographical contrast.

One of the main problems in imaging topographical structures by scanning electron microscopy (SEM) is the way in which the topography of the surface is reflected in the contrast appearing in backscattered electron (BSE) and secondary electron (SE) profiles. This contrast is caused either by the number of electrons backscattered in dependence on the topographical surface structure near the position of the incident electron beam, or by the numher of secondary electrons released by the incident and backscattered electrons. A further important aspect is the fact that BSE can *repenetrate* into the target. Such electrons produce additional SE and can be backscattered once more, i.e. these electrons modify the image contrast (Reimer et al. [1]). This fact has to be considered carefully when solving metrological problems, since target structures located some electron ranges away from the position of the incident electron beam can affect the measurement. The complicated history which repenetrated BSE undergo can be investigated in an excellent manner by the Monte Carlo simulation of the electron-target interaction. Using the Monte Carlo simulation of BSE profiles Reimer et al. [1] treated this problem by inspecting surface edges in dependence on the take-off angle. Joy [2] included the SE emission by using an exponential law to describe the escape probability of SE in dependence on the depth

* To whom correspondence should be addressed

of their production. Since this analytical description of the SE escape applies only to an infinitely flat surface this approach seems to be inappropriate to be applied to structures of a more complicated topography. In order to describe the SE emission from a complicated topography appropriately and more accurately it is necessary to trace each SE until it will be absorbed or emitted, as it was done by Kotera et al. [3]. Since such a simulation including all the interactions of the travelling SE is very time-consuming, instead of considering these interactions explicitly we calculated the SE emission by assuming a straight propagation of the SE generated along the trajectory of the primary electron (PE). First results of our simulation procedure applied to wall and trench structures are presented and discussed in the light of the influence of the surface profile on the BSE and SE profiles.

Simulation Model

The procedure starts with the calculation of the trajectories of PE using a Monte Carlo simulation procedure with a multiple scattering model based on the screened Rutherford scattering cross section (Werner and Heydenreich [4]). For a given primary energy the path-length of PE is divided into 50 segments. For each segment the number of released SE is calculated by dividing the energy loss of the PE along the segment by the mean energy to generate a SE (70 eV for Si as used by Joy [2]). Along the segments of the PE trajectories a definite number of points having the same distance in between, are defined as isotropic sources of the SE production (see Fig. 1). The number of point sources per segment and the number of simulated SE trajectories per source can arbitrarily be modified. We have found that for an energy of 5 keV a satisfactory option is given by three sources per segment and four simulated SE trajectories per source. It is assumed that the SE are emitted by the sources isotropically on straight paths until they are absorbed or until they have escaped from the target (see Fig. 1). In this simple model, the angular scattering

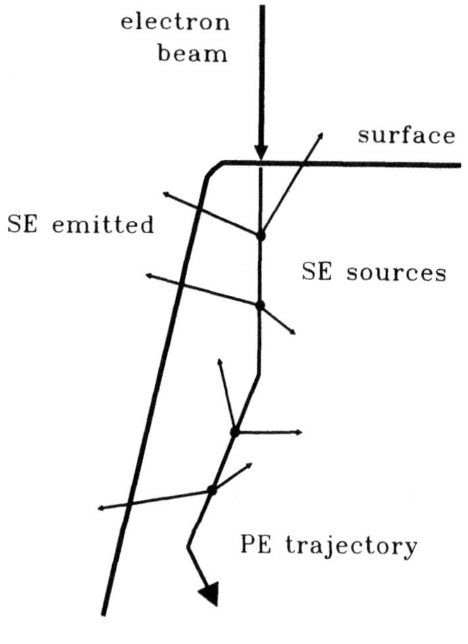

Fig. 1. Scheme of the simulation of the SE emission from two isotropic SE sources per segment of the incident primary electron and two released SE per source

events which the SE suffer on their paths and the condition for the SE to overcome the surface barrier are neglected. The mean path length of SE is chosen in such a way that the escape probability calculated by using the presented model for the SE emission from an infinitely flat horizontal plane fits the exponential escape probability applied by Joy [2]. For silicon and a primary energy of 5 keV a mean path-length of 6 nm is fitted. The actual path length the SE will travel within the target is considered to be distributed according to an exponential distribution, i.e. the mean path length has to be multiplied by the negative logarithm of an equidistributed random number. If a BSE repenetrates into the target the simulation of the electron trajectory and of the SE emission is carried on until the repenetrated electron is absorbed or backscattered once more. Each simulated profile shown in the figures results from 100 beam positions having equal distances of 50 nm in between. In each beam position the histories of 1000 incident electrons are simulated while all BSE finally escaping from the target and all SE emitted from the surface, are considered to be collected. For taking into account the distribution of the primary electrons within the incident electron beam the simulated profiles are convoluted with a Gaussian distribution. In the present simulations a beam diameter of 40 nm is used which is defined by $d = 2\sqrt{2}\,\sigma$, with σ denoting the standard deviation of the distribution.

Results and Discussion

Figures 2a and b show two-dimensional surface structures—i.e. walls and trenches, respectively—chosen to investigate the influence of both the surface shape and the BSE repenetration on the image contrast. The surface shapes are described by the well-known Fermi function which is mirrored at the central vertical line of both the diagrams. Both half width and height of the silicon structures are 2 μm. All calculations were performed at a primary electron beam energy of 5 keV with the electron range being about 0.5 μm. Comparing the BSE profiles of Figs. 2c and d, reveals the effect of surface details which are more than one electron range distant from the position of the incident electron beam. All the edges marked by either (A) or (B) in the surface structures indicated should have the same contrast provided that for the trench the edges would not partly absorb the electrons backscattered inside the trench. This absorption of BSE repenetrated into the trench edges causes a drop of the profile peaks in Fig. 2d unlike those in Fig. 2c. But this drop, however, is not uniformly distributed over the peaks. That part of a peak created by electrons backscattered near the bottom of a trench edge is more strongly affected than the remaining part, since these electrons are more strongly shadowed by the opposite trench edge. This leads to a shift of the peak centre with respect to the position of the corresponding profile peak of the wall.

Comparing the profiles in Figs. 2c and e, discloses an effect of the repenetration of BSE into the target. With the scanning electron beam approaching the foot of the wall (from left) the rate of the BSE decreases (Fig. 2c, left), which is due to the absorption of an increasing number of backscattered electrons in the wall. Unlike this, the SE profile increases (Fig. 2e, left). The SE yield rises since a relatively large number of electrons backscattered from the flat surface penetrate into the wall and release SE.

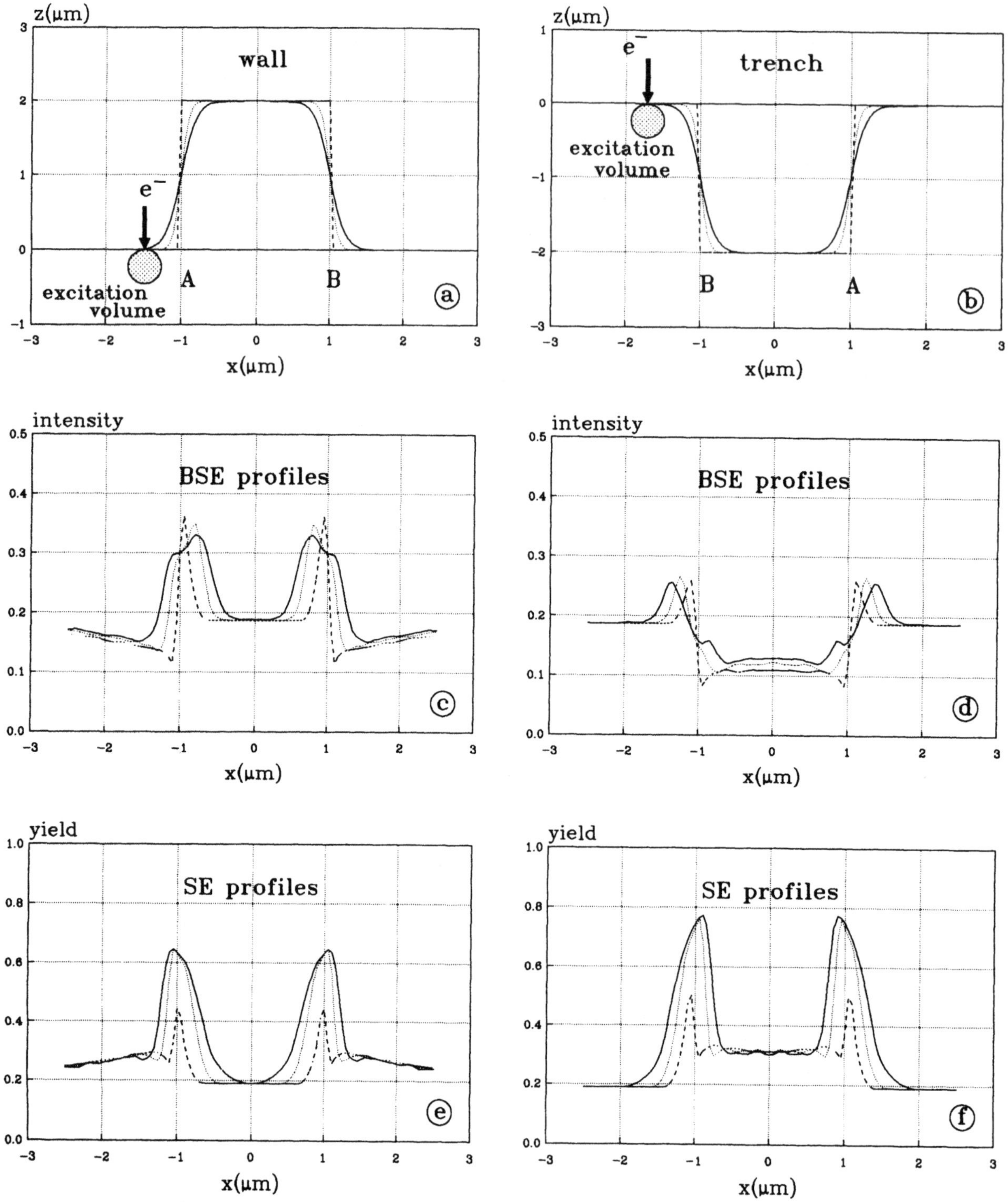

Fig. 2. Simulated BSE and SE profiles of wall (**c** and **e**) and trench structures (**d** and **f**) caused by a perpendicular incident electron beam (e^-) of 5 keV scanning the surface structures. The inclination of the structure edges (angle between edge and plane surface) of the walls (**a**) and trenches (**b**) is fixed at 80 (solid curves), 85 (dotted curves), and 90 degrees (dashed curves)

The positions of the peaks of the BSE profiles are approximately determined by the position of the upper end of the edges of the surface structures (cf. Figs. 2a and c, 2b and d). In these positions the primary electrons can escape from the target in the vertical direction as well as horizontally. The peaks of the SE profiles (Figs. 2e and f) show positions of the maxima which do not coincide with the maximum backscattering. The positions of the SE peaks are obviously determined by the high SE emission caused by primary electrons incident into the strongly declined surface of the edges and, mainly for the trenches, by the SE emission due to repenetrating BSE. Therefore, the SE peaks are situated near the middle of the edges (cf. Figs. 2a and e, 2b and f). This means that these peaks do almost not change their lateral positions with changing edge inclinations. Thus, the distance between the position of a BSE peak, almost indicating the upper end of an edge, and the position of the corresponding SE peak, roughly indicating the middle of an edge, is governed by the inclination of the edges.

Conclusions

Despite the simple character of the model in the present state, it enables assertions to be made as to the contrast arising from complicated topographic structures, which is prerequisite to the treating of metrological problems. The presented simulations show the important role of repenetrating BSE in contrast formation. The BSE and SE profiles calculated across both wall and trench structures of differently inclined edges show that relative to the positions of the BSE maxima, those of the SE maxima do not depend so strongly on the inclination of the edges. This fact has to be considered if walls and trenches are measured by SEM. Some modifications of the model with respect to the depth distribution of the SE escape probability as well as to the mean path length of the produced SE are necessary if the model should be used over the whole relevant energy range. A quantitative comparison of the simulated profiles with measured ones is required and in preparation.

References

[1] L. Reimer, M. Riepenhausen, M. Schierjott, *Scanning* **1986**, *8*, 164.
[2] D. C. Joy, *Scanning Microscopy* **1988**, *2*, 57.
[3] M. Kotera, T. Kishida, T. Fujiwara, H. Suga, D. W. Wittry, in: *12th Int. Congress X-ray Optics Microanal. Vol. 1*, Cracow 1989, p. 231.
[4] U. Werner, J. Heydenreich, *Ultramicroscopy* **1984**, *15*, 17.

Mikrochim. Acta (1992) [Suppl.] 12: 131–137

An Electron Scattering Model Applied to the Determination of Film Thicknesses Using Electron Probe Microanalysis

Hans-Jürgen August

Institut für Angewandte und Technische Physik, Technische Universität Wien,
Wiedner Hauptstrasse 8–10, A-1040 Wien, Austria
Present address: Siemens Österreich AG, Entwicklung Kommunikationstechnik, Hainburger Strasse 33, A-1030 Wien, Austria

Abstract. An electron scattering model, which has already been shown to be reliable in predicting X-ray intensities and depth distributions of X-ray generation in case of homogeneous samples, has been further developed and applied to layered structures. The accuracy of the model is checked by comparing calculated thin film intensity ratios (k-ratios) with those found experimentally. The deviations found are small and generally not exceed 6%. Furthermore, corresponding depth distributions have been calculated and are discussed from a physical point of view. The results prove that the electron scattering model copes with any number of layers and represents a time-saving alternative to Monte Carlo algorithms.

Key words: Electron probe microanalysis, electron scattering, film thickness.

Due to the increasing demand for electron probe techniques which enable in-depth analysis, publications on this subject are too manifold to be discussed in a single paper. Since the first investigations of Cockett and Davies in 1963 [1], much work has been done [e.g. 2, 3], culminating in the already quite sophisticated analytical model of Pouchou and Pichoir [4, 5, 6], which was the first to cope with multilayer structures.

Also Monte Carlo simulations have contributed to a better understanding of X-ray generation in thin films. The first important work on this subject was presented in 1974 by Kyser and Murata [7] and recently first results of the Monte Carlo approach quantifying both X-ray intensities and depth distribution functions have been published by Ammann and Karduck [8] and Karduck and Ammann [9]. These results show that the depth distributions in complex structures can no longer be described realistically using simple mathematical functions.

The development of a new electron scattering model [10], which has proved successful in predicting depth distributions in homogeneous samples [11], offers the possibility of investigating electron scattering and X-ray generation phenomena on a fundamental physical level. The present paper deals with the generalization of this model to samples consisting of any number of layers and with results from the corresponding calculations.

Theory

The principle of the electron scattering model, which is based on the theory of "multiple reflection" first presented by Cosslett and Thomas [12], has already been presented and discussed in detail elsewhere [11]. To cope with any number of layers, the model has been developed further. In this respect crucial developments are, (1) the employment of the theory for calculating backscattering coefficients for layered structures [13] and, (2) the introduction of effective mass depths. The calculation of the latter quantity shall be explained by means of an example: The effective mass thickness ρD^* of an Au film of real mass thickness ρD on a substrate of Al with regard to the electron transmission coefficient τ is given by the condition $\tau(\text{Au}, \rho D) = \tau(\text{Al}, \rho D^*)$. It is obvious that by applying subsequently this condition to more than one layer the model can be generalized to any number of layers.

Results and Discussion

Figure 1 shows a comparison of calculated values of the k-ratio of Al-K_α with experimental data of Reuter [14] for a film of Al on a substrate of B. Although the atomic number difference of the film and the substrate material is not dramatic, this example is interesting because of the low electron energies applied, viz. $E_0 = 4$ keV and 12 keV, and because the substrate consists of a very light element, so that the ratio of the backscattering coefficients of the film and the substrate material, respectively, is quite large, viz. 3.3 at $E_0 = 12$ keV. The deviations between measured and calculated data sets are very small, not exceeding approx. 3%

The results for a system with a very large atomic number difference between the film and the substrate material is given in Fig. 2. The calculated k-ratios for a Pt

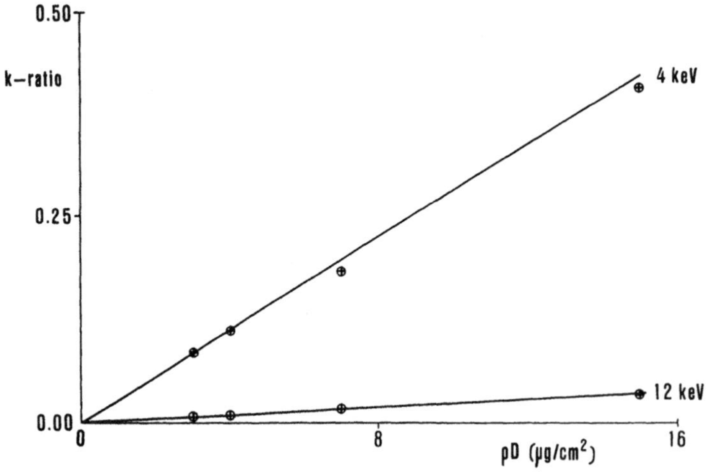

Fig. 1. Comparison of calculated values (solid line) of the k-ratio with those found experimentally by Reuter [14] (⊕). Film of Al on substrate of B. Primary electron energies $E_0 = 4$ keV (upper curve) and 12 keV (lower curve). Take-off angle $\psi = 52.5°$. ρD is mass thicknesses

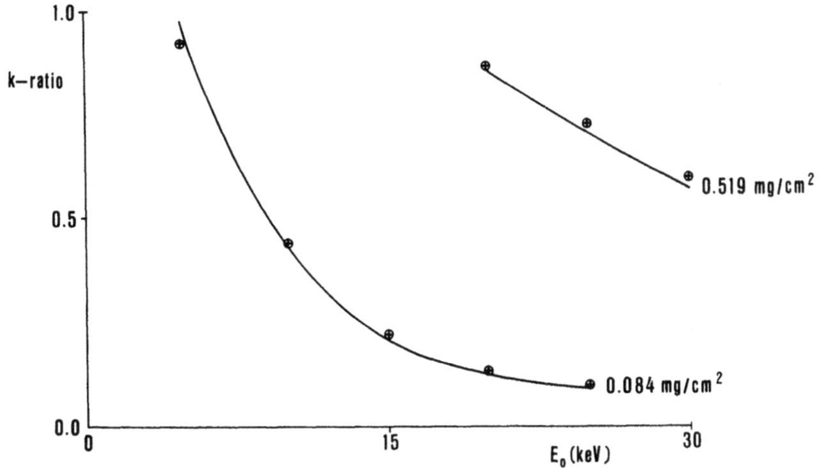

Fig. 2. Comparison of calculated values (Solid line) of the k-ratio with those found experimentally by Reuter et al. [15] (⊕). Film of Pt on substrate of Si. Mass thicknesses $\rho D = 0.519$ mg/cm^2 and 0.084 mg/cm^2. Take-off angle $\psi = 52.5°$

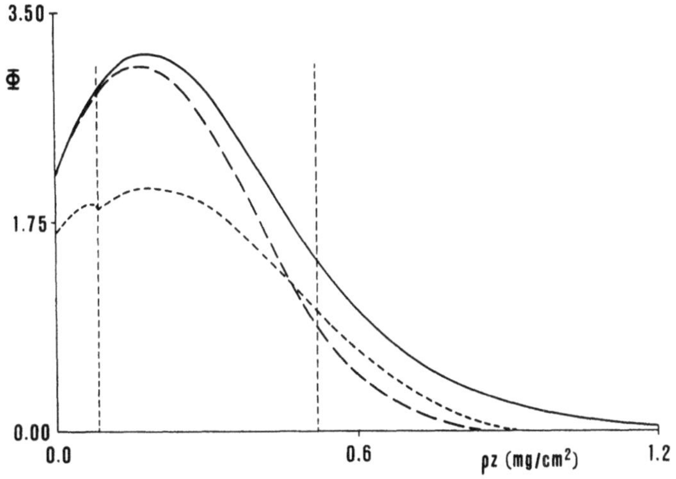

Fig. 3. Comparison of depth distributions for Pt-M_α radiation: Bulk sample of Pt (full curve), samples consisting of Pt-films ($\rho D = 0.084$ mg/cm^2, short-dashed line; $\rho D = 0.519$ mg/cm^2, long-dashed line) on Si. Primary electron energy $E_0 = 20$ keV

film on a substrate of Si correspond well with those found experimentally by Reuter et al. [15].

Figure 3 shows calculated depth distribution functions corresponding to the system defined in Fig. 2 for a primary electron energy $E_0 = 20$ keV. The curves were calculated assuming a tracer experiment, using an infinitesimally thin tracer of Pt which traverses the layered samples defined above. As can be seen, the influence of the layered structure is comparatively small in the case of the thicker layer (long-dashed line), while the reduction of the depth distribution curve in the case of the thinner layer (short-dashed line) is quite dramatic. The latter case is also interesting because of the unfamiliar shape of this curve, having two maxima, the first in the Pt layer, the second in the Si substrate. This can be explained by the fact that maxima of depth distribution curves are situated closer to the sample surface if the atomic number increases. The decrease of the curve in the Pt film in the vicinity of the interface is due to the fact that the amount of backscattered electrons decreases in this region because of the comparatively low backscattering coefficient of Si. In principle similar curves have also been reported recently by Ammann and Karduck [8].

While only k-ratios for X-rays from the film have been presented so far, Fig. 4 deals with X-rays emerging from a substrate of Si covered by a film of Ti. Deviations of the calculated curves from the data, measured by Bastin et al. [16], are very small for both mass thicknesses at all primary energies.

Figure 5 shows results for a system consisting of a Cu film on W, the W-L_α line having been measured by Pouchou and Pichoir [17]. Although in this case the

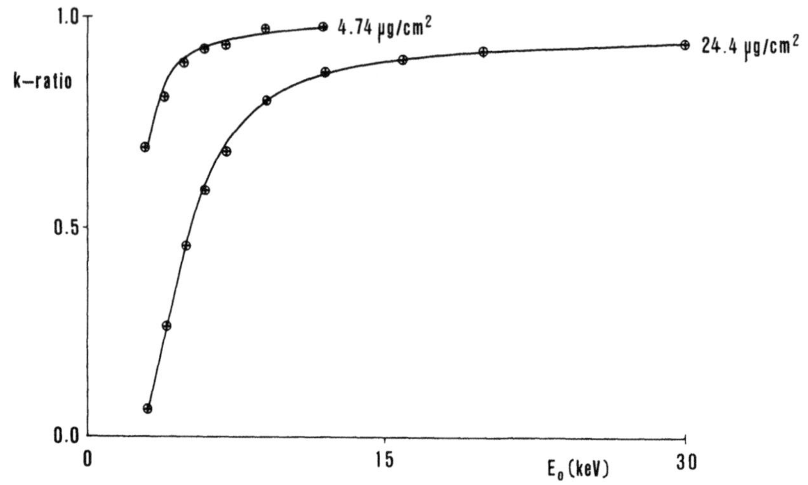

Fig. 4. Comparison of calculated values (Solid line) of the k-ratio of Si k_α with those found experimentally by Bastin et al. [16] (⊕). Film of Ti on substrate of Si. Mass thicknesses $\rho D = 4.74\ \mu g/cm^2$ and $24.4\ \mu g/cm^2$. Take-off angle $\psi = 40°$

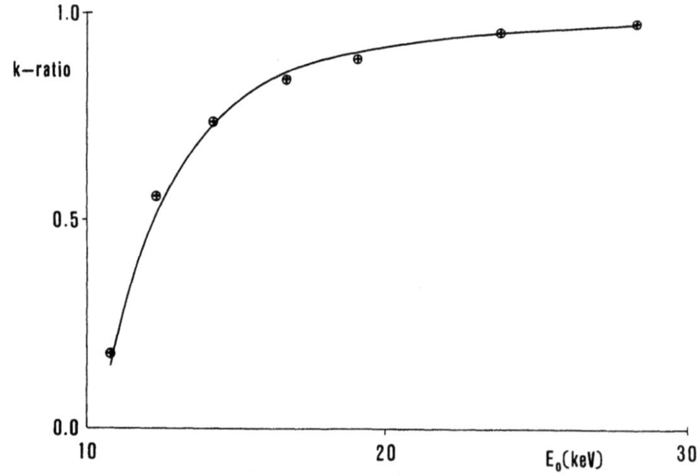

Fig. 5. Comparison of calculated values (Solid line) of the k-ratio of WL_α with those found experimentally by Pouchou and Pichoir [17] (⊕). Film of Cu on substrate of W. Mass thickness $\rho D = 42.7\ \mu g/cm^2$. Take-off angle $\psi = 18°$

deviations are somewhat larger, the correspondence of the data sets allows to determine the mass thickness with an accuracy of a few percent.

A more complicated system is dealt with in Fig. 6. In this case a substrate of Al is covered by three films consisting of, from top to bottom, Al, CuO, and Cu, respectively. Both the Al-K_α and the Cu-L_α line have been measured by Pouchou and Pichoir [4]. As can be seen, the X-ray k-ratios of even this complex structure are reproduced well by the electron scattering model. The correspondence with the experimental data is somewhat better in the case of the Al-K_α data, but the accuracy with regard to the reproduction of the Cu-L_α k-ratios is quite good.

The corresponding depth distribution curve at $E_0 = 10$ keV for Al-K_α radiation is shown in Fig. 7 and compared with the curve obtained for a bulk specimen of Al.

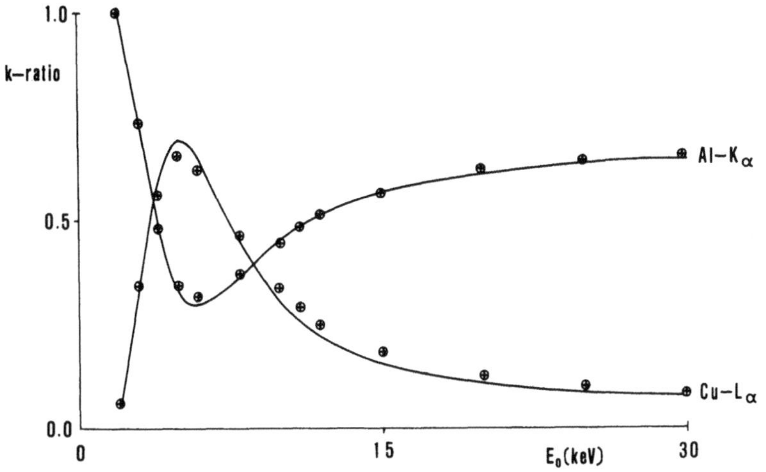

Fig. 6. Comparison of calculated values (Solid line) of the k-ratio with those found experimentally by Pouchou and Pichoir [4] (⊕). Film of Al ($\rho D = 13.76\ \mu g/cm^2$) on film of CuO ($\rho D = 5.44\ \mu g/cm^2$) on film of Cu ($\rho D = 51.96\ \mu g/cm^2$) on substrate of Al. Take-off angle $\psi = 40°$

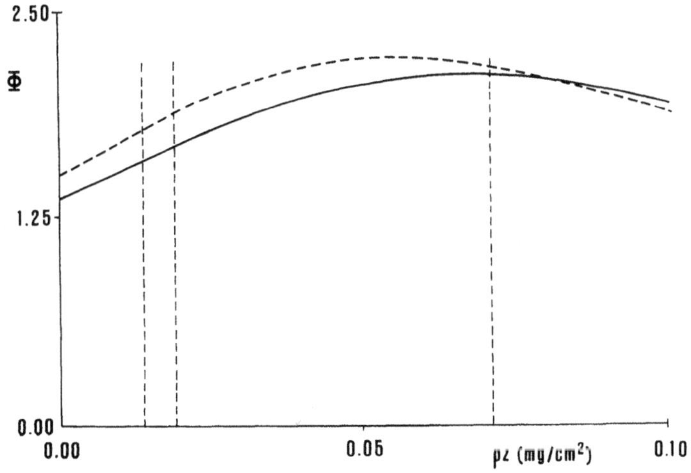

Fig. 7. Comparison of depth distribution curves for Al-K_α radiation. Bulk sample of Al (Solid curve) and sample (dashed line) consisting of a film of Al ($\rho D = 13.76\ \mu g/cm^2$) on a film of CuO ($\rho D = 5.44\ \mu g/cm^2$) on a film of Cu ($\rho D = 51.96\ \mu g/cm^2$) on a substrate of Al. Primary electron energy $E_0 = 10$ keV

Due to the larger backscattering coefficient of Cu as compared to Al the amount of generated X-rays is increased in surface-near regions. At larger mass depths, however, the curve in the layered structure decreases more rapidly due to the comparatively lower transmission coefficient of Cu.

Conclusion

The further development of the electron scattering model has been proved successful in calculating X-ray intensity ratios for samples consisting of even more than one layer. The accuracy of the model has been checked by comparison with experimental data obtained from a variety of samples covering a wide range of elements. Also with regard to the primary electron energy a wide range of values has been investigated, viz. between 2 keV and 30 keV.

Contrary to methods applying simple weighing procedures to obtain depth distribution functions for layered samples, the electron scattering model supplies the investigator also with physically plausible depth distribution functions, which of course can no longer be described using simple mathematical functions. According to the new model the resulting depth distributions are consistent also at the interfaces, which corresponds to basic physical considerations.

With regard to the performances, this model can be compared with Monte Carlo simulations, but it additionally offers the advantage of a considerable (factor approx. 100–200) saving of computing time. Although not shown explicitely in this paper, it is clear that—if an adequate database of experimental results is available—the composition of multilayer samples can also be obtained by iterative methods based on the new model. For this purpose several measurements carried out at different acceleration energies would be required. The electron scattering model presented in this paper thus offers itself as a fast alternative for the evaluation of measured data to obtain both film thicknesses and compositions.

Acknowledgement. The author gratefully acknowledges the support given by the Austrian Fonds zur Förderung der wissenschaftlichen Forschung (Projekt P7336 Phy).

References

[1] H. Cockett, C. D. Davies, *J. Appl. Phys.* **1963**, *14*, 813.
[2] H. Packwood, G. Remond, J. D. Brown, in: Proc. 11th ICXOM (J. D. Brown, R. H. Packwood, eds.), London, Ontario, 1986, p. 274.
[3] H.-J. August, J. Wernisch, *Scanning* **1987**, *9*, 145.
[4] J.-L. Pouchou, F. Pichoir, *Rech. Aérospatiale* **1984**, 349.
[5] J.-L. Pouchou, in: 1st EMAS-Workshop EMAS, Antwerp 1989, p. 127.
[6] J.-L. Pouchou, in: *Microanalyse par Sonde Electronique, Aspects Quantitatifs*, ANRT, Paris, 1989, section K.
[7] D. F. Kyser, K. Murata, *IBM J. Res. Dev.* **1974**, *18*, 352.
[8] N. Ammann, P. Karduck, in: *Proc. XIIth Int. Cong. for Electron Microscopy*, San Francisco Press, San Francisco, 1990, p. 214.
[9] P. Karduck, N. Ammann, in: *Proc. XIIth Int. Cong. for Electron Microscopy*, San Francisco Press, San Francisco, 1990, p. 14.

[10] H.-J. August, J. Wernisch, *X-ray Spectrom.* **1991**, *20*, 131.
[11] H.-J. August, J. Wernisch, *X-ray Spectrom.* **1991**, *20*, 141.
[12] V. E. Cosslett, R. N. Thomas, *Brit. J. Appl. Phys.* **1965**, *16*, 779.
[13] H.-J. August, J. Wernisch, *J. Microsc.* **1990**, *157*, 247.
[14] W. Reuter, in: *Proc. 6th ICXOM*, Osaka, 1971, (G. Shinoda, K. Kohra, T. Ichinokawa, eds.) University of Tokyo Press, Tokyo, 1972, p. 121.
[15] W. Reuter, J. D. Kuptsis, A. Lurio, D. F. Kyser, *J. Phys. D* **1978**, *11*, 2633.
[16] G. F. Bastin, H. J. M. Heijligers, J. M. Dijkstra, in: *Proc. XIIth Int. Cong. for Electron Microscopy*, San Francisco Press, San Francisco, 1990, p. 216.
[17] J.-L. Pouchou, F. Pichoir, *J. de Physique* **1984**, *C2*, 47.

Mikrochim. Acta (1992) [Suppl.] 12: 139–146

Calculation of Depth Distribution Functions for Characteristic and for Continuous Radiation

Hans-Jürgen August*

Institut für Angewandte und Technische Physik, Technische Universität Wien, Wiedner Hauptstrasse 8–10, A-1040 Wien, Austria

Abstract. Starting from empirically determined quantities of electron scattering, a model has been developed which allows the calculation of depth distribution functions for characteristic and continuous radiation. It is shown that both the total generated intensities of characteristic radiation and the distributions in depth are predicted accurately. The deviations in these values with regard to the quantification of the intensities are generally less than 5% when compared with those given by the model of Pouchou and Pichoir. The calculation of depth distribution curves for continuous radiation reveals that the differences between characteristic and continuous curves are larger than commonly expected. These differences depend on the overvoltage and on the atomic number.

Key words: Depth distribution function, continuous radiation, characteristic radiation.

The importance of element-characteristic X-ray depth distribution functions in electron probe microanalysis (EPMA) was already recognized in the early work of Castaing and Descamps [1] in 1955. Since these distribution curves are crucial quantities for the determination of, in principle, all correction factors applied in EPMA, increasing attention has been directed to this subject.

Apart from experimental strategies for obtaining characteristic depth distributions employing either the tracer method proposed by Castaing and Descamps [1] or the wedge specimen method of Schmitz et al. [2], also Monte Carlo methods [e.g. 3–5] and non-statistical mathematical models [e.g. 6–12] have been applied. The main advantage of the latter over Monte Carlo methods is the considerable saving of computing time, which is important especially if iterative calculation procedures are to be applied.

With respect to curves for continuous radiation only little work has been carried out so far. This is also due to experimental problems arising, because neither of the methods mentioned above can be considered as suitable for the determination of

* Present address: Siemens AG Österreich, Entwicklung Kommunikationstechnik, Hainburgerstrasse 33, A-1030 Wien, Austria

continuous radiation curves. A limited number of curves was presented by Reed [13] and Statham [14] as a result of Monte Carlo calculations. On the other hand, the shape of depth distributions for continuous radiation is important for a further development of the peak-to-background method and for any mathematical method for background subtraction.

The present paper gives an outline of the basics of the electron scattering model and presents a comparison of calculated characteristic and continuous radiation curves with those found experimentally for characteristic radiation.

Theory

The depth distribution function $\Phi(\rho z)$ is defined as the ratio of primary radiation generated in a certain mass depth ρz to the value of the cross section of radiation generation $Q(U_0)$, where U_0 equals E_0/E_x, E_0 being the primary electron energy and E_x being the edge energy E_c in the case of characteristic radiation and the radiation energy E_v in the case of continuous radiation.

The mathematical formulation of this definition can be represented as:

$$\Phi(\rho z) = \frac{\int_{U_0}^{1} \int_{0}^{\pi/2} \frac{dN^2(U, \alpha, \rho z)}{dU\, d\alpha} \frac{1}{\cos(\alpha)} Q(U) \sin(\alpha)\, d\alpha\, dU}{Q(U_0)}, \tag{1}$$

where U is the overvoltage ($= E/E_x$), and N is the number of electrons crossing an imaginary horizontal plane in the sample. α denotes the angle between the surface normal and the direction of the electrons.

The electron scattering model presented here can be understood as an elaboration of the ideas presented by Cosslett and Thomas [6] about 25 years ago. This model of "multiple reflection" considers the electrons to be scattered back and forth between two imaginary planes parallel to the sample surface and situated at a certain mass depth ρz. The electrons thus traverse an infinitesimally thin layer situated in between the imaginary planes, see Fig. 1. The aim of the model is to quantify the

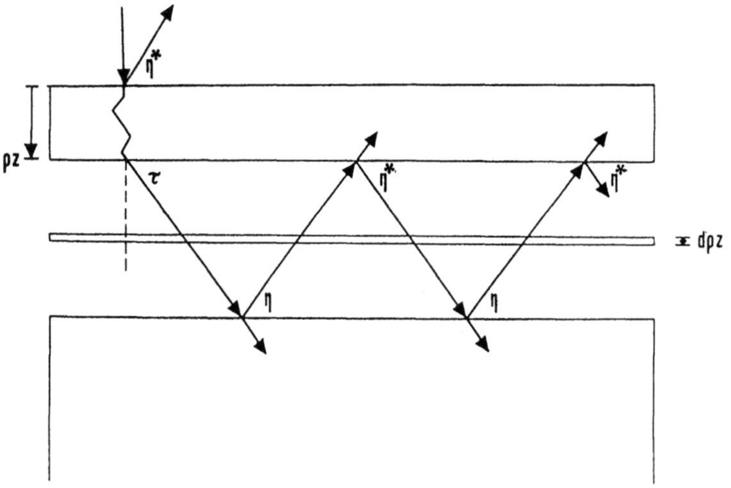

Fig. 1. Principle of the electron scattering model. The electron trajectories are calculated using electron backscattering (η) and transmission (τ) coefficients. The amount of radiation generated in the infinitesimally thin layer between ρz and $\rho z + d\rho z$ is to be quantified

radiation generation in this infinitesimally thin layer for several mass depths ρz, thus obtaining the depth distribution curves $\Phi(\rho z)$.

As can be seen from Fig. 1, the fundamental quantities which are required for the calculations are the backscattering coefficients for bulk $\eta(E, \alpha)$ and for layered samples $\eta^*(E, \alpha, \rho z)$, which of course depend on the electron energy E and on the incidence angle α, the electron transmission coefficient $\tau(E, \rho z)$ as a function of the primary electron energy E_0, the corresponding angular and energy distributions of transmitted and backscattered electrons, and the appropriate cross section for radiation generation.

A comprehensive discussion of this model including the presentation of the mathematical formalism is given elsewhere [10], while this paper is focused on the comparison of resulting depth distribution curves and on the discussion of the respective characteristics; the cross-sections employed in the respective calculations were that of Pouchou and Pichoir [15] for characteristic radiation and that of Kramers [16] for continuous radiation.

The units used are keV for energies and mg/cm^2 for mass thicknesses.

Results and Discussion

Figure 2 shows the ratios of the total generated intensities of characteristic radiation according to Pouchou and Pichoir [15] to those obtained using the new model. As can be seen, the best correspondences are obtained for Al at low overvoltages, while the differences at large overvoltages in the case of Au are larger. Nevertheless, the largest discrepancies amount to less than 7%, which means that both models yield

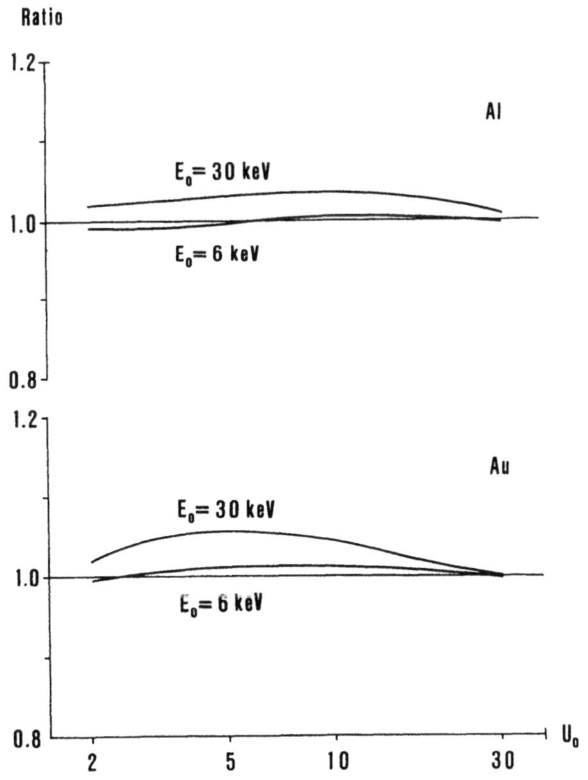

Fig. 2. Comparison of intensity ratios (Pouchou and Pichoir's values/new values) for Al and Au targets as a function of the overvoltage

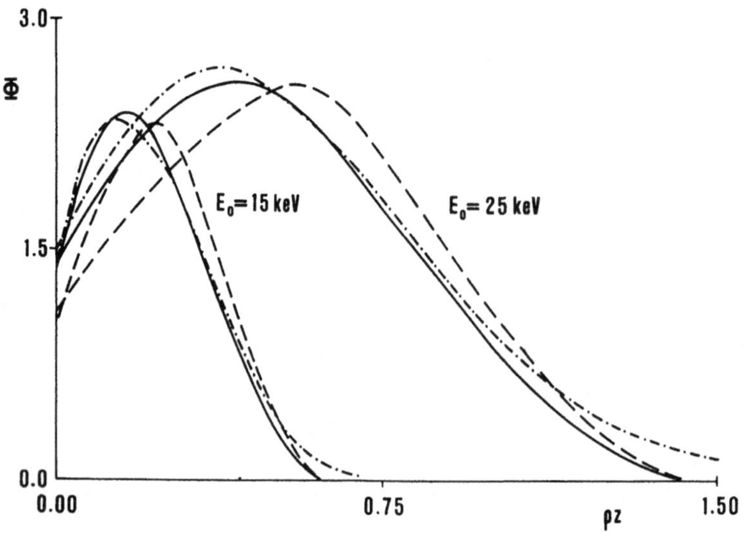

Fig. 3. Comparison of depth distribution functions in Al at $E_0 = 25$ keV and $E_0 = 15$ keV. Calculated (solid lines) and measured (dot-dashed lines) [20] curves for characteristic radiation (Mg-tracer, K_α-line), and calculated curves (dashed lines) for continuous radiation (with $E_v = 1.254$ keV). Mass depth units in mg/cm²

quite similar results. The model of Pouchou and Pichoir was chosen for comparison because it is known to be realistic and reliable [17–19].

In the following, measured curves for characteristic radiation will be compared with those calculated using the new model and with curves calculated for continuous radiation. To allow an easier comparison the latter are normalized so as to include the same area as the calculated curves for characteristic radiation. Continuous radiation curves are calculated for a radiation energy corresponding to the edge energy taken for the calculation of the characteristic curves.

A comparison of calculated curves with that obtained experimentally by Castaing and Hénoc [20] for characteristic radiation is shown in Fig. 3. The correspondence of the calculated curves for characteristic radiation with those measured is good for both $E_0 = 15$ keV and $E_0 = 25$ keV. With regard to the comparison with the continuous radiation curves some characteristic features of the differences can be observed: In the surface-near region the continuous curves are smaller, and their maximum lies deeper in the sample. As can be expected, the maximum range of X-ray production is the same for both types of radiation. This is because the lowest electron energy large enough for radiation generation is the same in both cases. This means that characteristic radiation curves cannot be transformed into continuous radiation distributions by simply changing an electron scattering parameter, e.g. Lenard's constant, as proposed in literature [13, 14], because this alters mainly the range of radiation production.

Depth distribution curves in Al are shown in Fig. 4. Again the correspondence between the curves measured by Sewell et al. [21] and the calculated ones is good, especially in case of $E_0 = 20$ keV. With regard to the continuous curves the same characteristics as discussed above can be observed.

Figure 5 shows curves in a somewhat heavier material, viz. Ti. Also in this case the correspondence of the characteristic curves is regarded good, although the

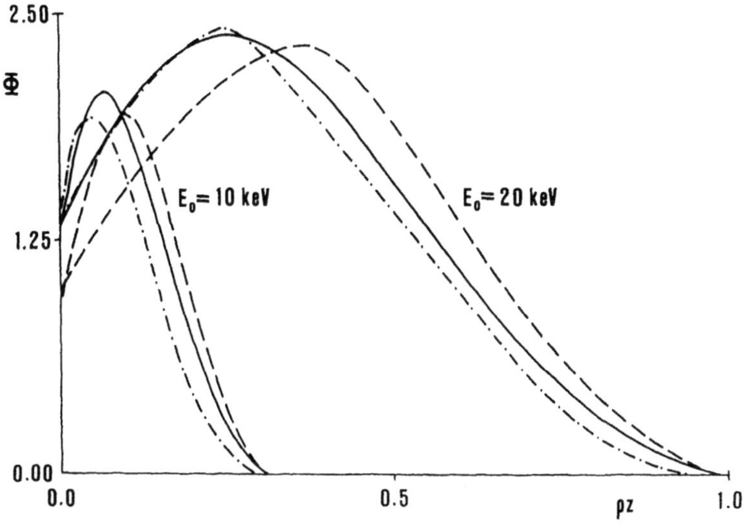

Fig. 4. Comparison of depth distribution functions in Al at $E_0 = 20$ keV and $E_0 = 10$ keV. Calculated (solid lines) and measured (dot-dashed lines) [21] curves for characteristic radiation (Si-tracer, K_α-line), and calculated curves (dashed lines) for continuous radiation (with $E_\nu = 1.74$ keV). Mass depth units in mg/cm^2

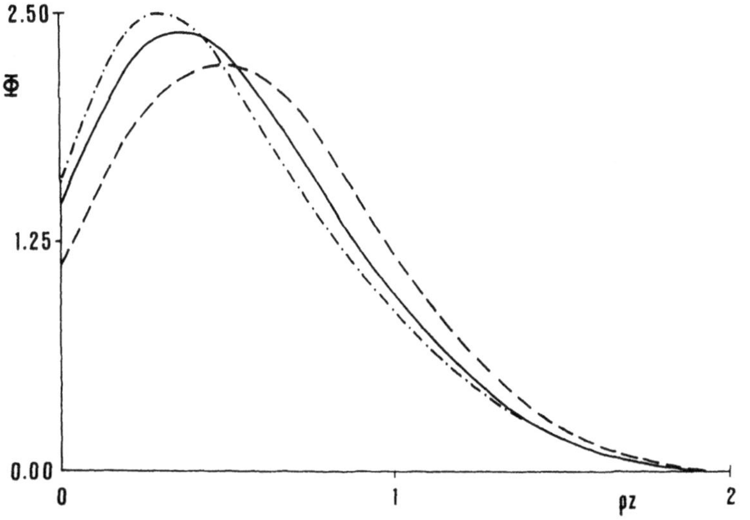

Fig. 5. Comparison of depth distribution functions in Ti at $E_0 = 29$ keV. Calculated (solid line) and measured (dot-dashed line) [22] curves for characteristic radiation (V-tracer, K_α-line), and calculated curve (dashed line) for continuous radiation (with $E_\nu = 4.95$ keV). Mass depth units in mg/cm^2

calculated curve yields a maximum lying somewhat deeper in the sample than that measured by Vignes and Dez [22]. The surface radiation value of the continuous curve again is smaller than that of the characteristic curves, the maximum situated at a larger mass depth. As already discussed in a previous publication [12], the differences between the characteristic and the continuous curves decrease with increasing atomic number and overvoltage. These observations are confirmed by the new results presented here.

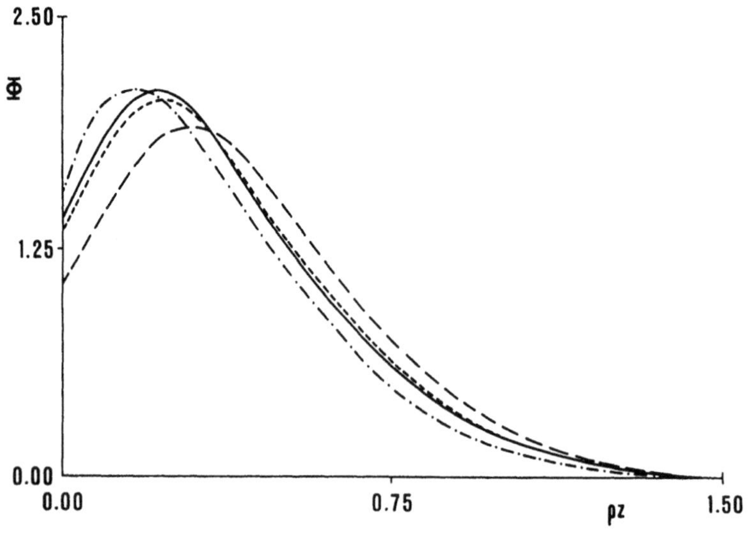

Fig. 6. Comparison of depth distribution functions in Cu at $E_0 = 25$ keV. Calculated (solid line) and measured (dot-dashed line) [23] curves for characteristic radiation (Zn-tracer, K_α-line). Calculated curve (long-dashed line) for continuous radiation (with $E_v = 8.639$ keV) and calculated curve for characteristic radiation with E_c replaced by E_v (short-dashed line). Mass depth units in mg/cm^2

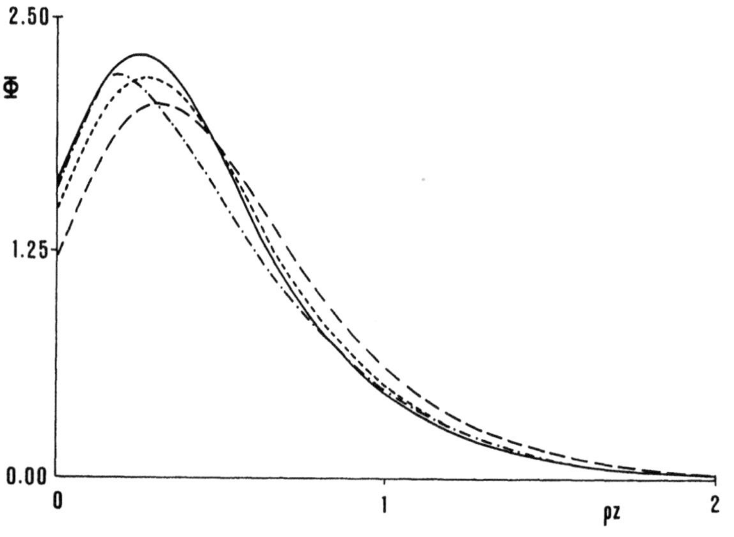

Fig. 7. Comparison of depth distribution functions in Au at $E_0 = 29$ keV. Calculated (solid line) and measured (dot-dashed line) [21] curves for characteristic radiation (Bi-tracer, L_α-line). Calculated curve (long-dashed line) for continuous radiation (with $E_v = 10.84$ keV) and calculated curve for characteristic radiation with E_c replaced by E_v (short-dashed line). Mass depth units in mg/cm^2

The same holds for the curves in a medium weight element, viz. Cu (see Fig . 6), and a heavy element, viz. Bi (see Fig. 7). The calculated characteristic radiation curves correspond well with the curves measured by Brown and Parobek [23] and by Sewell et al. [21], respectively, especially considering that the experimental curve shown in Fig. 7 seems to be too small, giving a too low value for the total intensity. Additional curves in Figs. 6 and 7 show the effect of substituting E_v for E_c when calculating the curve for characteristic radiation. To facilitate comparison, these curves again are normalized to the area under the original characteristic radiation curves. Although the difference between these energy values is very large, E_v being 8.64 keV and 10.84 keV and E_c being 9.66 keV and 13.42 keV, respectively, resulting in large changes of the overvoltage ratio applied, the alterations of the shape of the curves are not dramatic.

Conclusion

The calculated curves for characteristic radiation are close to those found experimentally both with regard to the total generated radiation intensity and with regard to the shape of the corresponding curves. The only systematic deviation is a slight overestimation of the depth at which the maximum of the curve can be found. This deviation increases with increasing atomic number of the matrix element. Nevertheless, it can be concluded that the electron scattering phenomena generally are quantified with high accuracy, so that the model offers itself to be extended in order to investigate other phenomena connected to electron beam microanalysis.

The first of these further developments, the calculation of depth distribution functions for continuous radiation, reveals characteristic deviations between the characteristic and the continuous radiation curves. The bremsstrahlung distributions always have comparatively lower surface values and have maximum values situated deeper in the sample. Especially in the case of heavier elements these maxima are less markedly peaked. As can be expected from physical considerations, the ultimate range of X-ray production is the same for both curves. These findings illustrate that small modifications of the formulae for the calculation of characteristic curves in order to obtain those for continuous radiation, e.g. the reduction of the value of Lenard's constant or the substitution of E_v for E_c, cannot be regarded as satisfactory. Even in cases of large differences between E_v and E_c the effect on the shape of the depth distribution function is comparatively small. A more realistic transformation algorithm, which will certainly depend on the atomic number of the target and the overvoltage ratio, is still to be established.

Acknowledgement. The author gratefully acknowledges the support of the Austrian Fonds zur Förderung der wissenschaftlichen Forschung (Projekt P7336 Phy).

References

[1] R. Castaing, J. Descamps, *J. Phys. Radium* **1955**, *16*, 304.
[2] U. Schmitz, P. L. Ryder, W, Pitsch, in: *Proc. 5th ICXOM Tübingen 1968* (G. Möllenstedt, K. H. Gaukler, eds.), Springer, Berlin, Heidelberg, New York, 1969, p 104.
[3] H. E. Bishop, *Proc. Phys. Soc.* **1965**, *85*, 855.
[4] J.-L. Pouchou, F. Pichoir, F. Girard, *J. Microsc. Spectrosc. Electron.* **1980**, *5*, 425.
[5] W. Rehbach, *Doctoral Thesis*, RWTH Aachen, 1987.
[6] V. E. Cosslett, R. N. Thomas, *Br. J. Appl. Phys.* **1965**, *16*, 779.
[7] D. B. Brown, D. B. Wittry, D. F. Kyser, *J. Appl. Phys.* **1969**, *40*, 1627.
[8] V. Lantto, *J. Phys. D* **1979**, *12*, 1181.
[9] R. H. Packwood, J. D. Brown, *X-ray Spectrom.* **1981**, *10*, 138.
[10] H.-J. August, J. Wernisch, *X-ray Spectrom.* **1991**, *20*, 131.
[11] H.-J. August, J. Wernisch, *X-ray Spectrom.* **1991**, *20*, 141.

[12] H.-J. August, J. Wernisch, *Scanning* **1991**, *13*, 207.
[13] S. J. B. Reed, *X-ray Spectrom.* **1975**, *4*, 14.
[14] P. J. Statham, *X-ray Spectrom.* **1976**, *5*, 154.
[15] J.-L. Pouchou, F. Pichoir, in: *Proc. 11th ICXOM* London, Ontario 1986 (J. D. Brown, R. H. Packwood, eds.), London, Ontario, 1986, p. 247.
[16] H. A. Kramers, *Philos. Mag.* **1923**, *46*, 836.
[17] H.-J. August, *Doctoral Thesis*, Technische Universität Wien, 1988.
[18] H.-J. August, R. Razka, J. Wernisch, *Scanning* **1988**, *10*, 107.
[19] G. F. Bastin, H. J. M. Heijligers, in: *Proc. 1st EMAS-Workshop Antwerp 1989* EMAS, Antwerp, 1989, p. 81.
[20] R. Castaing, J. Hénoc, in: *Proc. 4th ICXOM Orsay 1965* (R. Castaing, P. Deschamps, J. Philibert, eds.), Herrmann, Paris 1966, p. 120.
[21] D. A. Sewell, G. Love, V. D. Scott, *J. Phys. D* **1985**, *18*, 1233.
[22] A. Vignes, G. Dez, *J. Phys. D* **1968**, *1*, 1309.
[23] J. D. Brown, L. Parobek, *Adv. X-ray Anal.* **1973**, *16*, 198.

Mikrochim. Acta (1992) [Suppl.] 12: 147–152

A Method for In-Situ Calibration of Semiconductor Detectors

Johann Wernisch, Angela Schönthaler and Hans-Jürgen August*,**

Institut für Angewandte und Technische Physik, Technische Universität Wien, Wiedner Hauptstrasse 8–10, A-1040 Wien, Austria

Abstract. A Si(Li)-detector has been calibrated for low and medium X-ray energies ($E_v \lesssim 10$ keV) using the method of variation of the X-ray incidence angle, of which the theory is briefly outlined. If the thickness of the Au-contact layer is assumed to be known, the thicknesses of the Si-dead layer and of the Be-window have been determined from a least-squares fit to the experimental data, which were obtained at different X-ray energies. A mathematical simulation of the Bremsstrahlung spectrum, recorded by the detector, confirmed that the obtained results can be regarded as realistic and reliable.

Key words: Si(Li)-detector, calibration

Semiconductor detectors are widely used in connection with many analytical techniques, such as electron probe microanalysis (EPMA), X-ray fluorescence analysis (XRFA) or proton induced X-ray emission (PIXE), since they offer a low-cost possibility of recording in principle all energies which have analytical value, emitted by a sample.

To take advantage of this, it is necessary to know the detector efficiency, which depends mainly on the thicknesses of the layers through which the radiation has to penetrate before arriving at the detector's active zone, viz. the Be-window, the Au-contact layer, and the Si-dead layer. Additionally also other layers which are not genuine parts of the detector may appear, such as contamination layers (e.g. oil from the diffusion pump) or ice layers (due to the temperature gradient found near the detector, which is generally cooled by liquid nitrogen).

Apart from radioactive sources (see, e.g. [1]), also other radiation sources can be used for calibrating the detector, particularly the radiation used normally for the analytical investigations. This particle beam-induced radiation has already been used for the determination of detector efficiencies [1–5].

---

* To whom correspondence should be addressed

** Present address: Siemens Österreich AG, Entwicklung Kommunikationstechnik, Hainburger Strasse 33, A-1030 Wien, Austria

The second main aspect of detector calibration refers to the question of the comparison with detectors of known performance. If the features of the spectrum emitted by a given source are not accurately known, the use of a reference detector will be required. In principle the theoretical X-ray intensities emitted, e.g., from an electron-impact irradiated sample, can be calculated, but unfortunately these intensities cannot be quantified accurately enough.

Baker et al. [6] have presented an approach for calibrating the detector without the use of reference detectors. They used the detector under investigation as a reference spectrometer by simply varying the incidence angle of the X-rays to the detector. This model, which was designed originally for being employed together with a radioactive standard and secondary radiation sources, has been developed further for use in an electron microprobe or similar analytical instruments [7].

Theory

As discussed in detail previously by [7], the detected X-ray intensity I_{det} can be quantified in principle by:

$$I_{det} = I_o f(\chi) A \cos(\delta) \frac{1}{R^2} \frac{1}{4\pi} \exp\left[-\sum_i \left(\frac{\mu}{\rho}\right)_{E_v,i} \frac{\rho D_i}{\cos(\delta)} \right] \qquad (1)$$

where I_o is the primary X-ray intensity, generated in the sample, $f(\chi)$ denotes the correction for self-absorption in the sample, A is the active surface area of the detector, δ is the angle of X-ray incidence (being 0° in the case of perpendicular incidence), R is the distance between the X-ray source and the detector active zone, $(\mu/\rho)_{E_v,i}$ is the mass absorption coefficients for element i as absorber and X-rays having an energy E_v, and ρD_i is the corresponding mass thickness of one of the detector layers. The exponential expression represents a simplified quantification of the detector efficiency for X-ray energies below approx. 10 keV [7] according to Zaluzec [8].

The logarithmic representation of eq. 1 results, after rearrangement, in:

$$\cos(\delta) \ln\left[\frac{I_{det} R^2}{f(\chi)\cos(\delta)}\right] - \cos(\delta) \ln\left[\frac{I_o A}{4\pi}\right] = -\sum_i \left(\frac{\mu}{\rho}\right)_{E_v,i} \rho D_i \qquad (2)$$

Note that the second logarithmic term as well as the right-hand expression do not depend on the geometric situation. In the following it shall be assumed that two measurements at angles δ_j with j = 1, 2 are carried out.

Due to instrumental preconditions, e.g. the use of collimators, it may be possible that hardly quantifiable shadowing effects appear if the angle of radiation incidence δ_j is varied. The effect of these strictly geometrical, hence X-ray energy-independent phenomena is accounted for by a function $G(\delta_j)$ (see eq. (3)).

Denoting the first logarithmic term in eq. 2 by B_j, the second by C and the right-hand expression, which is the logarithm of the detector efficiency, by $D(E_v)$ we have:

$$\cos(\delta_j)B_j - \cos(\delta_j)C + \cos(\delta_j)G(\delta_j) = D(E_v) \qquad (3)$$

$D(E_v)$, being the logarithm of the detector efficiency $\varepsilon(E_v)$, can thus be obtained from eq. 3. It has been previously shown (see, e.g., [6]) that the mass thickness of the

Au-layer is the most difficult to be determined due to the small effect resulting from variations of this quantity. We have, therefore, assumed this mass thickness to be known, restricting the fitting procedure to the determination of the geometrical quantity G, and of the mass thicknesses of the Be-window and of the Si-dead layer, respectively.

Experimental

The instrument we used for our investigations is a Jeol JSM-T330A, equipped with a Link-detector, model 5508 (Si(Li), FWHM = 140 eV at 5.9 keV, detector area = 30 mm^2, thickness of the Be-window, as specified by the manufacturer = 8 μm).

The samples we used were pure element standard specimens embedded in sample holders made of brass or steel. Care was taken that the samples were conductive to avoid charging effects.

The two geometries used throughout our measurements are illustrated in Figs. 1 a and b. These two positions of detector and sample result in the largest difference of the X-ray incidence angle δ_i, which can be obtained in our instrument.

Spectrum processing was carried out using the software supplied by Link. Since for our purposes it is not important whether the radiation of a given energy E_v, is characteristic or continuous radiation we did not substract the background, thus considering all of the counts registered at a given channel.

Results

The results of our experimental investigations and the fitting procedure carried out to determine the mass thicknesses of the detector's layers are illustrated in Fig. 2.

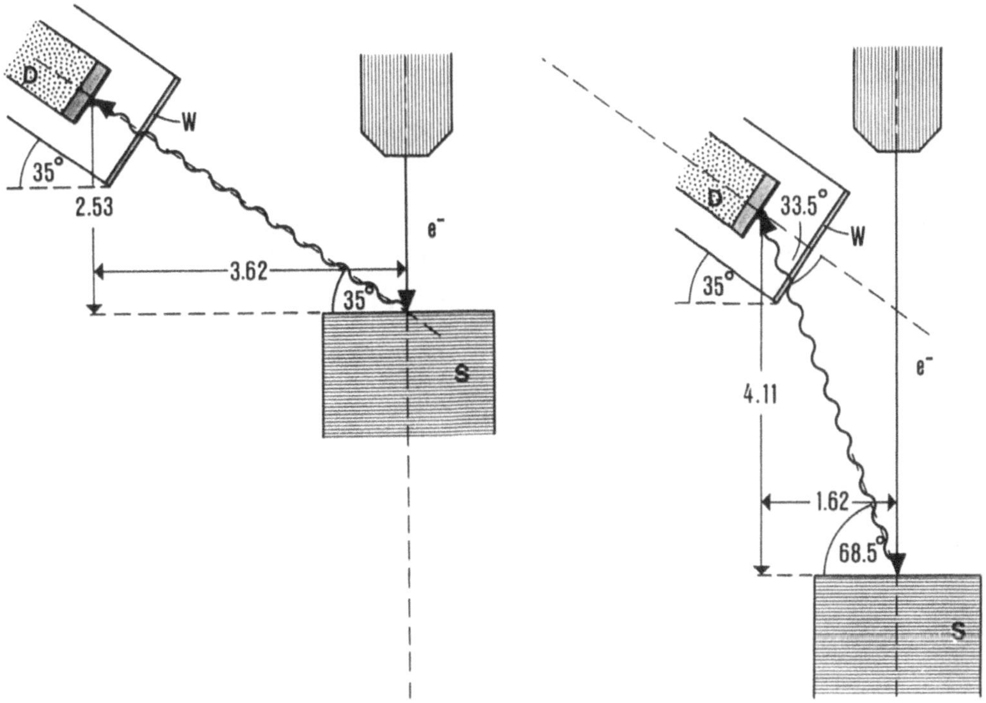

Fig. 1. (a, b) Geometries of the respective two measurements carried out at different X-ray incidence angles. D stands for the inner detector, W for the Be-window, and S for the electron beam irradiated sample. The values of the distances are given in cm

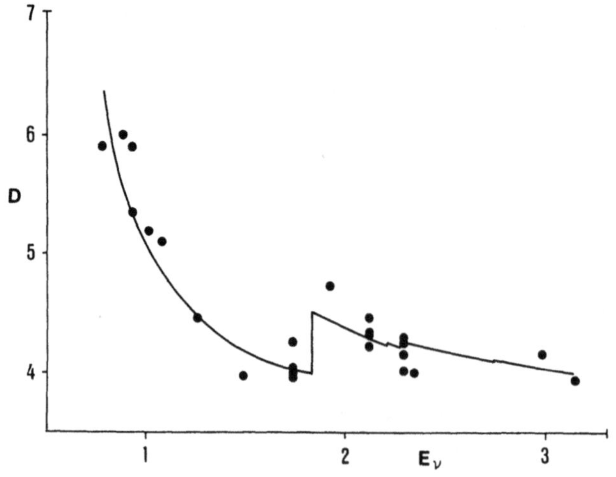

Fig. 2. Experimental results obtained for $D(E_\nu)$ (●) and fit to the data (curve)

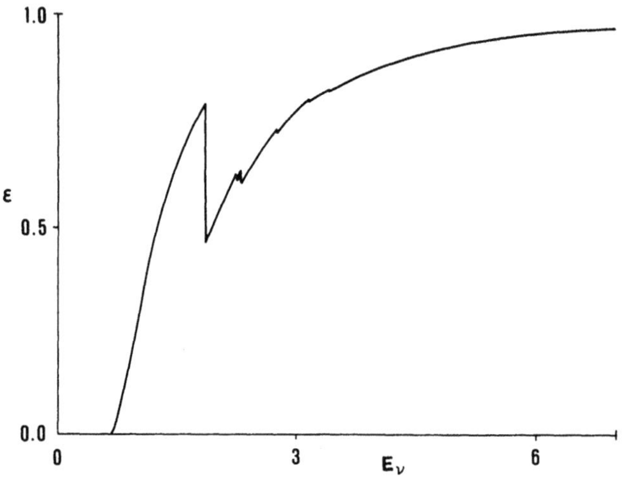

Fig. 3. Resulting detector efficiency curve. The curve is normalized to obtain unity for Cu-K_α X-rays

As mentioned above, we assumed the gold layer to have known mass thickness, which, according to literature, can be estimated as being between 28.95 and 38.6 $\mu g/cm^2$, i.e. 15 to 20 nm. In order to find out the appropriate value we carried out our fitting procedures assuming either 15 nm or 20 nm. The best 'root-mean-square' error was obtained using the lower value, although the effect resulting from the variation of the mass thickness of this layer is quite small. The mass thickness values, obtained for the other two layers, assuming the Au-layer to be 15 nm, were 1459 $\mu g/cm^2$ (=7843 nm) for the Be-window and 170.2 $\mu g/cm^2$ (=630.4 nm) for the Si-dead layer, respectively.

Fig. 3 shows the corresponding curve quantifying the dependence of the detector efficiency on the X-ray energy.

Discussion

The mass thickness of the Si-dead layer determined by the least-squares fitting procedure can be regarded as realistic. It reproduces the absorption edge jump which was found experimentally quite well. Of course the absorption edge jump will not be visible as a well-defined edge in measured spectra due to the limited resolution of the detector. It is, however, possible to simulate measured spectra. For this purpose Gaussian distributions are calculated of which the shape is determined on the one hand by the FWHM given by the resolution of the detector and on the other hand by the area given by the relative detector efficiency, which is calculated for the X-ray energy corresponding to the position of the maximum of the Gaussian distribution. To obtain the simulated spectrum, these Gaussian distributions placed, e.g., at every 10 eV, are convolved, which means that the respective contributions of each Gaussian distribution are summarized for each energy of the spectrum. For a simple simulation of the effects caused by the resolution of the detector and the absorption of emitted X-rays in the Si-dead layer we have calculated spectra in the way described above, assuming that the intensity of the continuous radiation emitted is constant in the small energy region of interest. Figs. 4 and 5 show comparisons of experimental spectra with calculated ones, assuming a FWHM value of 110 eV, which was determined from characteristic X-ray peaks in this energy region. It can be clearly seen that the calculations agree well with the spectra found

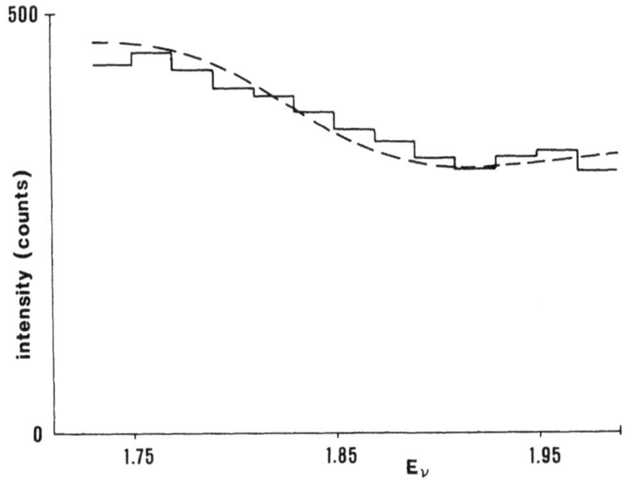

Fig. 4. Comparison of calculated (dashed curve) and measured (solid line) Bremsstrahlung spectra near the Si-K edge. Spectrum recorded for a Zn-target at an acceleration energy $E_o = 5$ keV

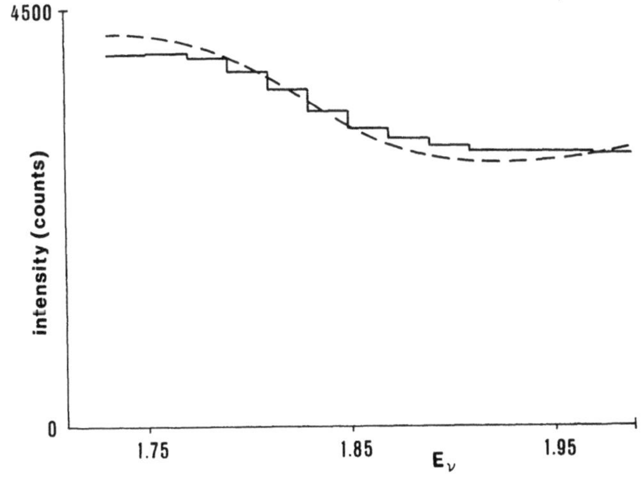

Fig. 5. Comparison of calculated (dashed curve) and measured (solid line) Bremsstrahlung spectra near the Si-K edge. Spectrum recorded for a La-target at an acceleration energy $E_o = 5$ keV

experimentally which is a further confirmation of the results obtained for the mass thickness of the Si-dead layer.

Acknowledgement. The authors gratefully acknowledge the support that has been given from the Austrian Fonds zur Förderung der wissenschaftlichen Forschung (Projekt P7336 Phy) and from the Jubiläumsfonds der Österreichischen Nationalbank (Projekt 3264).

References

[1] E. C. Montenegro, A. Oliver, F. Aldape, *Nucl. Instr. Meth.* **1985**, *B12*, 453.
[2] K. Shima, *Nucl. Instr., Meth.* **1979**, *165*, 21.
[3] D. M. Corallo, D. M. Creek, G. M. Murray, *J. Phys. E: Sci. Instr.* **1980** *13*, 623.
[4] D. G. W. Smith, *X-Ray Spectrom.* **1981** *10*, 78.
[5] W. Maenhaut, H. Raemdonck, *Nucl. Instr. Meth.* **1984** *B1*, 123.
[6] C. A. Baker, C. J. Batty, S. Sakamoto, *Nucl. Instr. Meth.* **1987** *A259*, 501.
[7] J. Wernisch, A. Schönthaler, H.-J. August, *Phys. Stat. Sol. (A)* **1990** *122*, 695.
[8] N. J. Zaluzec, in: *Introduction to Analytical Electron Microscopy* (J. J. Hren, J. I. Goldstein, D. C. Joy, eds.), Plenum, New York, 1979, p. 121.

Mikrochim. Acta (1992) [Suppl.] 12: 153–160

Background Anomalies in Electron Probe Microanalysis Caused by Total Reflection

Werner P. Rehbach* and Peter Karduck

Gemeinschaftslabor für Elektronenmikroskopie der RWTH Aachen, Ahornstrasse 55, D-W-5100 Aachen, Federal Republic of Germany

Abstract. In the present paper anomalies in the background at low spectrometer angles are investigated. It is found, that totally reflected low-energy characteristic X-rays cause this effect. The critical angles of total reflection are determined for a set of X-ray lines. It is shown, that total reflection may occur with all analyzing crystals which are used in electron probe microanalysis.

Key words: Electron probe microanalysis, soft X-rays, background, total reflection.

In wavelength dispersive analysis of soft X-rays anomalies in the background are found for several target elements. In the literature extremely high backgrounds in the wavelength range of the O Kα line are reported for C, Al, Si, Mo, and Zr [1]. This feature results in problems in the analysis of oxygen and nitrogen, as the anomaly depends strongly on the target material. As the same effect was found for for boron (Fig. 1), an investigation was started of the origin of the anomalies.

The Continuous Background

The background is that part of the signal which is not due to line emission [2]. It is caused by continuous radiation (Bremsstrahlung, i.e. deceleration radiation). This radiation is produced when an electron passes through the Coulomb field of the nucleus of an atom. The electron is deflected and decelerated and therefore it loses energy. This energy is emitted as an X-ray photon. The energy of the emitted photons varies continously from zero to the energy of the exciting electron.

Kramers [3] derived a theoretical equation for the intensity $I(\nu)$ of the generated continous radiation as a function of the observed frequency ν.

$$I(\nu)\,d\nu = h^2 kZ(\nu - \nu_0)\,d\nu, \tag{1}$$

with atomic number Z and frequency ν_0 for exciting energy E_0 (h is Planck's constant and k is a constant).

* To whom correspondence should be addressed

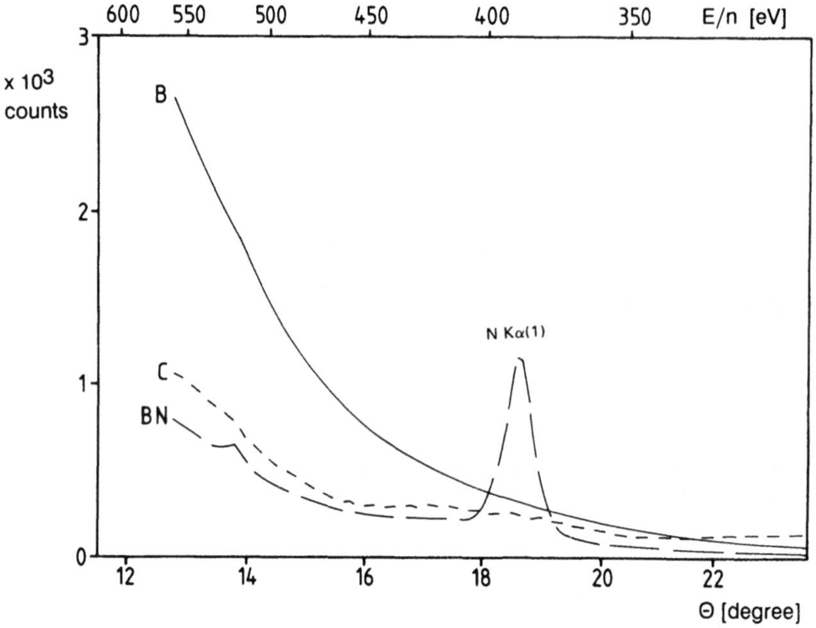

Fig. 1. Lead stearate crystal spectrometer scan on pure boron, boron nitride and vitreous carbon, discriminator settings for N Kα, $E_0 = 10$ keV

As the measured quantity in electron probe microanalysis is not the intensity but the number of photons N with the frequency ν, the equation must be divided by the electron energy $E = h\nu$:

$$N(\nu)\,d\nu = hkZ\left(\frac{\nu - \nu_0}{\nu}\right) d\nu. \tag{2}$$

Rewritten in units of wavelengths this results in:

$$N(\lambda)\,d\lambda = hckZ\left(\frac{1}{\lambda^2} - \frac{1}{\lambda\lambda_0}\right) d\lambda, \tag{3}$$

with a maximum at $\lambda = 2\lambda_0$.

In further investigations Rao-Sahib and Wittry [4] found a dependency with Z^n, where n varies between 1.12 and 1.38, depending on Z, λ and λ_0. In all cases the generated number of photons $N(\lambda)$ should increase with the atomic number Z.

Experimental

In order to elucidate the effect mentioned above, background measurements were carried out on a large number of pure elements and compounds. A JEOL 50A electron microprobe was used. The accelerating voltage was 10 kV and the lead stearate crystal was scanned from $\Theta = 12°$ to $\Theta = 24°$. ($n\lambda = 2.0$–4.0 nm). The lower limit of $\Theta = 12°$ is given by the mechanical construction of the spectrometer. The X-rays were detected by a gas flow proportional counter, the counter gas was 10 % argon with 90 % methane at a pressure of 450 HPa.

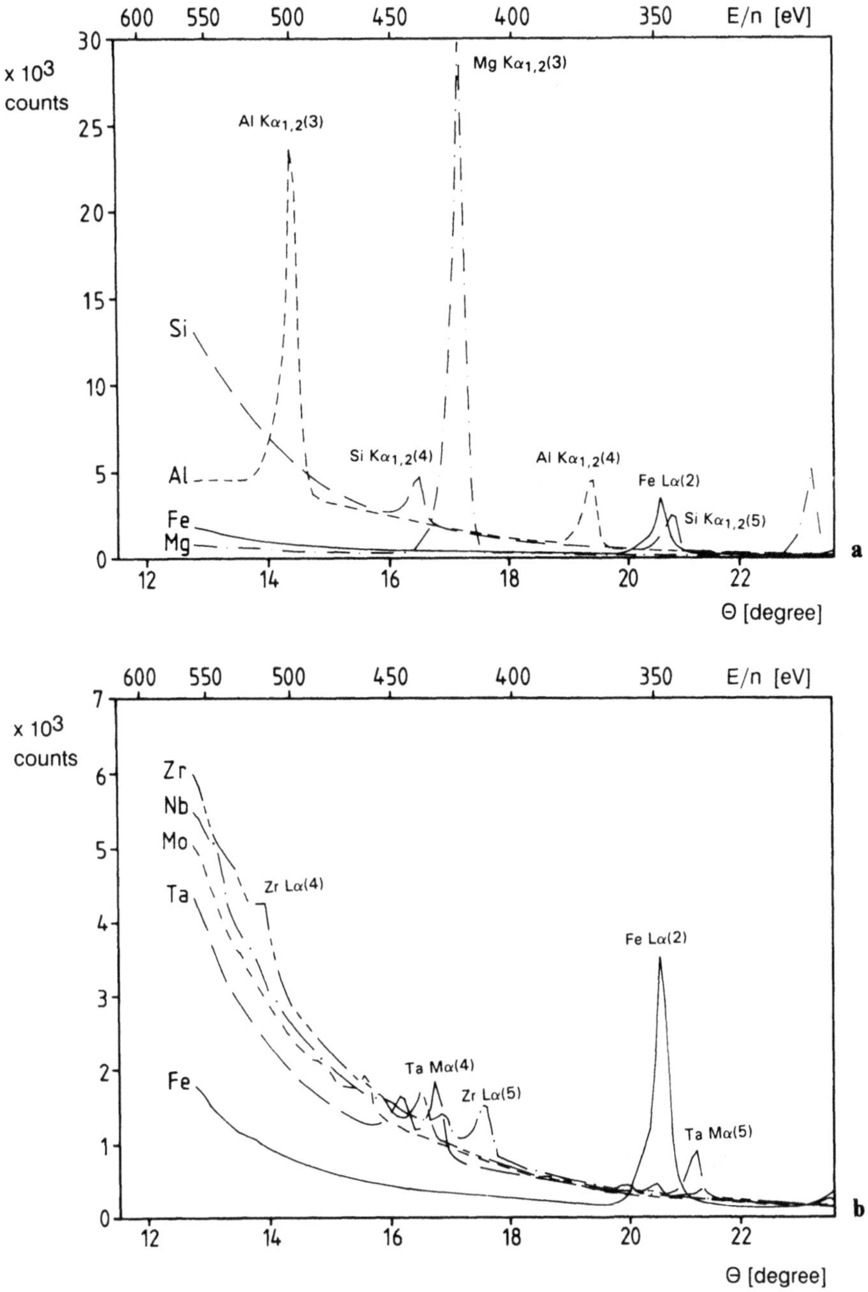

Fig. 2. **a** Lead stearate crystal spectrometer scan on Mg, Al, Si, and Fe, discriminator settings for O Kα, $E_0 = 10$ keV, **b** Lead stearate crystal spectrometer scan on Fe, Ta, Mo, Nb, and Zr, discriminator settings for O Kα, $E_0 = 10$ keV

Results and Discussion

Some examples are shown in Figs. 2a and 2b. Compared with Mg and Fe, the elements Al and Si have very high backgrounds especially for small values of Θ. A similar behavior was found for Zr, Nb and Mo, which showed high backgrounds compared with Ta. This phenomenon cannot be attributed to lines of higher order. Nor can backscattered electrons reaching the detector for small Θ-values explain

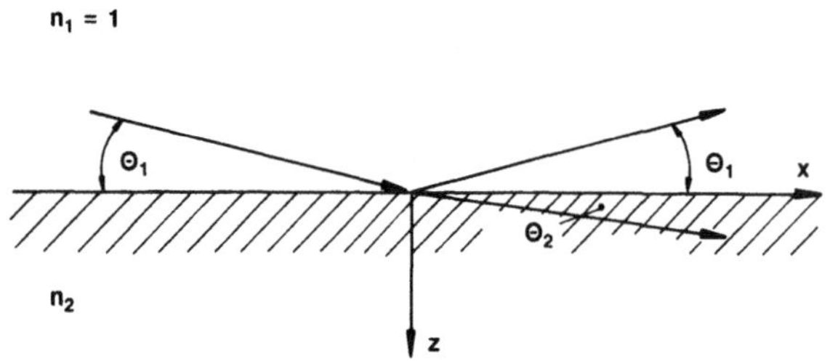

Fig. 3. Refraction and reflection of X-ray photons impinging from vacuum into a medium of refractive index n_2

the extraordinary behavior for specific elements. The backscatter coefficient η increases monotonously with Z and therefore the effect should also increase with Z in the same way, which is not found. Nor does different absorption of continuous radiation cause the effect. The fact, that the anomaly is the more pronounced the smaller Θ makes it probable that total reflection of X-rays at the surface of the analyzing crystal is the reason.

Total Reflection

Total reflection can occur if radiation arrives from an optically denser medium and impinges on an optically less dense medium. For X-rays, any medium is less dense than vacuum, so total reflection is generally possible (Fig. 3). According to [5], the refractive index for photons is

$$n = 1 - \frac{N_a r_0}{2\pi}\left(\frac{hc}{E}\right)^2 \frac{\rho}{A} f. \tag{4}$$

N_a is Avogadro's number, r_0 is the classical electron radius, E is the energy of the X-rays, ρ is the density, A is the atomic weight, and f is the atomic scattering factor. Far from the absorption edges f nearly equals Z. In the vicinity of absorption edges, dispersion and absorption must be considered which result in an atomic scattering factor

$$f = Z + f' + if'', \tag{5}$$

in which f' is a term related to dispersion and f'' is a term related to absorption.

The refractive index becomes complex and may be written as

$$n = 1 - \delta - i\beta, \tag{6}$$

where

$$\delta = \frac{N_a r_0}{2\pi}\left(\frac{hc}{E}\right)^2 \frac{\rho}{A}(Z + f'), \tag{7}$$

and

$$\beta = \frac{N_a r_0}{2\pi}\left(\frac{hc}{E}\right)^2 \frac{\rho}{A} f''. \tag{8}$$

For X-rays δ and β are positive and small, which result in refraction indices that are below unity, i.e. X-rays are deflected away from the normal when impinging from vacuum into material. According to Snell's law of deflection (Fig. 3):

$$n_1 \cos\Theta_1 = n_2 \cos\Theta_2 \qquad (n_1 = 1). \tag{9}$$

There exists a critical angle of total reflection Θ_{1c}, with $\Theta_{2c} = 0$ and therefore $\cos \Theta_{2c} = 1$.

$$\cos \Theta_{1c} = n_2 . \tag{10}$$

Far from absorption edges and expanding the cosine in a power series for small angles Θ this results in:

$$\Theta_{1c} = \sqrt{2\delta_2} \simeq \sqrt{\frac{N_a r_0}{\pi} \frac{\rho Z}{A}} \lambda . \tag{11}$$

Hence the critical angle is proportional to the wavelength of the photons. Taken absorption into account, no sharp limit of total reflection exists. For increasing Θ a decreasing reflectivity is to be found also for $\Theta < \Theta_{1c}$, and in the region $\Theta_{1c} < \Theta < 2\Theta_{1c}$ the reflectivity is not completely negligible (Fig. 4) [6].

The Θ-range of the crystal spectrometers is about 12° to 75°. Therefore X-ray lines with Θ_c higher than about 6° should show total reflection at the lower end of the spectrometer range. Some of such lines are tabulated in Table 1. Other elements

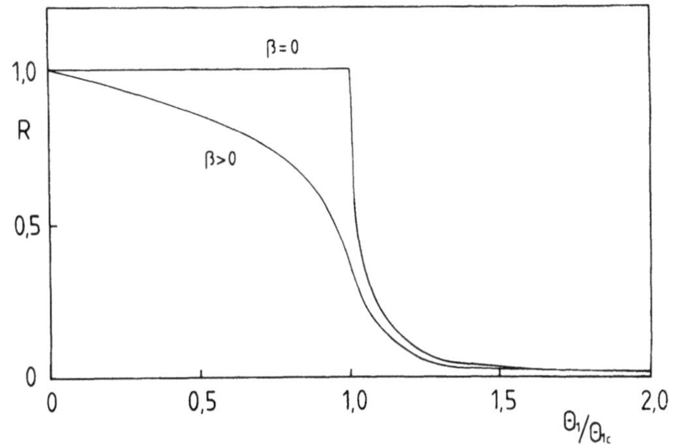

Fig. 4. Reflection coefficient R of X-ray photons versus Θ/Θ_c

Table 1. Energies and critical angles of total reflection for several low-energy lines

X-line	E [eV]	Θ_c [°]
Be Kα	108.5	15.18
B Kα	183.3	9.00
C Kα	277.0	5.95
Na $L_{2,3}M$	30.5	54.22
Mg $L_{2,3}M$	49.3	33.49
Al $L_{2,3}M$	72.4	22.83
Si $L_{2,3}M$	91.5	18.05
Zr Mζ	151.1	10.93
Nb Mζ	171.7	9.61
Mo Mζ	192.6	8.58

like Fe or Ta have no such lines and show subsequently no total reflection. They are presented in Fig. 2 and Fig. 6 for purposes of comparison only.

Totally reflected lines should produce pulse height distributions in the proportional counter similar to the distributions produced by the same lines at the Bragg-reflection positions. This assumption is proved in Fig. 5a for the B Kα line. The Bragg-reflected photons, just like those totally reflected at a lead stearate crystal, produce qualitatively similar pulse-height distributions. Photons reflected at a RAP crystal at the same angle also show this behavior. Thus it is demonstrated that the Bragg-reflected as well as the totally-reflected signals have the same nature, but are deflected by two different mechanisms.

It can also be shown for elements like Si, Zr, Mo, that the high background at small Θ-values is caused by total reflection of low-energy L- and M-lines (Fig. 5b).

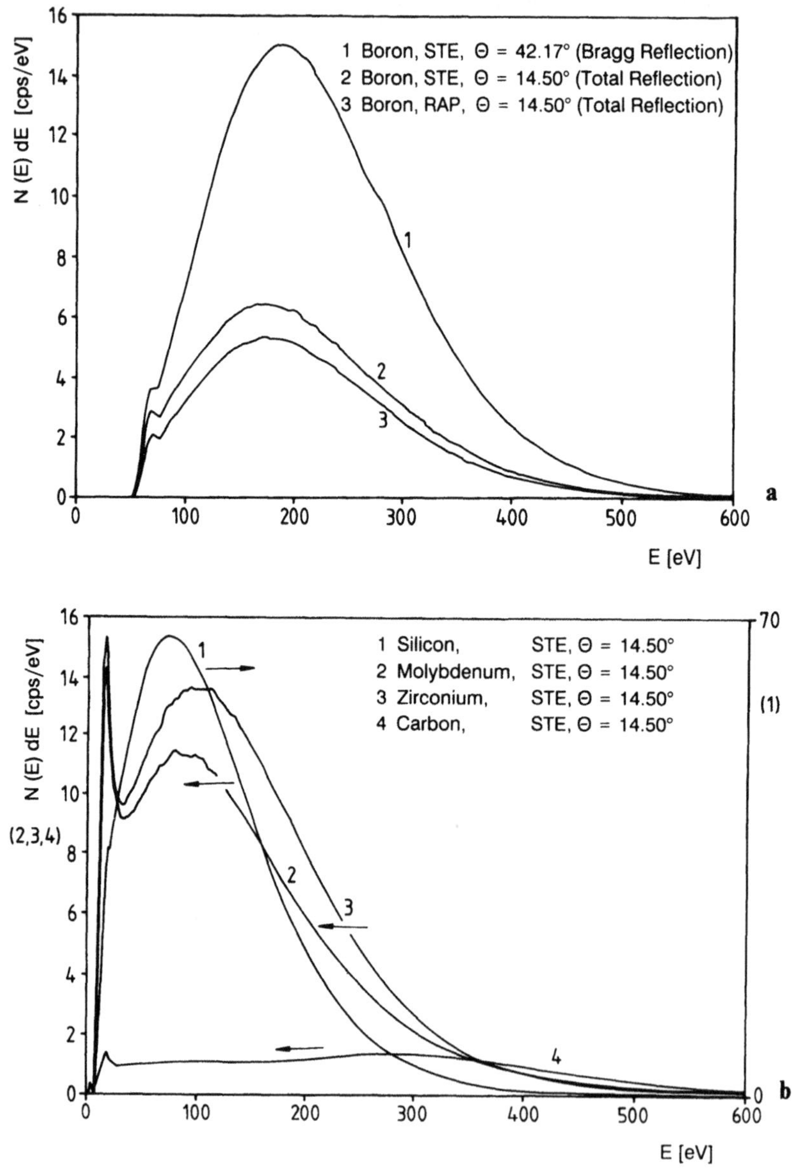

Fig. 5. a Pulse height distributions for boron, **b** Pulse height distributions for C, Si, Zr and Mo

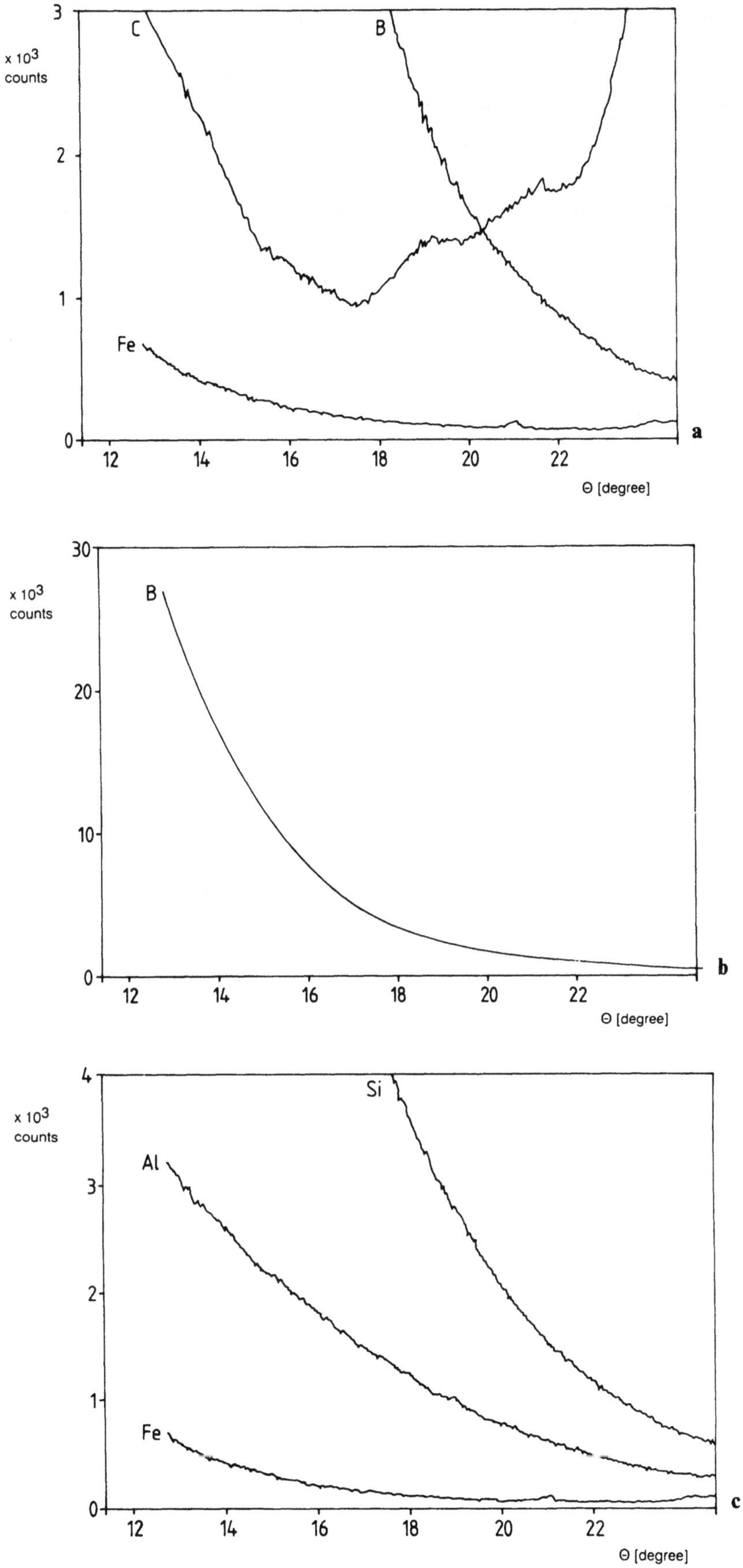

Fig. 6. **a** Spectrometer scan on boron, carbon and iron with a synthetic multilayer crystal (Ni–C, $2d = 9.5$ nm), $E_0 = 10$ keV, **b** Spectrometer scan on pure boron with a synthetic multilayer crystal (Ni–C, $2d = 9.5$ nm), $E_0 = 10$ keV, **c** Spectrometer scan on Al. Si, and Fe with a synthetic multilayer crystal (Ni–C, $2d = 9.5$ nm), $E_0 = 10$ keV

The effect vanishes in carbides, nitrides, and oxides because the severe absorption of the specific lines in the compounds reduces the emission to a large extent. It was observed, that even a very thin layer of an oxide, on Si e.g., results in a strong diminution of the background [1]. To avoid uncertainties in the evaluation of the background in the case of total reflection it is advisable to avoid the region of small Θ-angles. This may be achieved by using synthetic multilayer crystals. The question whether these crystals show the effect at all is answered by Figs. 6a to 6c. With an Ni–C crystal ($2d = 9.5$ nm) we found an extremely high background for boron and also very high values for Si and Al, compared to Fe.

Although they show the anomaly too there is an advantage in the use of synthetic multilayers. By suitable choice of the $2d$ values the region of total reflection is avoidable in most cases. If it is unavoidable to measure at small deflection angles, off-peak measurements of the background on the specimen are preferred instead of peak-measurements on a reference specimen. Only in this way the true background may be determined.

References

[1] J. Betzold, *Beitr. elektronenmikroskop. Direktabb. Oberfl.* **1986**, *19*, 7.

[2] K. F. J. Heinrich, *Electron Beam Microanalysis*, Van Nostrand, New York, 1981.

[3] H. A. Kramers, *Phil. Mag.* **1923**, *46*, 836.

[4] T. S. Rao-Sahib, D. B. Wittry, *J. Appl. Physics* **1974**, *45*, 5060.

[5] R. W. James, *The Optical Principles of Diffraction of X-Rays*, G. Bell, London, 1962.

[6] B. Lengeler, in: *Synchrotronstrahlung in der Festkörperforschung*, Kernforschungsanlage Jülich, Jülich, 1987.

Mikrochim. Acta (1992) [Suppl.] 12: 161–171

Automatic Analysis of Soft X-Ray Emission Spectra Obtained by EPMA

Ifan A. Slavic[1], Jasna I. Slavic[2], Ivan A. Grzetic[1], and Miodrag K. Pavicevic[1,*]

[1] Faculty of Mining and Geology, Univerisity of Belgrade, Djusina 7, P.O. Box 244, YU-11001 Belgrade, Yugoslavia

[2] Republic Statistics Institute, P.O. Box 313, Milana Rakica 5, YU-11000 Belgrade, Yugoslavia

Abstract. In this paper the so-called independent parameter (IP) fitting method within the standard representation (SR) is explained to the case of a non-linear model. It is shown that the SRIP-method applied to a non-linear model, converges steadily, if the model is properly chosen and thus, it can be used as a measure of the correctness of the model. This is accomplished by assuming the existence of at least one peak—described by a Gaussian formula—while the SRIP-method is applied in the iterative procedure until convergence is attained. While observing the residuals, the next peak is added to the previous model and the procedure continues. The results are illustrated by examples of oxygen soft X-ray emission spectra deconvolution for periclase, korund, and quartz.

Key words: Automated EPMA, soft X-ray, curve fitting.

The so-called independent parameter (IP) fitting method has been described by Slavic [1], for the cases of complex spectra described by sum of Gaussians. This approach gives better results than those achieved by the modified Gauss-Newton algorithm. Different parameter initialization approaches have already been investigated, in order to solve the complex spectrum bumps and extremely small peaks [2, 3, 4]. However, the problem of divergency, like in other fitting methods, has not satisfactorily been solved. But, it has recently been shown, that the IP linear fitting in the SR converges steadily [5]. By introducing the SR we made the simple data transformation—a special kind of scaling—of our nonlinear case, where a sum of Gaussians is reduced, upon linearization to the linear approximation problem. It is also evident that SRIP process keeps on converging steadily. The term "convergence" refers to the convergence of at least one out of three Gaussian-parameters of the j-th peak, Eq. (2′), when applying the IP fitting iterative process. (See theorem 1 in Appendix A (PASR) and increment Eq. (B14) in Appendix B, (NASR)). Rare "crumbs of divergence" (i.e. the 10^{-3}-rd parts of the neighbor converging sum of squared error- increments) will not disturb the main process. The advantage of

---

* To whom correspondence should be addressed

the IP method is, either in the R^2 or in the SR^2 space, seen also in the fact that the number of parameters may exceed the number of available data points [5]. This means, that one can fit 10 Gaussians (30 parameters) to 20 data pairs per sample (so-called "over-rank" parameter approximation), that certainly illustrates the essential difference between the IP's and Seidel's method of numerical analysis.

In our case we applied the SRIP method in a non-linear way when analyzing the soft X-ray emission spectra of oxygen.

Oxygen X-Ray Emission Spectra

General characteristics of the light element X-ray emission spectra include a relatively low peak-to-background ratio and a complex line or band structure. Therefore, the objective of this paper might also be formulated as a study that illustrates a possible application of the fitting program when investigating the hyperfine structure of X-ray spectra that reflect molecular states rather than atom states.

Recording of spectra was performed by a lead stearate crystal in a SEMQ—ARL instrument using acceleration voltage of 20 keV and 50 nA beam current [9].

The oxygen K_α emission spectra originate from the 2p electron level in oxygen which is at the same time the valence band and it is highly influenced by valence electrons of Mg, Al and Si in periclase, corund and quartz [6, 7, 8]. The influence of the valence electrons causes different line or band shape, nonuniform full-width-at-half-maximum (FWHM) and wavelength shift of the spectral peak maxima. The most outstanding consequence is a chemical shift of the band maximum. Application of the fitting program should reveal both the shape and energy position of the most intense peaks in the oxygen K emission band.

The whole procedure itself presents the advantage of EPMA in performing X-ray spectrometric studies in microvolumes. This can be of great importance for both material science or other disciplines dealing with small particles and phase boundaries.

Independent Parameter Method in Standard Representation

Let us denote the wavelengths by V_i and measured intensities by W_i, respectively; thus, any sample of our emission spectrum will be denoted onwards as the set of basic experimental data

$$(V_i, W_i) \in R^2, \qquad i \in N[0, m], \tag{1}$$

where $M = m + 1$ represents the number of data pairs in the R^2 space for which we seek a model. It is given as a sum of the J Gaussian functions as follows:

$$W_i(V_i) = \sum_{j=1}^{J} b_{3j-2} \exp\{-(V_i - b_{3j})^2/b_{3j-1}\} \pm E_i, \tag{2}$$

where J is the total number of peaks, j is the current peak and b's are the parameters in the j-th peak, ordered by their significance: intensity, FWHM and position. Significance means that the same relative change of the more significant parameter produces more model dislocations than the corresponding change of a less significant one.

To accept the model as an approximation, the errors E_i should satisfy the statistical fluctuation error condition:

$$|E_i| \leqslant +\sqrt{W_i} \text{ or } \sum_j E_i^2 \to \min.$$

The model (2) is a nonlinear one and is similar to models processed by Slavic [1–4]. To avoid problems of divergence we use here the method of so called *standard representation independent parameter* (SRIP) fitting, developed by Slavic and Slavic [5]. The rigorous mathematical theory for the linear case is given in Appendix A (PASR); the theory is then expanded here to the nonlinear case and is given in detail in Appendix B (NASR). The SRIP method implies the new model (Eq. 2′) whose parameters a's in SR space are uniquely related to the main model parameters b's from R space (Eq. 2). Inverting equations (A5a) and (A5b), after some simple mathematics, return the parameters from SR into R space.

It should be noted that the *standard representation* (SR) is a simple transformation, where (A4a) and (A4b) give the results in the SR space. The maximum ordinate in this space ($y_{max} \equiv 1.$) determines the highest peak from which the initial—starting parameters are taken. After very few iterative fitting steps by the algorithm (B14), the highest residual value determines the second peak and the procedure continues automatically.

In such a way the new peaks will be added to the model and, afterwards, the whole model will be fitted together by a higher number of iterations. The stopping criterion for adding the new, smaller peak is the appearance of divergence (for all parameters for such a peak), and the stopping criterion for the whole model is convergence. In such a way this method appears selfadjusting by solving the problem of complex spectra deconvolution automatically.

The SRIP method can be extended to cases where the number of data pairs is less than number of parameters—this can be easily proved by experiments for both linear and nonlinear cases.

Experimental Results and Concluding Remarks

Our approach is based on the following three items:

(a) the simple transformation of data known as *standard representation* (SR) can be defined. The model in SR, of Eq. (2), is:

$$y_i(x_i) = \sum_{j=1}^{J} a_{3j-2} \exp\{-(x_i - a_{3j})^2/a_{3j-1}^2\} \pm e_i, \tag{2'}$$

where the values of all the relevant data are simply derived from Eq. (A5a) and (A5b), respectively;

(b) the so-called *independent parameter* (IP) fitting method [1, 5] is applied. It can be proved that IP fitting in SR implies stable convergence, if the parameters are ordered by their significance; and

(c) a favorable Taylor's series expansion of a suitable nonlinear model reduces the case to a linear one; this means that the model parameters have to be ordered by their significance, e.g. in the sum of the model Gaussians the peaks are ordered by the intensities, while for each peak, according to Eq. (2′), the parameter order is: intensity, FWHM and position.

Illustration of the fitting is given in Figs. 1, 2 and 3 showing the deconvolution of the oxygen K_α emission spectra of MgO, SiO_2 and Al_2O_3. The complex band structure of molecular spectra is represented by the molecular orbital diagram for SiO_2 (Fig. 4; developed on the basis of the results in [8]). As can be seen molecular

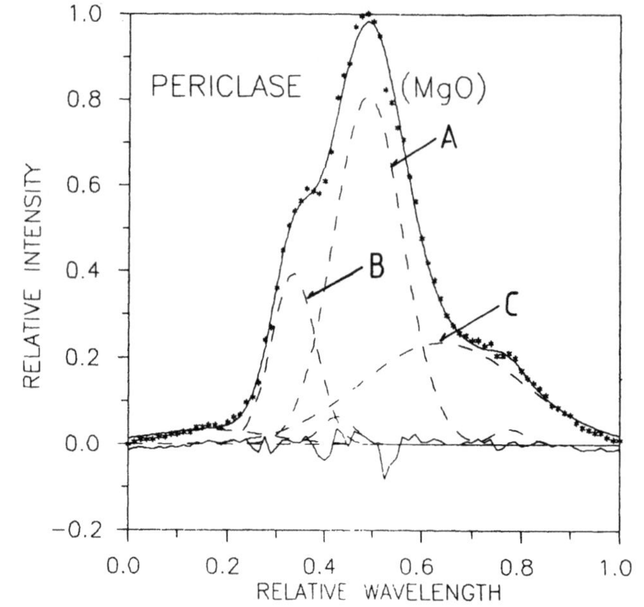

Fig. 1

Fig. 1, 2 and 3. O K_α emission spectra of periclase, quartz and corund (dots represent experimental data, solid lines represent the total model curve and residuum, while dashed lines represent the Gaussian components found). Relative wavelength adjustment (arbitrary units in SR to nm or keV): 0.0 = 2.37764 nm or 0.5214 keV and 1.0 = 2.36136 nm or 0.5250 keV

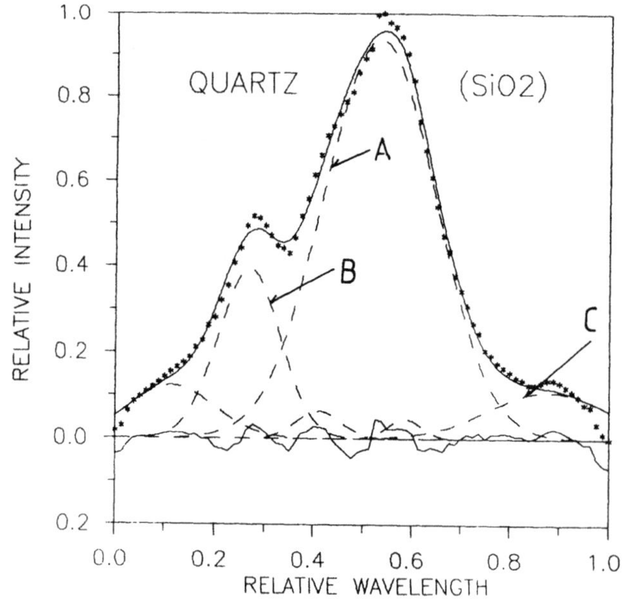

Fig. 2

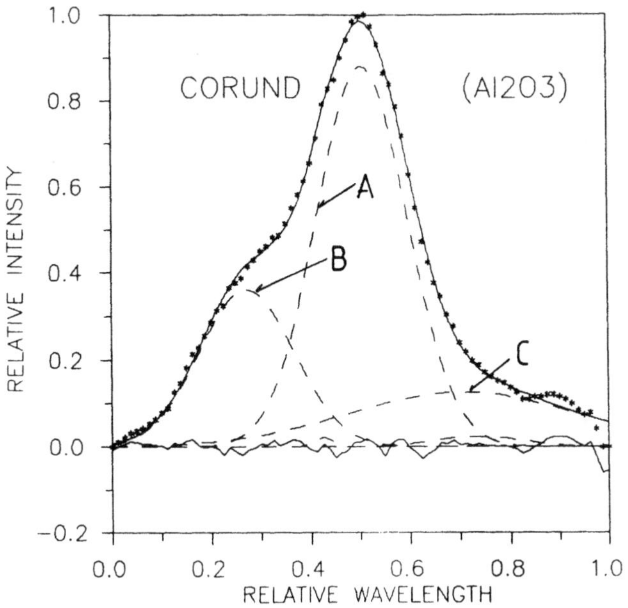

Fig. 3

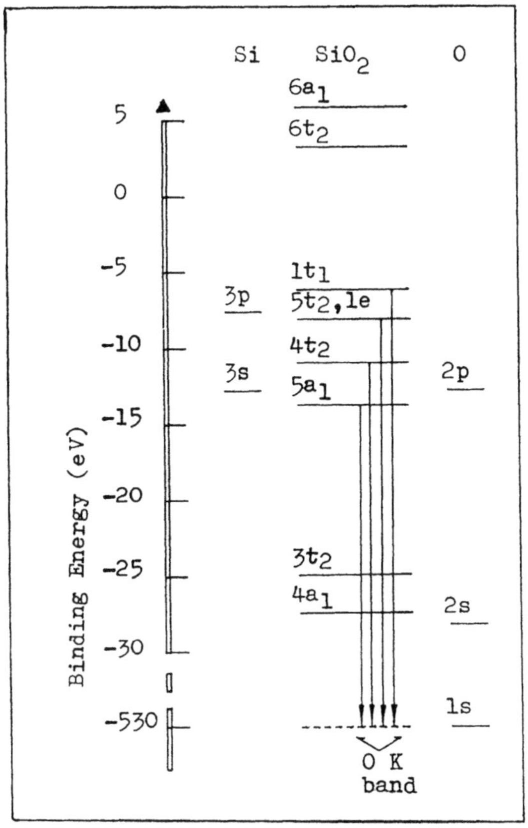

Fig. 4. Molecular orbital diagram for quartz

orbitals $1t_1$, $5t_2$ and $1e$, $4t_2$, $5a_1$ are energy comparable with oxygen 2p electrons. The O K_α emission spectra originate from these levels in SiO_2. When experimental spectra are examined, three dominant peaks for MgO, SiO_2 and Al_2O_3 can be recognized (Table 1).

Table 1. The constituent peaks of the oxygen K_α emission bands in periclase (MgO), quartz (SiO_2) and corund (Al_2O_3)

Mineral	The positions of the constituent peaks represented in arbitrary units in SR, nm and keV order by intensity		
	peak A	peak B	peak C
MgO	0.489 2.36964 nm 0.52321 keV	0.335 2.37219 nm 0.52264 keV	0.640 2.36723 nm 0.52374 keV
Al_2O_3	0.505 2.36942 nm 0.52325 keV	0.271 2.37323 nm 0.52241 keV	0.706 2.36615 nm 0.52398 keV
SiO_2	0.531 2.36900 nm 0.52335 keV	0.269 2.37326 nm 0.52241 keV	0.879 2.36333 nm 0.52460 keV

Table 2. The relative chemical shifts of O K_α lines of MgO and Al_2O_3 in comparison to SiO_2 (in nm and eV)

	peak A	peak B	peak C
MgO	+0.00064 nm −0.14 eV	−0.00107 nm +0.23 eV	+0.039 nm −0.084 eV
Al_2O_3	+0.00042 nm −0.10 eV	0.00 nm 0.0 eV	+0.00282 nm −0.62 eV

The relative position and energy separation between these peaks depend, for each compound, on crystal lattice parameters mainly governed by the metal associated with the oxide. In the comparison of the relative chemical shifts of the O K_α constituent lines within the emission band for MgO and Al_2O_3 to SiO_2 (Table 2), it is quite obvious that the influence of Mg, Al and Si on position of oxygen constituent lines is significant. The position of the main line (peak A) shifts to higher energies from MgO to Al_2O_3 and further in SiO_2 due to the increasing electronegativity of the metal (Mg: 1.2, Al: 1.5 and Si: 1.7).

Chemical shift of the emission line, together with change in band shape and in relative intensity of the constituent lines within the band, became quite evident when using the described program for automatic analysis of soft X-ray emission spectra. X-ray emission spectrometry performed by EPMA with a lead stearate crystal has some disadvantages in peak-to-background ratio in comparison with new types of crystals covering the same wavelength range. However, the problem of energy resolution in applying the new crystals, has not improved very much. In case of the O K_α band the constituent lines originate from molecular orbitals with different symmetries and energies having natural FWHM much greater than pure atomic orbitals. It is known from photoelectron spectroscopy that the natural FWHM for certain electron levels in the valence region could be $\geqslant 1$ eV. The complete energy range of all investigated spectra (Table 1) is about 2 eV. Under the presumption that the FWHM of each constituent line of the emission band is greater than 1 eV a better energy resolution can hardly be expected, as it does not depend on instrument capabilities, but is its natural property. The presented program cannot offer solutions for the problem of transition probabilities in molecular orbital region that differs from one compound to another.

Acknowledgement. This work was supported by the Serbian Republic Fund of Science.

References

[1] I. A. Slavic, *Nucl. Instrum. Methods* **1976**, *134*, 285.
[2] I. A. Slavic, *Nucl. Instrum. Methods* **1973**, *112*, 253.
[3] I. A. Slavic, S. P. Bingulac, *Nucl. Instrum. Methods* **1970**, *84*, 261.
[4] I. A. Slavic, S. P. Bingulac, *Data Processing in Experimental Physics* (Series of Selected Papers in Physics), Physical Society Japan, **1971**, pp. 101–108.
[5] I. A. Slavic, J. I. Slavic, *Standard Representation and Polynomial Recognition of Discharge versus Water Level Calibration Charts* (to appear).
[6] J. A. Tossell, *Seochim. Cosmochim. Acta* **1973**, *37*, 583.
[7] J. A. Tossell, *J. Spectrosc. Relat. Phenom.* **1976**, 1.
[8] J. A. Tossell, G. V. Gibbs, *Phys. Chem. Minerals* **1977**, *3*, 21.
[9] I. A. Grzetic, D. M. Golijanin, M. K. Pavicevic, *Sintering—Theory and Practice, Material Science Monographs, Vol. 14*, Elsevier, New York, Amsterdam, 1982, pp. 631–636.

Appendix A. Polynomial Approximation in Standard Representation (PASR)

If we have a discrete data set

$$(V_i, W_i) \in R^2, \qquad i \in N[0, m], \tag{A1}$$

for which we would like to find out the functional relation

$$W(V) \simeq F(V), \tag{A2}$$

we can obtain a new data set, i.e.

$$(x_i, y_i) \in SR^2 \equiv R^2[0, 1], \qquad i \in N[0, m], \tag{A3}$$

given by the following simple transformation:

$$x_i = \frac{V_i - V_{min}}{V_{max} - V_{min}}, \; V_{min} = \min_i \{V_i\}, \; V_{max} = \max_i \{V_i\}, \tag{A4a}$$

and

$$y_i = \frac{W_i - W_{min}}{W_{max} - W_{min}}, \; W_{min} = \min_i \{W_i\}, \; W_{max} = \max_i \{W_i\}. \tag{A4b}$$

Note that

$$0 \leqslant x_i \leqslant 1, \qquad 0 \leqslant y_i \leqslant 1, \qquad 0 \leqslant i \leqslant m. \tag{A4c}$$

The new data set (x_i, y_i) is said to be the *standard representation* (SR) of the starting data set (V_i, W_i); the transformation returning SR chart to the starting field of values is given by

$$V_i = V_{min} + (V_{max} - V_{min})x_i, \tag{A5a}$$

and

$$W_i = W_{min} + (W_{max} - W_{min})y_i. \quad \text{(A5b)}$$

The SR is characterized by some valuable properties that enable easy solving of both interpolating or fitting (smoothing approximation) problems. This is done by parameters disclosing of a linear or nonlinear functional form, which approximates successfully the given SR data, i.e.

$$y \simeq f(x). \quad \text{(A2')}$$

Hence, the searched function F, Eq. (A2), will be known, implicitly by (A2'), or explicitly expressed by simple returning transformations (A5).

IPF Polynomial Approximation in SR

The polynomial approximation appears to be the most simple one in further application. Therefore we consider the case when the SR (A3) can be mathematically identified by the n-th degree polynomial, i.e.

$$y_i(x_i) = P_n(x_i) \equiv \sum_{j=0}^{n} a_j x_i^j, \qquad 0 \leqslant i \leqslant m. \quad \text{(A6)}$$

In SR the natural constraints appear as follows:

$$P_n(x_0 = 0) \equiv a_0 = y_0, \quad \text{(A7)}$$

$$P_n(x_m = 1) \equiv a_0 + a_1 + \cdots a_n = y_m, \quad \text{(A8)}$$

and

$$I = \int_0^1 P_n(x)\,dx \equiv a_0 + a_1/2 + \cdots a_n/(n+1). \quad \text{(A9)}$$

If

$$M = m + 1 \quad \text{(A10)}$$

is the total number of available data points and $\Delta x \cong 1/M$, then

$$I = \lim_{M \to \infty} [\textstyle\sum y_i/M] \equiv \lim_{M \to \infty} [\bar{y}] \quad \text{(A9')}$$

The total number of polynomial parameters is given by

$$N = n + 1. \quad \text{(A11)}$$

Lemma 1. If the SR set (A3) is represented by the polynomial (A6), and if it can be found the best agreement of the constraint (A8), i.e. if it can be found the polynomial degree *n* for which

$$|a_0 + a_1 + \cdots a_n - y_m| = \min, \quad \text{(A8')}$$

then, this *n* is said to be the *optimum* degree (n_{opt}).

Proof. No matter how parameters of the polynomial (A6) have been found any number of those parameters make a sequence

$$a_0^{(n)} + a_1^{(n)} + \cdots + a_n^{(n)} - y_m \equiv \delta(n);$$

therefore it can be determined

$$\delta_{min}(n_{opt}) = \min_i |\delta(n)|.$$

Let us consider the following obviously chain of inequalities:

$$\sum (\partial Y_i/\partial a_j)^2 \leqslant \sum (\partial Y_i/\partial a_j)(\Delta Y_i/\Delta a_j) \leqslant \sum (\Delta Y_i/\Delta a_j)^2 , \tag{A12}$$

where, in our case

$$Y_i \equiv P_n(x_i),\ \Delta Y_i \equiv \varepsilon_i = y_i - P_n(x_i). \tag{A13}$$

From (A12) we derive the independent parameter estimator (IPE) for the (k + 1)-st iterative step of the j-th parameter optimization

$$\Delta a_j^{(k+1)} = D_j \sum_j x_i^j \varepsilon_i^{(k)}. \tag{A14}$$

Here the "weighting value" is given by

$$D_j = 1 \Big/ \left[\sum_i x_i^{2j}\right] \tag{A15}$$

and should not be computed to (A15) at each step, except at the very first one.

Theorem 1. The simple iterative IPF method, for the SR polynomial, given by

$$a_j^{(k+1)} = a_j^{(k)} + \Delta a_j^{(k+1)}, \qquad 0 \leqslant j \leqslant n, \tag{A16}$$

with the IPE (A14) and (A15), is steady converging.

Proof. Sufficiency. We shall prove the sufficient condition: |IPE| < 1. Let us consider (A12) as

$$\sum x_i^{2j} \leqslant (1/\Delta a_j) \sum x_i^j \varepsilon_i \leqslant (1/\Delta a_j)^2 \sum \varepsilon_i^2 . \tag{A12'}$$

Let it be

$$|\varepsilon_i|/|\Delta a_j| = r_{ij} \leqslant r_{i,j}, \qquad r = \max\{r_{ij}\}. \tag{A17}$$

Hence

$$\sum x_i^{2j} \leqslant r \sum x_i^j \leqslant \sum r^2 . \tag{A12''}$$

It follows from (A4c) that:

$$0 < \sum x_i^n < \sum x_i^{n-1} < \cdots < \sum x_i^2 < \sum x_i < \sum \equiv M \tag{A4d}$$

for the nontrivial case $M > 1$. This can be fulfilled only for

$$r \geqslant 1 \quad \text{and therefore} \quad |\Delta a_j| \leqslant |\varepsilon_i|. \tag{A17'}$$

Taking into account the condition $|\varepsilon_i| < y_i$ as well as (A4c), the final conclusion is that

$$|\mathrm{IPE}| \equiv |\Delta a_j| < 1 . \tag{A17''}$$

Necessity. Let $r < 1$, say $r = 1/M$; by (A12'') it follows that

$$\sum x_i^{2j} \leqslant (1/M) \sum x_i^j \leqslant 1/M, \quad \text{i.e.} \quad \sum x_i^{2j} \leqslant (1/M),$$

what is in contradiction with the chain inequalities (A4d).

Appendix B. Nonlinear Approximation in Standard Representation (NASR)

A mathematical model F(x) approximating a set of data (x_i, y_i) satisfies the relation

$$y_i \equiv y(x_i) = F(x_i) + \varepsilon_i , \tag{B1}$$

where ε_i are some errors appearing due to any possible reason. The model F(x) can be accepted generally as a good approximation under the following condition

$$|\varepsilon_i| \ll y_i , \tag{B2a}$$

e.g.

$$|\varepsilon_{\max}| < \varepsilon_{\text{allowed}}, \tag{B2b}$$

or

$$q(n) \equiv \sum \varepsilon_i^2(n) \to \min. \tag{B2c}$$

The number of data pairs M is given by (A10) and number of model parameters N by (A11); accordingly, one can define the model approximation generally as

(A) *analytical form* if:

$$\varepsilon_i \equiv 0, 0 \leqslant i \leqslant m, \quad \text{and} \quad N < M, \tag{B3}$$

(I) *interpolating form* if:

$$\varepsilon_i \equiv 0, 0 \leqslant i \leqslant m, \quad \text{and} \quad N = M, \tag{B4}$$

(F) *fitting form* if:

$$\varepsilon_i \neq 0, \quad \text{and} \quad N < M, \tag{B5}$$

and (O) *over-rank parameter form* if:

$$\varepsilon_i \neq 0, \quad \text{and} \quad N \geqslant M. \tag{B6}$$

One can yield the over-rank polynomial parameters only by the so-called independent parameter fitting (IPF) iterative optimization method. The IPF method appears also especially useful in SR, where it is steadily converging; the property has been proved in Appendix A for the linear model fitting in SR.

SR and Nonlinear Model Approximation

Even if a model for exact solution of a problem exists, and is denoted under "analytical form" (B3), this rarely could be found in practice, mainly because of the statistical fluctuations of the data. The functional connection between experimental data variables is almost given by a fitted form of a model which may combine linear and nonlinear part [1].

Depending on aim, it might be more convenient to search either linear or nonlinear model approximation; e.g. pure linear model is convenient for differentiating and integrating purposes; the nonlinear model can extract important investigating constituent (e.g. the Gaussian peaks in X-ray, γ-ray or optical spectra analysis [2]), or it can be more efficient in the smoothing—with a smaller number of parameters—than a linear model.

Let it be, in SR, the general mathematical model (including linear and nonlinear part) given by

$$Y_i = f(x_i; b_1, \ldots, b_N), \tag{B7}$$

where N unknown parameters have to he estimated. Let they be ordered by their significance (this means that the same relative change of the b_1 parameter produces more model dislocation than the corresponding change of b_2 parameter, etc.).

We write a Taylor series expansion for the model value in the vicinity of the experimental i-point ($Y_{i(\theta x)}$) at (k + 1)-st step for the iterative procedure as

$$Y_{i(\theta x)} = Y_i^{(k)} + [\partial Y_i^{(k)}/\partial b_1]\Delta b_1^{(k+1)} + \cdots + [\partial Y_i^{(k)}/\partial b_N]\Delta b_N^{(k+1)}. \tag{B8}$$

This linearization leads, in matrix notation, to the following system of equations

$$U^{(k)}\Delta B^{(k+1)} = E^{(k)}, \tag{B9}$$

and, after normalization (without weighting matrix)

$$H^{(k)}\Delta B^{(k+1)} = \tilde{U}^{(k)}E^{(k)}, \tag{B10}$$

where

$$H^{(k)} = \tilde{U}^{(k)}U^{(k)}, \tag{B11}$$

where $\Delta B = \{\Delta b_j\}$ is the N-dimensional parameter increment vector to be found, $U = \{u_{ij} \equiv \partial Y_i/\partial b_j\}$ $(i = 0,\ldots,m; j = 1, 2,\ldots, N)$ denotes the (M, N) matrix of the partial N-parameter derivatives for M points; $\tilde{U}$ indicates the transposed matrix and $E = \{\varepsilon_i \equiv Y_{i(\theta x)} - Y_i\}$ is the M-dimensional error vector.

Further, the fitting is usually realized by the modified (damped) Gauss-Newton method (MGN), as the solution of eq. (B10), i.e.

$$\Delta B^{(k+1)} = p^{(k)}[H^{(k)}]^{-1}\tilde{U}^{(k)}E^{(k)}, \tag{B12}$$

where p is an arbitrary constant influencing the speed and convergency of the iterative process; this constant is given in [1] as

$$p^{(k)} = \begin{cases} 1, \text{if converging,} \\ 0.5\, p^{(k-1)}, \text{if diverging.} \end{cases} \tag{B13}$$

Note that the MGN method has the simpler form in SR than in R, omitting the weighting diagonal matrix D [1].

So, we assume the following:

(a) the chosen mathematical model suits to the general data pattern;
(b) initial parameter vector $B^{(0)}$ is sufficiently good, and
(c) the model parameters are ordered by their significance.

Under these assumptions the convergency is guaranteed ($p \equiv 1$) and the solution can be found by the IPF method; therefore, we calculate the parameter increments simply by

$$\Delta b_j^{(k+1)} = [\sum (\partial Y_i^{(k)}/\partial b_j)\varepsilon_i^{(k)}]/\sum (\partial Y_i^{(k)}/\partial b_j)^2, j = 1,\ldots, N. \tag{B14}$$

Mikrochim. Acta (1992) [Suppl.] 12: 173–177

The Scanning Very-Low-Energy Electron Microscope (SVLEEM)

Ilona Müllerová* and Michal Lenc

Institute of Scientific Instruments, Czechoslovak Academy of Sciences, Královopolská 147, 612 64 Brno, Czechoslovakia

Abstract. The idea of the scanning very low energy electron microscope is discussed. The use of the immersion objective lens (the specimen as a cathode immersed in the electrostatic field of the cathode lens and in the magnetic field of the single pole-piece lens) is proposed to achieve a higher resolution and the optical properties of the lens are calculated. Preliminary results with the cathode lens in a classical SEM are presented and the new configuration in an ultra-high-vacuum (UHV) SEM is described.

Key words: Scanning electron microscopy, low-energy electron microscopy, immersion electron lenses.

Low energy electron diffraction (LEED) represents one of the classical tools of surface science. There is a very tight connection, both in theory and experiment, between very low energy electron diffraction (VLEED) and low energy electron microscopy (LEEM) [1]. But it was only recently, that Telieps and Bauer [2] were able to demonstrate the first pictures obtained with LEEM, although this type of microscope was invented more than two decades ago. In theory, the separation of the illuminating and the imaging part of a LEEM enables the application of all techniques used in transmission electron microscopy [3].

In principle, the imaging carried out by scanning the surface using a very low energy electron beam is possible too, but the attainable resolution is often poor. The scanning system would allow a simultaneous and much simpler detection and analysis in comparison with direct imaging. Recently Ichinokawa [4] made an important step forward to improve the spatial resolution of LEED. He used the optical column of the UHV scanning electron microscope (SEM) equipped with a field emission gun. The primary beam energy varied between 250 eV and 1 keV and correspondingly, spatial resolutions of 50 nm and 15 nm respectively were obtained.

To achieve good optical properties at low incident electron energies, several authors have proposed the combination of magnetic and electrostatic retarding field lenses (e.g. [5]). Frosien et al. [6] used a compound magnetic and electrostatic

---

* To whom correspondence should be addressed

retarding field as the objective lens for their low voltage SEM. The spherical and chromatic aberration coefficients (C_{si} resp. C_{ci}) with respect to the image plane were $C_{si} = 59$ mm and $C_{ci} = 15$ mm if only a magnetic field of the objective lens was applied and $C_{si} = 3.7$ mm and $C_{ci} = 1.8$ mm if an electrostatic retarding field was added to the magnetic field.

Experimental Set Up of the VLVSEM

We made our first experiment in the Tesla BS340 SEM with a poor vacuum, a thermionic gun and a large specimen chamber, so that we were able to implement all modifications. The set up used is shown in Fig. 1. The high energy electrons ($E_p = 10$ keV $-$ 40 keV) in the primary beam are focused onto the specimen and decelerated to the final very low energy which is given by the adjustable potential difference $E = e(U_{ac} - U_{sp})$. U_{ac} is the potential of the cathode of the electron gun and U_{sp} the potential of the specimen. The secondary electrons are re-accelerated to the PIN structure semiconductor detector. In such a simple way we were able to obtain a picture with an energy of the electrons impinging onto the specimen as low as $E = 10$ eV, and a result is shown in Fig. 2.

To achieve a higher resolution we proposed a new arrangement [7], schematically shown in Fig. 3. It is based on the Tesla BS350 UHV SEM with a field emission gun and a TF-W/100-Zr cathode. In Fig. 3 only the pole-piece of the objective lens of this microscope is shown. The single pole-piece lens with the cathode lens, the magnetic shielding and the crossed field deflector were additionally installed from the bottom of the microscope. The primary beam electrons are decelerated to the

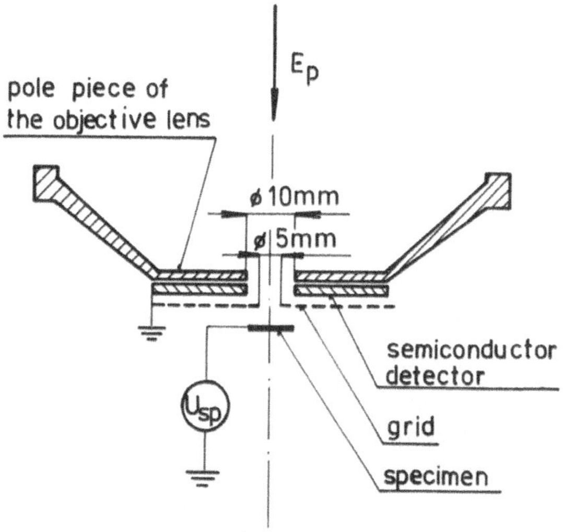

Fig. 1. Low voltage SEM based on the SEM BS340 with the electrostatic lens near the specimen. E_p is primary electron beam energy, U_{sp} is the potential of the specimen

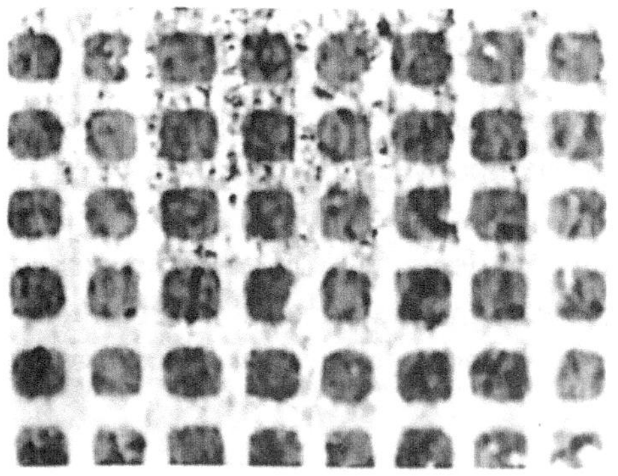

Fig. 2. A polished aluminium disc covered with a 50 nm thick layer of gold, which has been vacuum deposited through a grid with a mesh width of 64 μm. From the picture we can estimate a resolution of 3 μm. The calculated value is by more than one order lower

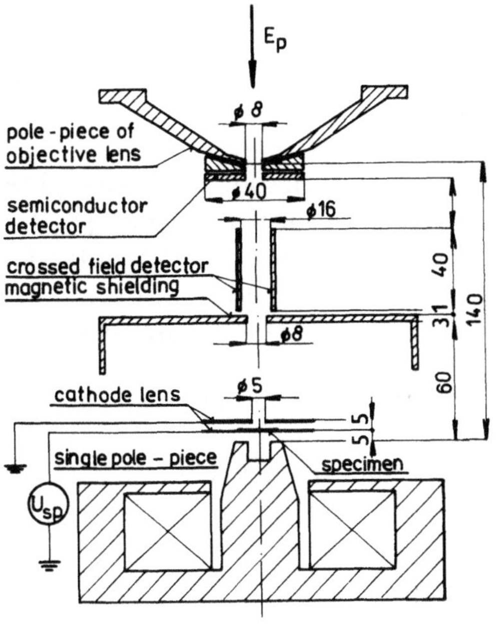

Fig. 3. The experimental set up of SVLEEM with the cathode lens and single pole-piece lens (all dimensions indicated in mm)

specimen by the cathode lens, as described in connection with Fig. 1, and the reflected (RE) and secondary (SE) electrons are accelerated by the cathode lens to the detector. The immersion objective lens (the specimen as the cathode immersed in the electrostatic field of the cathode lens and in the magnetic field of the single pole-piece lens) focuses both the primary beam to the specimen and RE and SE to the detector.

The Optical Properties of Our Set-Up

The cathode lens may be visualized as a uniform electric field of length d (d is the cathode-to-anode distance), terminated by a thin diverging lens of focal length $f = -4d$. The magnification of the cathode lens (M^c) is given by:

$$M^c = 2E_p^{1/2}/(3E_p^{1/2} - E^{1/2}), \tag{1}$$

for small incident electron energies ($E \ll E_p$), $M^c = 2/3$. The angular magnification M_a^c is obtained from:

$$M_a^c M^c = (E/E_p)^{1/2}. \tag{2}$$

The spherical and chromatic aberration coefficients, C_s^c and C_c^c resp., of the cathode lens are:

$$C_s^c = C_c^c = [(E_p - E)^2 E\, d]/[E_p + (E_p E)^{1/2}]^3, \tag{3}$$

the limiting cases being $C_s^c = C_c^c = 0$ for $E = E_p$ and $C_s^c = C_c^c = (E/E_p)d$ for $E \ll E_p$.

The aberration coefficients of the cathode lens and the magnetic lens are added according to:

$$C_s = C_s^c + \frac{1}{(M^c)^4}\frac{E^{3/2}}{E_p^{3/2}} C_s^m, \tag{4}$$

$$C_c = C_c^c + \frac{1}{(M^c)^2}\frac{E^{3/2}}{E_p^{3/2}} C_s^m, \tag{5}$$

where C_c^m and C_c^m are the aberration coefficients of the magnetic lens.

The radii of aberration discs are

$$r_s = [C_s(0.6\ \lambda_E)^3]^{1/4}, \tag{6}$$

$$r_c = [C_c(0.6\ \lambda_E)(\Delta E/E)]^{1/2}. \tag{7}$$

For a fixed primary beam energy E_p, r_s and r_c are increasing for decreasing values of the incident beam energy E. This is, in other words, Scherzer's statement [8]. On the other hand for a fixed incident beam energy E, r_s and r_c are decreasing for increasing values of the primary beam energy E_p. This is the reason why the cathode lens is so useful for low energy electron microscopy. Munro et al. [9] derived under simplifying assumptions that in the case of a scanning system not only on-axis aberration coefficients (C_s and C_c) are reduced by the retarding field, but the deflection aberration coefficients as well.

The possibilities of our new experimental set-up are well illustrated in the following comparison. For the geometry shown in Fig. 1 we calculated C_s = 0.0171 mm and C_c = 0.0063 mm for the immersion ratio E_p/E = 1000 and C_s = 53.4 mm and C_c = 19.27 mm for E_p/E = 1. For the set up with the single pole-piece lens shown in Fig. 3 we get C_s = 3.94 mm and C_c = 8.31 mm for E_p/E = 1 and C_s = 0.0051 mm and C_c = 0.0052 mm for E_p/E = 1000. For these values the resolution limit of 25 nm for the energy E = 10 eV and the immersion ratio of 1000 can be achieved.

In spite of the strong influence of the electrostatic part of the immersion lens on the angular magnification, its optical strength is small. When the incident electron

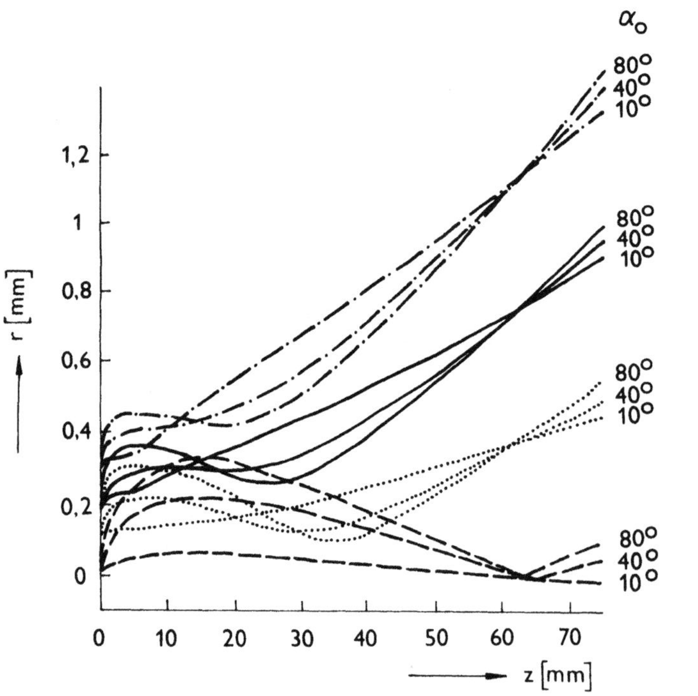

Fig. 4. The trajectories (r) of elastically reflected electrons (E = 10 eV, E_p = 10 keV), r_0 = 0 mm ———, r_0 = 0.1 mm --------, r_0 = 0.2 mm ———, r_0 = 0.4 mm —·—·—·—, α_0 = 10°, 40°, 80°. The object plane (with the Gaussian image plane at the specimen surface z_i = 0 mm) for primary electrons is positioned at z_0 = 65 mm

energy E is varied from 10 keV to 10 eV by changing the specimen potential U_{sp}, the excitation of the single pole-piece lens has to be changed from 2030A-t (*Amper-turns*) to 1760A-t only.

Detection of RE and SE

For a high value of the immersion ratio E_p/E, the RE and SE move close to the optical axis with an energy near to that of the primary beam. In the first stage of the experiment, RE and SE are deflected to the off-axis semiconductor detector. The crossed field deflector (dipole field Wien filter) is adjusted so that it does not influence the first order optical properties of the primary beam.

Future Work

After adding an energy analyzer and a multisegment detector, we plan in the next stages of the experiment to form the image in elastically or inelastically scattered electrons (our ultimate goal: in Auger electrons) or to form the image with electrons scattered into the different directions (our ultimate goal: in diffracted beams). There are many problems to be solved which concern the transport of the RE or SE, the choice of the energy analyzer etc. An important precondition is satisfied, however: the immersion lens transports fairly well a broad range of angles and a large field of view, which is illustrated in Fig. 4.

References

[1] E. Bauer, in: *Studies in Surface Science and Catalysis 36*, J. Koukal, ed., Elsevier, Amsterdam, 1988, p. 26.
[2] W. Telieps, E. Bauer, *Ultramicroscopy* **1985**, *17*, 57.
[3] M. Lenc, *Čs. Čas. Fyz.* **1975**, *A25*, 28 (in Czech).
[4] T. Ichinokawa, in: *Proc. XIIth ICXOM* (S. Jasienska, L. J. Maksymowicz, eds.), Acad. Min. and Metallurgy, Cracow, 1989, 25.
[5] R. S. Paden, W. C. Nixon, *J. Phys. E: Sci. Instr.* **1968**, *1*, 1073.
[6] J. Frosien, E. Plies, K. Anger, *J. Vac. Sci. Technol.* **1989**, *B7*, 1874.
[7] I. Müllerová, M. Lenc, in: *Proc. XIIth ICEM* (L. D. Peachey, D. B. Williams, eds.), San Francisco Press, San Francisco, 1990, p. 398.
[8] O. Scherzer, *Ultramicroscopy*, **1982**, *9*, 385.
[9] E. Munro, J. Orloff, R. Rutherford, J. Wallmark, *J. Vac. Sci. Technol.* **1988**, *B6*, 1971.

Mikrochim. Acta (1992) [Suppl.] 12: 179–186

To the Backscattering Contrast in Scanning Auger Microscopy

Luděk Frank

Institute of Scientific Instruments, Czechoslovak Academy of Sciences, Královopolská 147, 612 64 Brno, Czechoslovakia

Abstract. The backscattering contrast was studied using silicon-metal-film sandwiches in which the incomplete metal layer consists of a regular array of strips. For various combinations of metals and thin film materials, data concerning contrasts of the strips in Auger emission are presented. For different configurations the Auger image contrast varies in the magnitude and even contrast reversal is observed in the dependence on the primary energy. It even varies for the different constituents of the surface films.

Key words: Auger electron microscopy, backscattering contrast, ratio techniques

The mean free path length of Auger electrons, amounting to several atomic layer distances, is considered approximately equal to the information depth of the analysis. Only the Auger electrons emitted within this depth can be detected outside the specimen. Nevertheless, chemically heterogeneous structures present below this surface layer up to the depth corresponding to the range of primary electrons, cause variations in the total electron emission from the specimen surface. This is due to variations in backscattering from these structures. Consequently, the number of electrons returning through the surface layer changes thereby influencing the excitation level of the Auger emission. For a specimen with a chemically homogeneous surface layer but with a heterogeneous subsurface layer, differences in the Auger electron signal can be observed in microanalyses performed at various points or spurious contrast may arise in the Auger image.

This effect was studied by Kirschner [1]. He reports that with a structure consisting of islands of an Au layer on a Si substrate, covered with an Al film, the enhancement of Auger signals of Al and of the contaminants (O, Ar) from the Au island, amounts to tens of percent relative to that from the Si surroundings and increases with the primary electron energy within the range 1 to 10 keV.

It has also been found that the contrast cannot be successfully corrected using ratioing, i.e. by dividing the net Auger signal by the background signal, taken at an energy slightly higher than the peak energy [2] or measured at 2 keV [3]. The development of methods for the suppression of the backscattering contrast requires a more detailed experimental investigation using different chemical systems in which the energy distribution of the background can be recorded along with the Auger signals. This paper presents the first part of these data, namely those for the Auger image contrast.

Experimental

Specimens suitable for the study of backscattering contrast were prepared by electron-beam lithography using exposed resist masks on silicon (see Fig. 1). Cu strips ($x = x' = 8$ or 16 μm) were produced by chemical metallizing, a method based on differences in the electrochemical potential between silicon and metal. The strips were at least partially immersed into the Si substrate. By evaporating Bi, islands were created in the trenches etched by the radio-frequency plasma sputtering (CF_4 + O_2(15%) plasma, planar reactor), through the mask into a certain depth. The thickness of Bi was the same as the measured depth of the trenches (i.e. $l_2 \approx 0$). If not mentioned otherwise, the thickness of Cu amounted to l = 200 nm (l_2 = 5 to 60 nm) while thickness of Bi amounted to l = 120 nm.

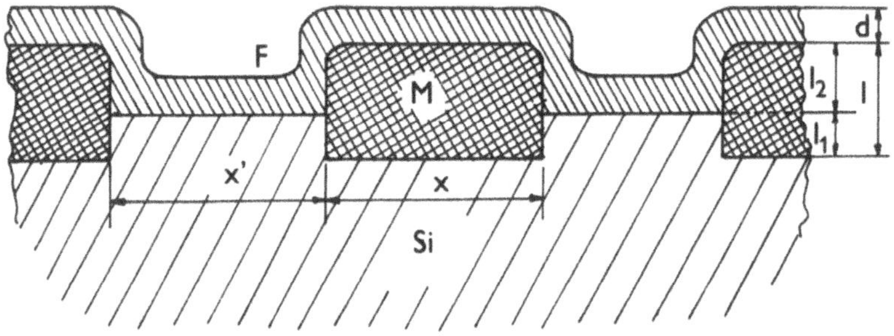

Fig. 1. The cross-section of the specimen showing the strips of metal M on the Si substrate, covered with a film F (see text for further explanation)

The uppermost films were sputtered in the Leybold-Heraeus Z-550 equipment with a planar magnetron. Al_2O_3, SiO_2, Si_3N_4 and TiN_x ($x \simeq 1$) films were reactively sputtered from a metal or Si target in an appropriate atmosphere, other films were sputtered using targets consisting of the same material. Thicknesses within the range d = 5 to 50 nm were used.

The experiments were carried out in the Perkin-Elmer PHI 595, with MACS V6D Surface Analysis Software. The cylindrical mirror analyzer in the co-axial configuration was used for electron detection. The standard take-off angle is 42° which in combination with the specimen holder inclination of 30° gives the measured intensity as a weighted average within the emission angle range of 12° to 72°. The measurements were carried out in the pulse counting mode at peak energy and at an energy slightly higher than the peak energy. The difference of both signals is registered in the device but the signals A (peak) and B (background) can be extracted off-line from the files and treated separately in addition to $P = A - B$.

The specimens were carefully cleaned in-situ using a standard built-in argon ion source at 4 keV and a dose just below the dose causing changes in the backscattering contrast. The chemical homogeneity of the surface was checked by the point analyses in the strip and substrate sites.

Quantitative data on signal ratios were extracted from the line-scans measured perpendicularly to the strips over several (usually three) structure periods (see Fig. 2a). The edge effect was mostly easily recognizable in the neighbourhood of the signal transitions at metal/silicon edges. Its influence on the results has been avoided by extracting the data from central parts of the strips only. From the error bars shown in Fig. 2b and c, the degree of uncertainity in the data presented can be derived. The fluctuations of the signal ratio $\varepsilon = S_1/S_2$ (where S_1 and S_2 are the signals from the strips and from the substrate, respectively) were calculated from the standard deviations $s(S_i)$ of the averages of groups of line-scan points [4].

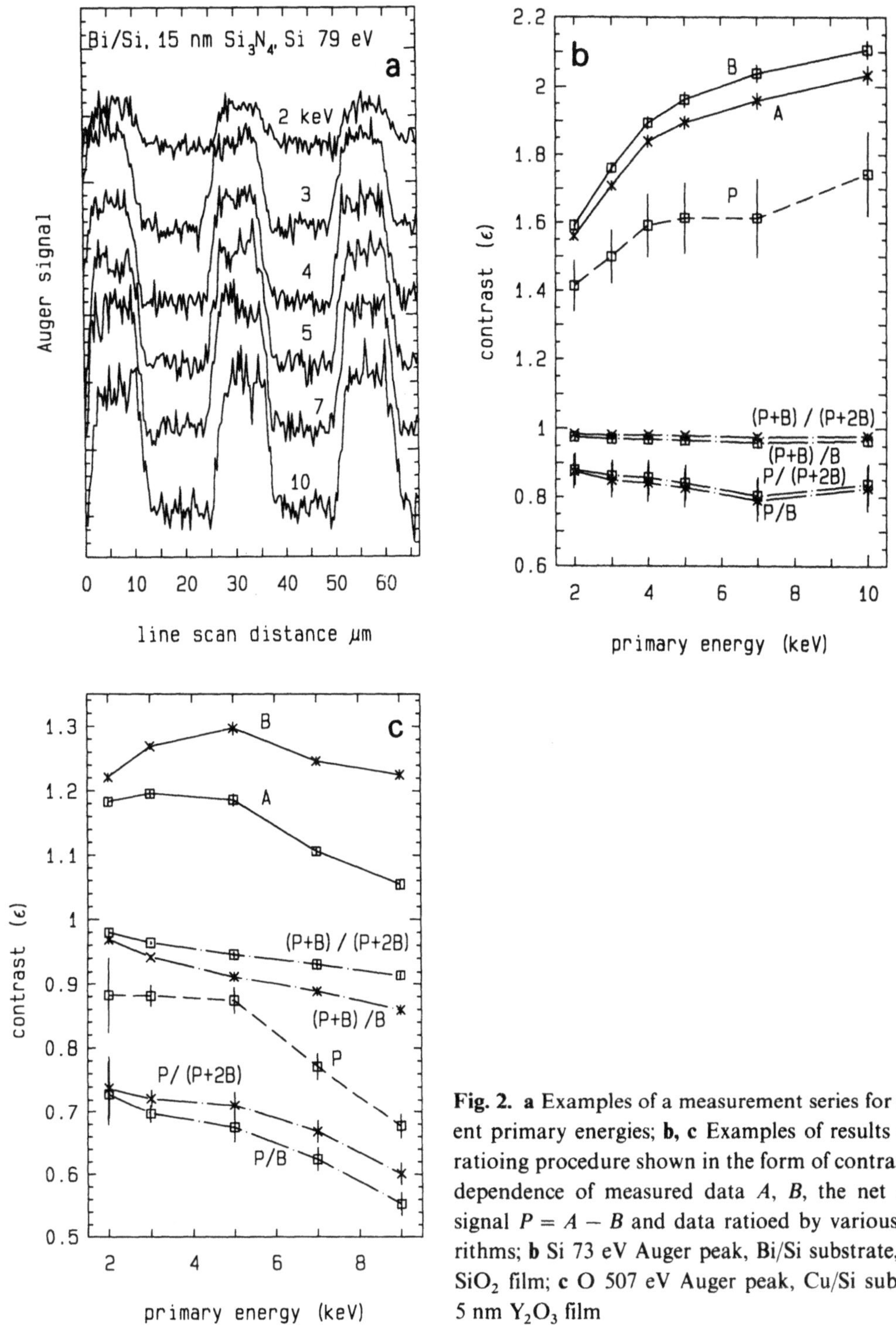

Fig. 2. **a** Examples of a measurement series for different primary energies; **b, c** Examples of results of the ratioing procedure shown in the form of contrast $\varepsilon(E)$ dependence of measured data A, B, the net Auger signal $P = A - B$ and data ratioed by various algorithms; **b** Si 73 eV Auger peak, Bi/Si substrate, 5 nm SiO_2 film; **c** O 507 eV Auger peak, Cu/Si substrate, 5 nm Y_2O_3 film

The preparation of a metal substrate in the form of a film only, troubles the interpretation of the measurement due to effects localized at the interface below the metal film. To evaluate this influence, the electron scattering process was simulated using programs provided by Joy (described in ref. [6]) for Cu and Bi bulk sample. It has been found that, e.g., the depth of 200 nm in Cu is reached by a considerable part of the primary electrons with primary energy higher than 9 keV. Nevertheless, these electrons have a very limited chance to be backscattered and escape from the specimen or contribute

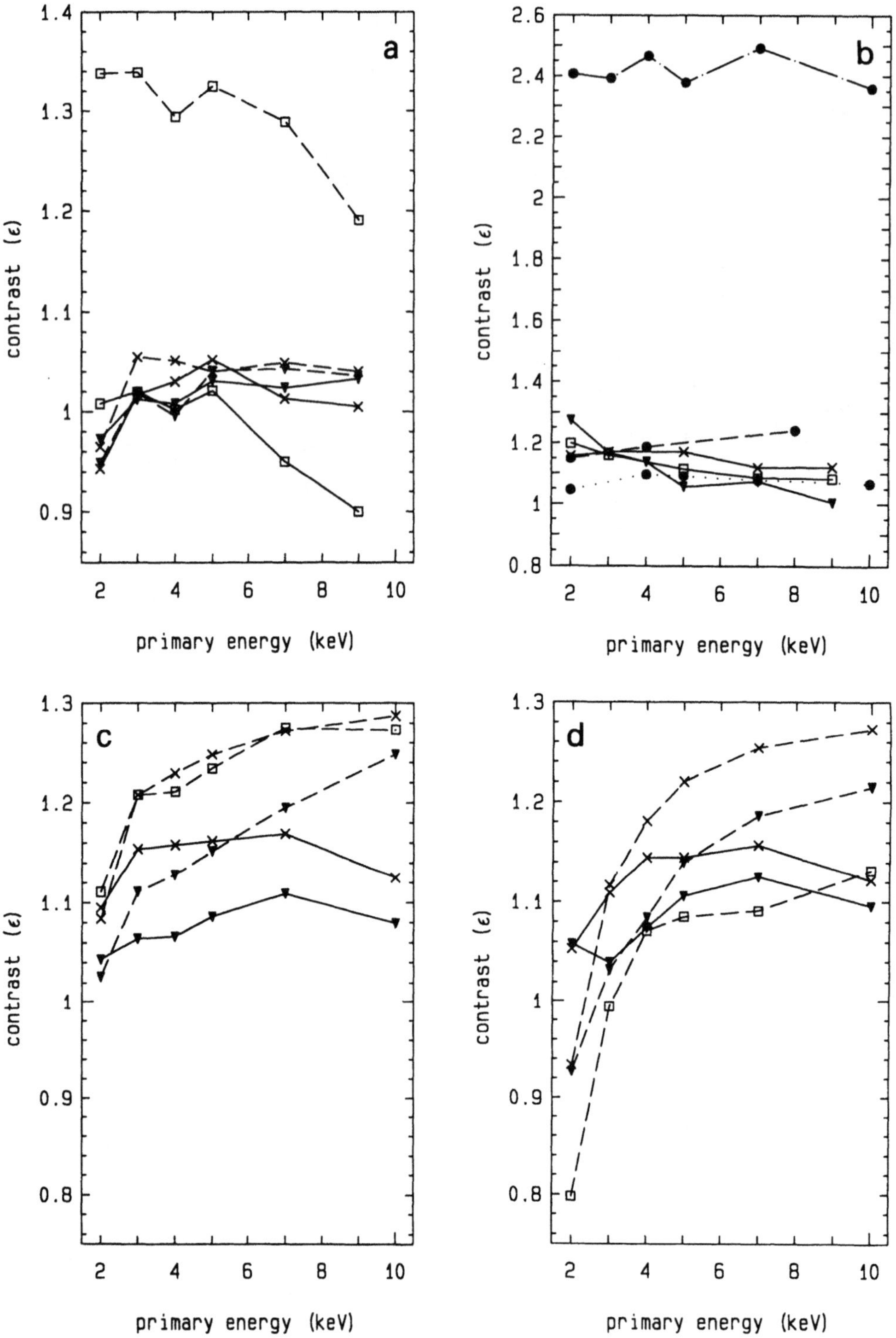

Fig. 3

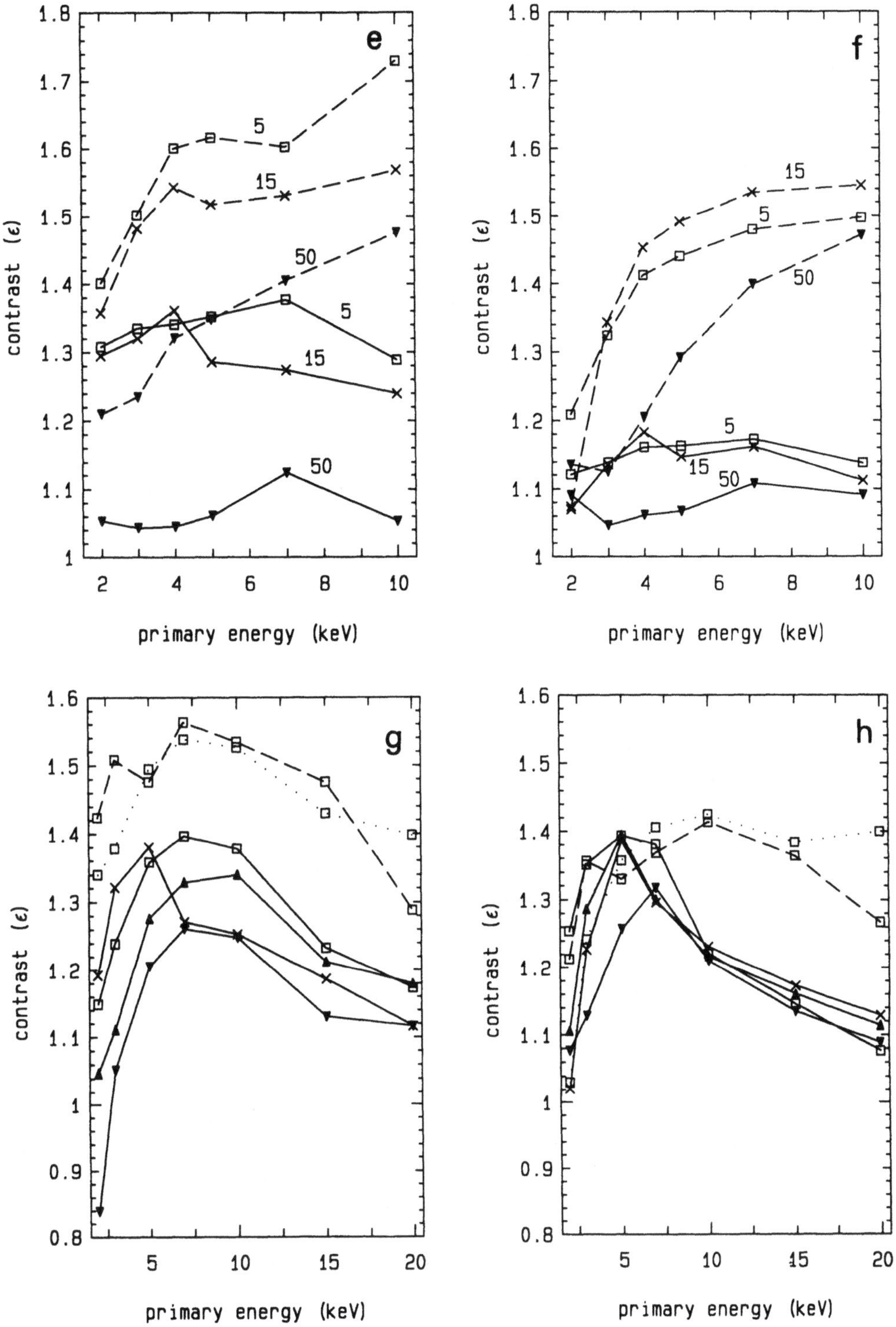

Fig. 3

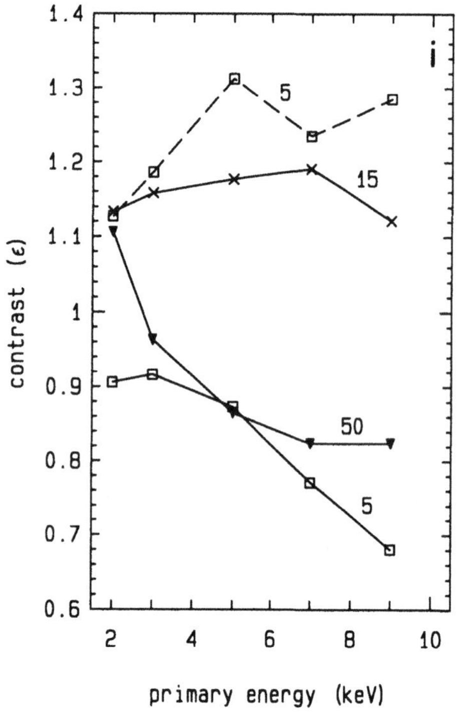

Fig. 3. Contrast (ε) data plotted against the primary energy, for various combinations of the metal/Si structures (Cu/Si or Bi/Si) and surface films of SiO_2, Si_3N_4, Y_2O_3, ITO (0.9 InO_2 + 0.1 SnO_2), Al_2O_3, TiN_x and Al. Curve labels indicate the film thickness in nm. **a** Cu/Si substrate, ITO film, solid line: In 379 eV Auger peak, dashed line: O 509 eV Auger peak, film thickness: □ 5 nm, × 15 nm, ▾ 50 nm; **b** solid line: Cu/Si substrate, O 503 eV peak, Al_2O_3 film (□ 5 nm, × 10 nm, ▾ 50 nm); dashed line: Bi/Si, Ti 415 eV, 10 nm TiN_x; dash-dotted line: Bi/Si, O 503 eV, 12 nm Al; dotted line: Cu/Si, Al 64 eV, 12 nm Al; **c** Si 79 eV peak, Si_3N_4 film, solid line: Cu/Si, dashed line: Bi/Si, film thickness: □ 5 nm, × 15 nm, ▾ 50 nm; **d** N 376 eV peak, otherwise see (**c**); **e** Si 73 eV peak, SiO_2 film, solid line: Cu/Si, dashed line: Bi/Si; **f** O 503 eV peak, otherwise see (**e**); **g** Bi/Si substrate, Si 73 eV peak, SiO_2 film; Bi thickness: solid line 75 nm, dashed line 200 nm, dotted line 600 nm; SiO_2 thickness: □ 7 nm, ▴ 13 nm, × 20 nm, ▾ 44 nm; **h** O 503 eV peak, otherwise see (**g**); **i** O 508 eV peak, Y_2O_3 film, solid line: Cu/Si, dashed line: Bi/Si

to emission of radiation. A relevant evaluation of the influence of the metal-substrate interface seems to be a comparison of the backscattering coefficients η of the bulk metal and of the film of a given thickness, both for a specific energy. Both η values for the 200 nm Cu film and the bulk Cu do not differ below 10 keV. For Bi films, a difference appears above 10 keV for the 75 nm film, above 12 keV for 120 nm, above 17 keV for 200 nm and above 20 keV for 600 nm. The corresponding thresholds are observable (Fig. 3h): the curve for 75 nm Bi departs downwards at 10 keV and the curve for 200 nm at 20 keV, from the 600 nm curve. Other phenomena connected with the metal-substrate interface are neither expected nor observed.

Discussion

From Fig. 2, it follows that mainly the net Auger data, $P = A - B$, suffer from significant noise. To identify its origin the variance $s(S_i)$ was compared with the statistical noise contribution $\sqrt{S_i}$. Factor $K = s(S_i)/\sqrt{S_i}$ was found within the range of 0.8 to 2.15 for both A and B data, namely $K = 1.22 \pm 0.31$. Because $s(S_i)$

is the sample standard deviation of sample size $n \approx 30$ (number of line-scan points taken for estimation of each of the signals S_i) drawn from an infinite population with variance s (which is best estimated by $s(S_i)$), the variance of $s(S_i)$ is $s(S_i)/\sqrt{2n}$ for large n values [5]. Accordingly, $s(K)/K \approx 1/\sqrt{2n}$ for pure statistical noise without topographical contribution. As this limit is exceeded only twice and the K value is not far from unity, and because no ratioing type (i.e. P/B and $P/(P + 2B)$) changes significantly the relative fluctuations of the signal (which has to occur in the case of topography) we can conclude that our results are not considerably influenced by a surface roughness.

Figs. 2b,c show effects of the ratioing, using the algorithms proposed by El Gomati et al. [2]. The two examples used differ expressively in the $\varepsilon(E)$ behaviour for the net Auger signal $P = A - B$. In one of the examples $\varepsilon > 1$ and increases while in the other $\varepsilon < 1$ and decreases. Nevertheless, the ratioed curves are very similar in both cases and no one approaches the constant function $\varepsilon(E) \equiv 1$, which corresponds to a corrected contrast. Deviating not far from this are the curves $(P + B)/B$ and $(P + B)/(P + 2B)$, but they show no true Auger contrast, which is hidden inside a high "black level" background. A subtraction of this background would transform the curves into the P/B and $P/(P + B)$ curves being different from unity. One can conclude that the existing algorithms are not successful in correction of the backscattering contrast.

The contrasts ε of the Auger line-scans are given in Figs. 3a–3i. Their most important properties can be summarized as follows:

a) ε values vary, as a function of the material and thickness of the surface layer, within a broad range, falling even below 1, i.e. the metal strips are dark in some cases.
b) Dependences of ε on the primary electron energy have not the same character. These are slightly increasing, constant or even decreasing for the Cu strips and increasing for the Bi strips. Above 10 keV the dependence is decreasing.
c) The influence of the surface film thickness on ε is not unambiguous. In most cases ε reaches its maximum at 15 nm thickness.
d) The increase in the Bi layer thickness enlarges ε for Auger electrons of oxygen from SiO_2 but for Si electrons the same is not true for all SiO_2 thicknesses.
e) Differences in the behaviour of ε for both constituents of the film are quite significant in some cases, e.g. for Y_2O_3 and ITO (0.9 InO_2 + 0.1 SnO_2).

Conclusion

It can be concluded that none of the ratioing algorithms proposed so far is successful in suppression of the Auger backscattering contrast. For a broader variety of materials and geometrical parameters of the sandwich structure, the behaviour of this contrast is much more complex than a simple atomic number contrast of the substrate, increasing with the primary energy as shown in [1].

The experimental data presented can contribute to the development of an algorithm for suppression of the backscattering contrast in routine Auger electron microscopy. Success in the search for an empirical formula like those mentioned above, is not probable. An efficient algorithm will likely incorporate the electron backscattering factor corresponding to the in-depth interface. It would be, nevertheless, desirable to extract all the necessary information from the spectral measurement in the course of a single experiment. No attempts to do this seem to have been published so far.

The data presented can also be applied to verification of computer models of the electron scattering and Auger electron generation processes in solids.

Acknowledgements. The author is indebted to Professor E. Bauer, the Institute of Physics of the Technical University Clausthal, for placing the Multiprobe Laboratory at his disposal and for valuable discussions. Thanks are due to F. Matějka, J. Matějková, J. Dupák and J. Sobota from the Institute of Scientific Instruments Brno for specimen preparation. The provision of the backscattering simulation programs by Professor D. C. Joy, University of Tennessee, is gratefully acknowledged.

Reference

[1] J. Kirschner, *Scanning Electron Microsc.* **1976**, *I*, 215.

[2] M. M. El Gomati, J. A. D. Matthew, M. Prutton, *Appl. Surface Sci.* **1985**, *24*, 147.

[3] H. E. Bishop, *Scanning Electron Microsc.* **1983**, *III*, 1083.

[4] W. T. Eadie, D. Drijard, F. E. James, M. Roos, B. Sadoulet, *Statistical Methods in Experimental Physics, 1st Ed.*, North-Holland Publ., Amsterdam, 1971, p. 27.

[5] I. V. Dunin-Barkovskyi, N. V. Smirnoff, *The Probability Theory and Mathematical Statistics in Technology*, GITTL, Moscow, 1955, p. 227 (in Russian).

[6] Z. Czyżewski, D. C. Joy, *J. Microsc.* **1989**, *156*, 285.

Mikrochim. Acta (1992) [Suppl.] 12: 187–190

Design Consideration Regarding the Use of an Accelerator Mass Spectrometer in Ion Microanalysis

K. M. Subotic[1] and Miodrag K. Pavicevic[2,*]

[1] Boris Kidrich Institute, Belgrade, Yugoslavia
[2] Faculty of Mining and Geology, University Laboratory for Electron Microanalysis, University of Belgrade, Belgrade, Yugoslavia

Abstract. In this paper the limitations of conventional mass spectrometry are studied together with accelerator based methods which could help circumvent these limitations. In particular, cyclotron-based accelerator mass spectrometry (AMS) is discussed with an emphasis on the performances of superconducting minicyclotrons, designed for use as AMS instruments. If ionic and laser micro-excitation are used in the ion source, the system would have a considerably better analytical sensitivity and isotope resolution compared to standard SIMS and LMMS techniques.

Key words: Ion source, superconducting minicyclotrons, accelerator mass spectrometry, high mass resolution.

The application of ion microbeam methods in solving geochemical and cosmophysical problems offers information on sample composition, structure, trace elements, distribution of elements and isotopes, isotopic ratio, concentration profiles etc. However, in the majority of cases this information is limited by the detection limit, thus representing the main obstacle for a more profound and better interpretation of the results obtained. When separation of elements and isotopes is carried out with conventional systems for mass spectrometry, one of the main problems is to distinguish the isobaric molecules from the ions, i.e. separate particles whose A/q ratio's (A, particle mass and q, ionic charge) are approximately identical. With AMS (accelerator mass spectrometry) which separates elements i.e. isotopes on the basis of classical spectrometry, extended by nuclear acceleration techniques and by stripping and detection methods, an extremely high degree of sensitivity as well as a very low background level can be obtained together with solving the problem of isobaric effects. In the case that the particle path length is extended, an increased sensitivity of AMS is expected. Our idea is to change the linear AMS into a cyclotron that can

* To whom correspondence would be addressed

enable extention of a charged particle trajectory within a detection system to practically infinite values and consequently can detect extremely small differences in A/q ratio [1].

Cyclotron AMS

The principle of the cyclotron based AMS is determined by the phase slip equation:

$$\sin \Phi = 2\pi \mathrm{hn}\, df/f,$$

where Φ is the phase slip and df/f is the frequency error. Mass resolution R reads $R = m/\delta m = \pi hn$, where m is mass, h is harmonic number and n is the number of the particle turns executed. The maximum intensity ratio of the in-phase component I and the out-of-phase beam component I_0 is however determined by:

$$I/I_0 = [N/N_0]_{nat},$$

where N/N_0 is the natural isotope ratio.

Cyclotrons are devices with an inherently high mass-resolution. When tuned for a given atomic mass, they are detuned for molecules with the same atomic mass number but different true masses, due to the accelerator phase slip. An AMS dedicated cyclotron is characterized by a high resolution $R = \pi hn$ number value. This requirement does not necessarily imply high energy needs. In a sufficiently strong magnetic field, the extraction radii may be retained in the 10–20 cm range. Especially, if built in the superconducting version in persistent mode, the electron magnet does not need a power supply while in operation. The only power used would be to operate the RF accelerating system and the ion source. At high bending power of superconducting cyclotrons, ions may be accelerated up to 25–50 times higher energies than in cyclotrons with the same polar radii at room temperature. In a mini cyclotron the particle energy is so low that the isochronous field is almost constant. Putting sectors on a mini-sized uniform magnet will not improve axial focusing significantly. The flat magnetic field configuration of high homogeneity thus suggests itself as a reasonable design approach. Such a dipole field producing constant homogeneity at all field levels could be realized in an air-core-type superconducting version by two symmetrically, relative to the medium plane, positioned superconducting coil sections. The demand for high particle-turn-numbers means very low radio frequency (RF) power requirements. Typical voltage requirements are in the range of 100–500 V, which means less demanding design criteria for the system and relatively cheap commercial components in power supply circuits to be used.

The Working Principle of an Ion Microbeam Analysis Instrument with Superconducting Mini Cyclotron

Figure 1 shows the working principle of an instrument for ion microbeam analysis with a superconducting mini cyclotron. The ion source (1) produces either positive or negative ions from different gases which are accelerated from several to twenty keV and then used to bombard the specimen (2).

The ion-source optics system enables the beam diameter to vary from 1 to 500 μm on a sample and produces, through bombardment, in a sample the ion

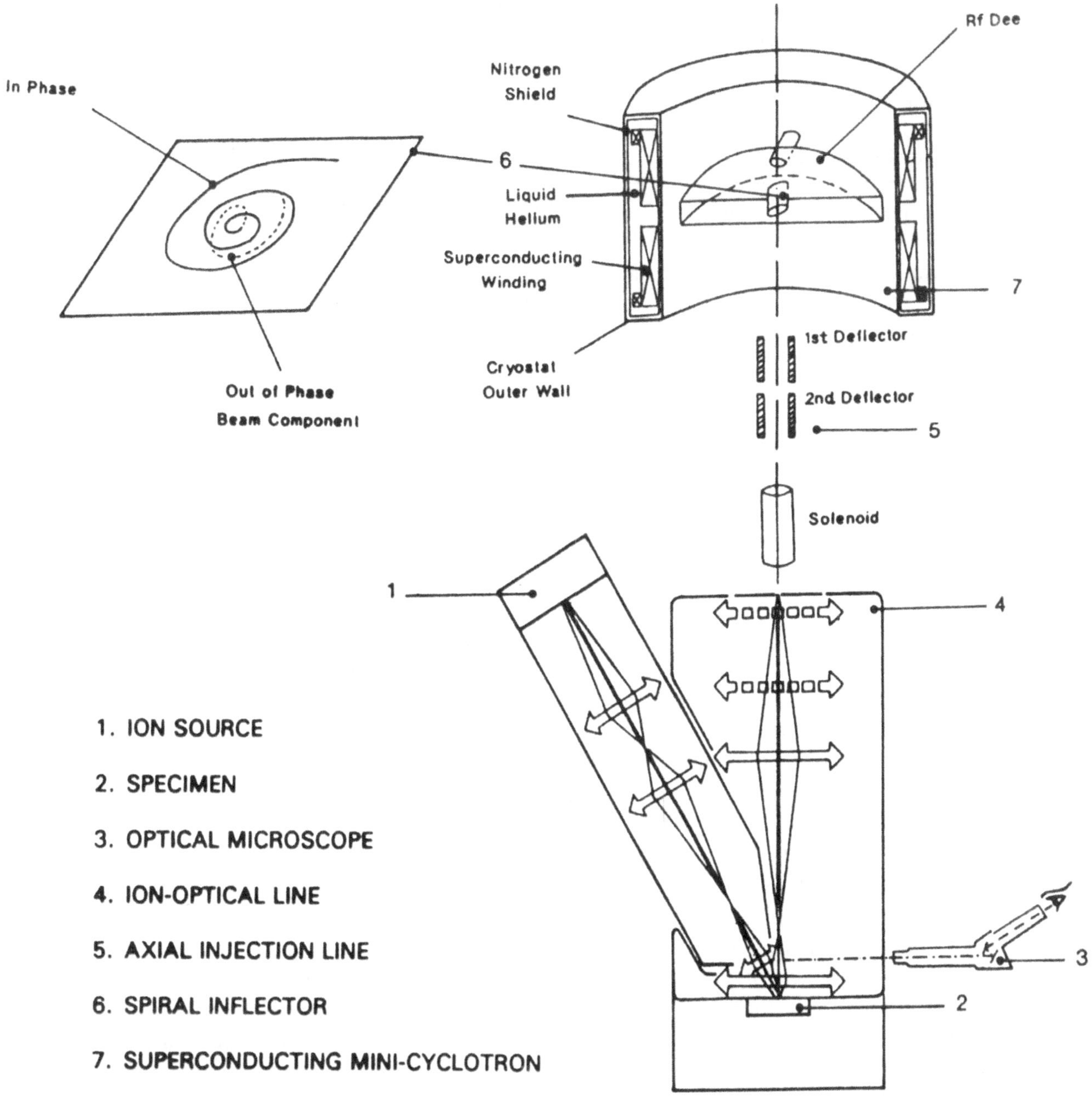

Fig. 1. The working principle of an instrument for ion microbeam analysis with a superconducting mini-cyloctron

current ranging from several up to 50 mA/cm^2. The selection of spots to be analysed is controlled by an optical microscope, (3). The ion-optical system (4) transfers the sputtering ions through linear injection (5) and spiral inflector (6) into a superconducting mini cyclotron (7). A short account of the basic performances of the instrument is given below. Evidently high device resolution can be achieved by making the nh factor as high as possible. However, the highest harmonic number available is determined by coupling effects between transversal and longitudial motion, while the maximum number of turns is determined by clearance requirements at beam injection and extraction. In this situation the device resolution may be increased by

Table 1. Calculated analysing performances of an AMS with superconducting mini-cyclotron

Cyclotron resolution	100,000
Secondary beam current	$2\mu A = 1.2 \times 10^{13}$p/s
Cyclotron transparency	10%
Detecting limit	$D_0 = 10^{-12}$ at 3% statistics in 10^3 s
Detection limit	$D_1 = 10^{-16}$ at 20% statistics in 24 h

the use of stronger magnetic fields in order to increase the number of particle turns executed. The maximum field strength is limited in the conventional design approach. Furthermore the field homogeneity reproducibility deteriorates with the change of field strength of iron core magnets. If a wide spectrum of nuclei has to be detected, conventionally designed AMS mini-cyclotrons should be operated at constant injection and extraction rigidity, and constant orbit geometry for all nuclear species. This implies respectively smaller injection energies and smaller final energies for heavier nuclei. Additionally the isobar problem may not be solved at such low cyclotron rigidities. Introducing the superconducting air core design with a flat magnetic field for an AMS mini-cyclotron, high field strengths with full reproducibility of field homogeneity are possible. Flexible magnetic field settings make the cyclotron rigidity at injection and extraction radius adjustable. Operating a high field strength, the in-phase beam component trajectory is resolved at much inner radii and the rest of the radial space may be used to accelerate the in-phase beam component to sufficiently high energies at which they may be detected and identified. The unavoidable 'isobar' problem, which is the limiting factor in the use of conventional AMS may be solved, however, in superconducting mini-cyclotrons as a result of much higher resolutions available (at fixed radius frequency the cyclotron resolution increases proportionally with field strength) [2].

The cyclotron focusing limit determines the number of particles that can be transmitted through the device according to the maximum current I_0 which corresponds to the cyclotron space charge limit:

$$I_0 = \varepsilon_0 A v^2 d\Phi / 2\pi dE/q$$

where ε_0 = permissivity, A = full beam aperture, v is vertical oscillation frequency, $d\Phi$ = beam phase width, dE is energy again per turn and q = ion charge [1]. Based on detailed calculations, the most important parameters of a future instrument, are summarized in Table 1.

References

[1] K. M. Subotic, *J. Phys G. Nucl. Part. Phys.* **1991**, *17*, 363.

[2] K. M. Subotic, D. Novkovic, M. S. Stojanovic, Lj. S. Milinkovic, *Nucl. Instr. Meth. Phys. Res.* **1991**, *B50*, 267.

Mikrochim. Acta (1992) [Suppl.] 12: 191–195

Accurate Estimation of Uncertainties in Quantitative Electron Energy-Loss Spectrometry

Jan J. Y. Van Puymbroeck[1], Wim A. Jacob[2], and Pierre J. M. Van Espen[1,*]

Departments of [1]Chemistry and [2]Medicine, University of Antwerp, Universiteitsplein 1, B-2610 Antwerpen, Belgium

Abstract. To determine the background under an ionization edge, the pre-edge background intensities are extrapolated using a function which describes the intensity decrease as a function of the energy loss. When intensities and energy losses are logarithmically transformed, a linear relationship is found. The uncertainty of the background, which determines to a large extent the minimal detectable signal and the uncertainty of the calculated concentration, depend on the uncertainties of the slope and intercept of this linear relationship. The uncertainties in slope and intercept must be estimated by a weighted linear regression procedure if the variances of the intensities are non-uniform. If the variances of the normal distributed intensities are determined with few degrees of freedom, they are not suited as weight factors in the least squares minimalisation, however they can be replaced by smoothed values. These smoothed values are calculated with variance functions which describe the relation between the variance and the intensity. The results of the weighted least squares estimation with the smoothed variances did not differ much from the results of an unweighted regression. The weighted regression with the experimental variances led to erroneous results.

Key words: Electron energy-loss spectrometry, background extrapolation, least squares estimation, variance functions.

Electron energy-loss spectrometry (EELS) in a transmission electron microscope is a submicron analysis technique particularly appropriate for the quantitative determination of low atomic number elements. In EELS the inelastic processes between the beam electrons and the electrons of the sample atoms are studied. The energy distribution of the originally monochromatic electrons is recorded after they are transmitted through a thin sample. A small part of these fast electrons ionise the sample atoms by exciting, among others, their inner-shell electrons. The interactions for different shells are observed in the high energy-loss part of the spectrum as edges on a gradually decreasing background of non-characteristic energy-loss electrons

---

* To whom correspondence should be addressed

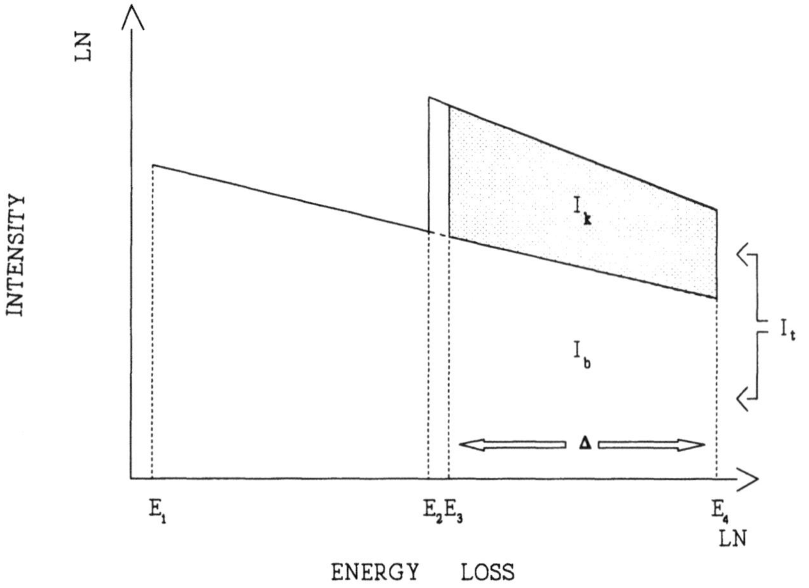

Fig. 1. Schematic representation of an EELS signal. The represented energy-loss range is 150 to 200 eV wide (see text for further explanation).

and electrons of the tails of preceding edges. The net-intensity integral of the edge is proportional to the concentration of the element in the analyzed area of the sample. A schematic representation of an EELS edge is shown in Fig. 1.

The net-intensity I_k is the difference between the total intensity I_t and the background I_b in the energy window Δ. I_b can only be determined from the background which precedes the edge. It is assumed that the background gradually decreases as a power function of energy loss E_i

$$Y(E_i) = A \cdot E_i^R \tag{1}$$

and that this relationship remains valid in the region between E_1 and E_4. If the intensities Y_i and E_i are logarithmically transformed, the parameters A and R can be calculated using linear regression. The background between E_3 and E_4 is then obtained by extrapolation.

Since the minimal detectable signal and the uncertainty of the elemental concentration depend on the uncertainty of the net intensity s_k, an accurate estimation of s_k is crucial. According to error propagation rules, s_k depends on s_b, the error of I_b. Since I_b was obtained by extrapolating the background intensities using the parameters A and R, s_b is a function of the variances of A and R [1]:

$$s_b^2 = s_A^2 \cdot \left(\frac{\partial I_b}{\partial A}\right)^2 + s_R^2 \cdot \left(\frac{\partial I_b}{\partial R}\right)^2 + \cdots \tag{2}$$

Various authors pointed out that a weighted least squares estimation (WLSE) is preferable over an unweighted estimation (ULSE) for an accurate estimation of s_A^2 and s_R^2 for two reasons [2, 3, 4]: (a) the variances s_i^2 of the intensities Y_i are non-uniform (heteroscedastic) because they are the result of a counting experiment and have a Poisson distribution and (b) the logarithmic transformation introduces a bias and makes a transformation of s_i^2 necessary.

In our experiments, the intensities Y_i follow a normal distribution. The variances s_i^2 are heteroscedastic. Their magnitude depends on the intensity Y_i. Based on the considerations given above, it is expected that WLSE must be used. Since s_i^2 is obtained with few degrees of freedom, the estimation of the underlying population variance σ_i^2 by the sample variance s_i^2 is not optimal. Because s_i^2/σ_i^2 has a χ^2 distribution, there will be a large sampling error for the s_i^2 which makes them unsuitable as weights in the WLSE. We tried to improve the WLSE by replacing the s_i^2 by values originating from a variance function which expresses the relationship between s_i^2 and Y_i.

Experimental

The data were obtained with a ZEISS CEM902 microscope (Carl Zeiss, Germany). Spectra are acquired by sequential recording of the intensities in successive energy loss channels. Each channel has a width of approximately 1 eV. The intensity Y_i at each energy loss is the average from n measurements. In practice, the operator selects a small value for n to keep acquisition times acceptable and radiation damage low. The variance s_i^2 is calculated and stored together with Y_i.

The data used in this paper, originate from a BN crystallite sample. The intensities which were recorded between 100 and 600 eV energy loss, are averages of 3 measurements ($n = 3$).

Results and Discussion

First, the non-uniformity of the variances of the experimental data is checked. This is done using Bartlett's test [5]. Because this test is criticised for its inefficiency in the presence of low degrees of freedom [6], we use also Cochran's test [7]. Both tests always indicated heteroscedastic variances.

From literature we found polynomial (eq. 3), linear (eq. 4) parabolic (eq. 5) and multiplicative (eq. 6) models as possible relationships [6–8].

$$s^2(Y_i) = a_0^n + a_1^n \cdot Y_i + a_2^n \cdot Y_i^2, \tag{3}$$

$$s^2(Y_i) = a_0^l + a_1^l \cdot Y_i, \tag{4}$$

$$s^2(Y_i) = a_0^b + a_1^b \cdot Y_i^2, \tag{5}$$

$$s^2(Y_i) = a_0^m \cdot Y_i^{a_1^m}. \tag{6}$$

Their purpose is to smooth the s_i^2 values and eliminate extreme deviant values. The s_i^2 values are then replaced by $s^2(Y_i)$ values as weight factors in the WLSE. The coefficients of the models are estimated by a least squares fitting. The data of the pre-edge background regions and the entire spectrum was used to fit the models. Both procedures gave essentially the same results. They were also fitted using the logarithmically transformed intensities Y_i' and variances $s_i'^2$.

$$Y_i' = \ln(Y_i), \qquad s_i'^2 = \frac{s_i^2}{Y_i^2}. \tag{7}$$

The best model is considered the model with the lowest residual variance [6].

The coefficients for the models of equations 3 to 6 are determined for the total intensity range between 100 and 600 eV energy loss of a spectrum of BN. The coefficients and their residual variances are given in Table 1. The polynomial model shows a local minimum. The linear and polynomial models both generate negative

Table 1. Coefficients and residual variances of the least squares fit of the variance of the intensity as a function of the intensity using different models (equations 3 to 6)

Model	Coefficients a_0	a_1	a_2	Residual variance ($\times 10^9$)
Polynomial	15810	-5.13	$2.63 \cdot 10^{-4}$	8.60[1]
Linear	-28522	11.28	—	3.55[2]
Parabolic	163	$1.91 \cdot 10^{-4}$	—	3.09
Multiplicative	$6.19 \cdot 10^{-6}$	2.31	—	2.95

[1] local minimum
[2] negative variances

Table 2. Least squares estimated power law parameters (eq. 1) and their estimated errors for different weighing strategies

	Weighing strategy	A ($\times 10^{13}$)	s_A ($\times 10^{11}$)	R	s_R
Boron	ULSE	5.47	1.39	-4.484	0.040
	WLSE s_i^2	7.02	0.49	-4.532	0.014
	WLSE parabolic model (eq. 5)	5.63	1.78	-4.490	0.050
	WLSE multiplicative model (eq. 6)	5.20	0.99	-4.474	0.030
Nitrogen	ULSE	2.47	0.69	-4.029	0.092
	WLSE s_i^2	1.75	0.18	-3.972	0.036
	WLSE parabolic model (eq. 5)	2.40	1.11	-4.024	0.154
	WLSE multiplicative model (eq. 6)	2.50	0.38	-4.031	0.051

variances. The parabolic and multiplicative models do not show this unacceptable behaviour. The multiplicative model gives the lowest residual variance.

In general, the polynomial and linear models generate often negative values at the extremes of the data. This behaviour occurs less when fitting the $s_i'^2$ values. If negative values are obtained, the models are inappropriate to generate weight factors for the WLSE. The multiplicative model shows in most cases the lowest residual variance.

The results of the background fitting for boron and nitrogen using ULSE and WLSE using respectively the experimental sample variances s_i^2, the parabolic and multiplicative variance models as weight factors, are given in Table 2. The results of the ULSE are very similar to the WLSE using both variance models. The errors of A and R are underestimated when the WLSE is performed with the experimental s_i^2 values. The similarity between the ULSE and the WLSE with the smoothed variance suggests that the logarithmic transformation of equation 7 reduces to a large extent the non-uniformity of the variances making weighting less necessary. From the variance fitting results in Table 1, it can be explained that the non-uniformity is less pronounced after log-log transformation. Inspection of the coefficients of the multiplicative model, which describes the s_i^2 values most successfully, shows that a_0^m is a small constant while a_1^m is close to 2, therefore after the logarithmic transformation of equation 7, the $s_i'^2$ values will almost be constant.

Our results suggest that although the experimental s_i^2 values are heteroscedastic, the ULSE can accurately estimate the errors, s_A and s_R, on the power law parameters A and R from equation 1. This is caused by the logarithmic transformation which reduces the non-uniformity. The laborious WLSE procedure with variance smoothing becomes unnecessary. The WLSE with the experimental s_i^2 values as weight factors leads to erroneous results by underestimating the errors on A and R.

References

[1] R. F. Egerton, *Electron Energy-Loss Spectroscopy in the Electron Microscope*, Plenum, New York, 1986, p. 259.
[2] T. Pun, J. R. Ellis, M. Eden, *J. Microsc.* **1985**, *137*, 93.
[3] D. R. Liu, L. M. Brown, *J. Microsc.* **1987**, *147*, 37.
[4] P. Trebbia, *Ultramicroscopy* **1988**, *24*, 399.
[5] D. L. Massart, A. Dijkstra, L. Kaufman, *Evaluation and Optimization of Laboratory Methods and Analytical Procedures*, Elsevier, Amsterdam, 1978, p. 105.
[6] D. Rodbard, R. H. Lenox, H. L. Wray, D. Ramseth, *Clin. Chem.* **1976**, *22*, 350.
[7] H. Bubert, R. Klockenkämper, *Fresenius Z. Anal. Chem.* **1983**, *316*, 186.
[8] L. M. Schwartz, *Anal. Chem.* **1979**, *51*, 723.

Mikrochim. Acta (1992) [Suppl.] 12: 197–204

An EELS System for a TEM/STEM—Performance and Its Use in Materials Science

Reinhard Schneider* and Wolfgang Rechner

Institute of Solid State Physics and Electron Microscopy, Weinberg 2, D-O-4050 Halle/Saale, Federal Republic of Germany

Abstract. A transmission/scanning transmission electron microscope (TEM/STEM) of the type JEOL JEM 100S was complemented by a self-constructed system for electron energy loss spectroscopy (EELS). The spectrometer has an energy resolution of about 1.5 eV. For controlling the microscope and the spectrometer an external computer control was designed. A digital image acquisition system was attached to the microscope providing energy filtering microscopy by EELS. Some examples demonstrate the potential applicability of the system to problems in materials science.

Key words: Analytical electron microscopy, electron energy loss spectroscopy (EELS), computer control, digital image recording.

TEM/STEMs are employed frequently in combination with analytical methods [1] and the combination with EELS is particularly useful if light elements are to be detected. EELS yields information on specimen composition, structural properties and electronic properties.

For the JEM 100S microscope a 90° magnetic sector analyzer was developed and mounted below the camera chamber in place of the STEM detector which is now at the side of the column. The spectrometer can be used for EELS in the TEM/STEM imaging and diffraction modes at primary energies of 40 to 100 keV.

Magnetic Sector and Spectrometer Design

The main part of the magnetic sector and spectrometer consists of a magnetic prism (Fig. 1). The imaging properties have been calculated on basis of the transfer-matrix method and for the aberration-free electron-optical design of sector fields [2]. Advantageous imaging properties are attained by inclining the entrance and the exit pole-piece boundaries of the magnet to the optical axis, thus yielding quadrupole field components [3]. This enables a microscope crossover to be imaged at a

* To whom correspondence should be addressed

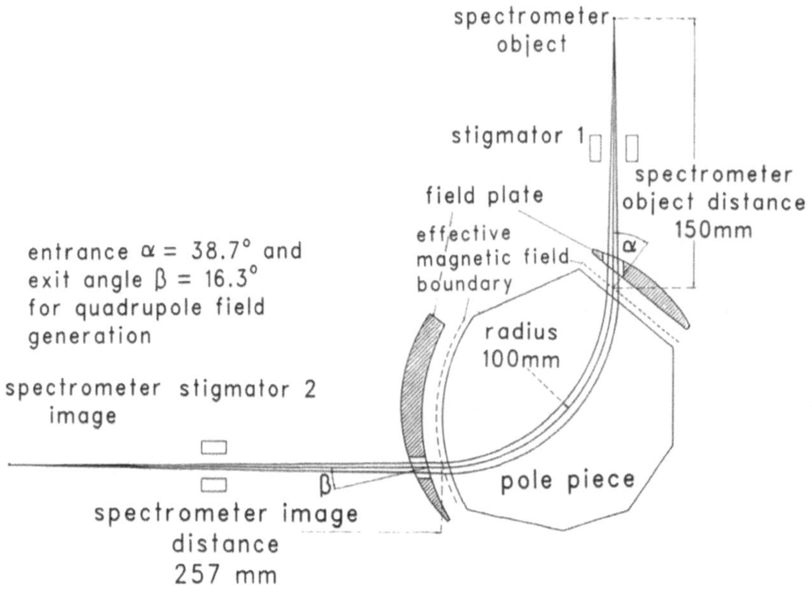

Fig. 1. Schematic presentation of the magnetic prism

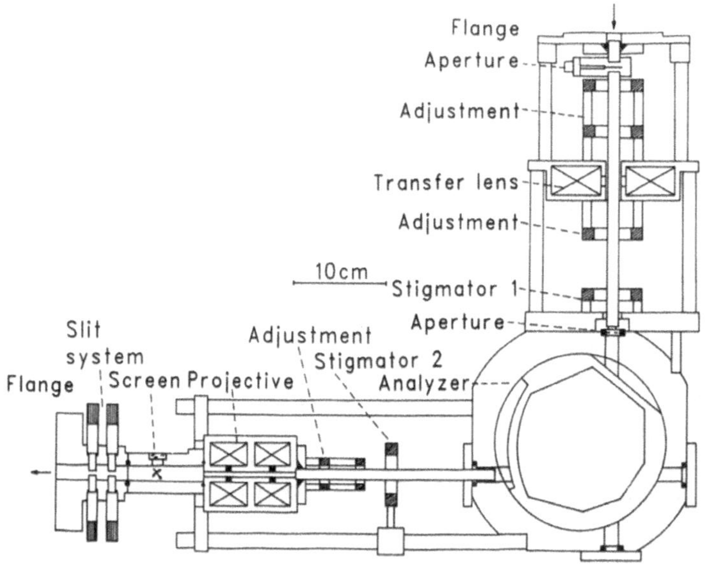

Fig. 2. Layout of the spectrometer system

point or on a line (length <0.01 cm). The image plane of the spectrometer is the dispersion plane.

Pre- and post-spectrometer optics (round lenses, stigmators, alignment units) are provided to adapt the operating modes (see Fig. 2). A transfer lens is situated between the microscope column and the spectrometer. The lens transfers a TEM/STEM-crossover to the object plane of the spectrometer. The aberrations are corrected for by introducing hexapole fields in the form of stigmators and a curved magnetic exit boundary. Corrections have been made for the spherical aberration in the dispersion direction and for the dispersion plane tilt. The dispersion of the prism is about 4 μm/eV at a primary energy of 100 keV. With the help of a rotation-free lens the spectrum can be magnified up to 80 times. The energy loss electrons are selected using a double-slit. The selected electrons are detected serially by a scintillator/photomultiplier arrangement working in the counting mode. The counts are recorded by a multichannel analyzer. The spectrometer system has an energy resolution of 1.5 eV for a thermionic tungsten cathode (Fig. 3a,b). This is particularly the case if the electronics for compensating 50 Hz parasitic voltages are used [4]. The performance of the EELS system presented, is nearly comparable with that of commercial equipment like the GATAN model 607 magnetic prism, but its electron optics offers more possibilities of influencing the electron beam. This is advantageous in operating the system in the different imaging and diffraction modes of the JEM 100S microscope.

The TEM/STEM is not only combined with EELS but also with energy-dispersive X-ray spectroscopy and convergent-beam electron diffraction (CBED). To reduce the contamination, a special anticontamination unit is used [4]. Because

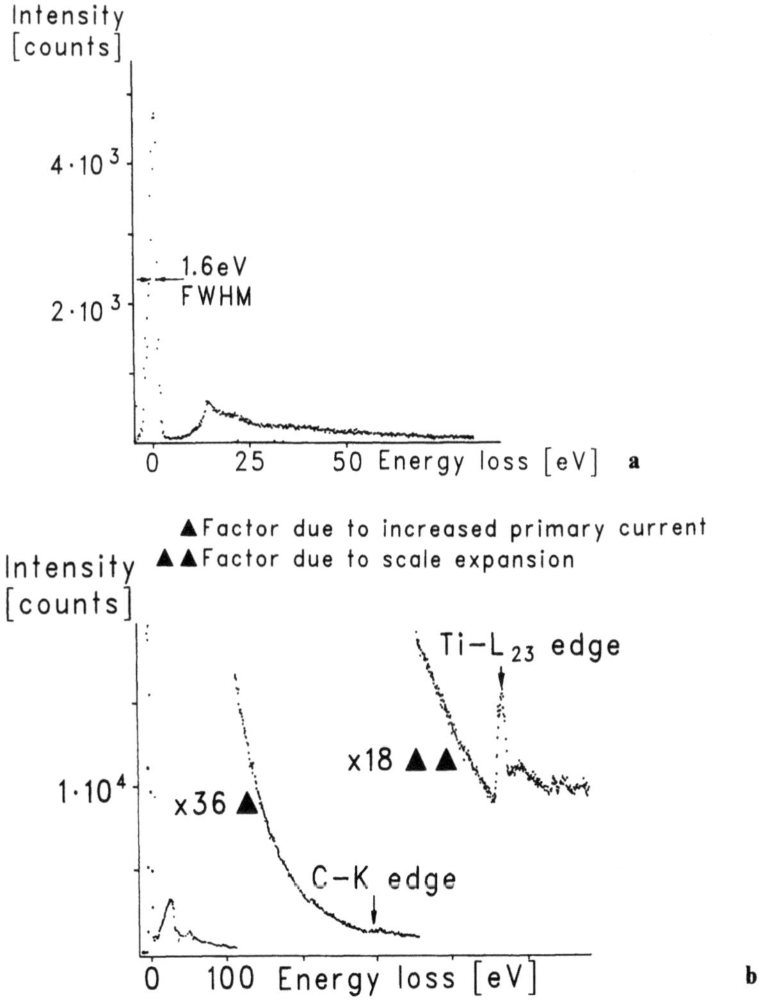

Fig. 3. Electron energy loss (EEL) spectra (60 kV accelerating voltage). **a** Al, **b** Ti-L_{23}-edge

of instability effects, specimen damage and contamination, the operating time should be restricted to approx. 15 min.

Remote Control of Microscope Electronics and Acquisition

A fast and optimum setting of all working parameters is required. To solve this problem the microscope control system has been modified. A computer is applied to control the column or to record and process data arising from the signal detectors. For the mode-control a system has been designed, which allows one to set currents for all lenses, the magnetic prism and most of the alignment elements in addition to the high voltage (minimum step width 0.1 eV) [5]. The arrangement of the electronics is illustrated in Fig. 4. It consists of digital-to-analogue converter (DAC)-interfaces and components of data transfer and decoding in addition to the microprocessòr. The DAC output voltages are applied to the power circuits of the electron-optical elements as references. The hardware is contained in a box on the microscope console. Two different types of circuit boards have been developed, each equipped with two 12-bit DACs. Type (a) meets medium stability requirements, for instance for controlling alignment coils. Type (b) can be used for the lens setup because of its very high stability of about 10^{-6}. Separate electronics were constructed for controlling the high voltage in connection with EELS. This board consists of connected 12-bit and 16-bit DACs that have a particularly high stability.

To carry out EELS, a spectrum is recorded by increasing the high voltage of the microscope, thus scanning the spectrum across the energy selection slit. The advantage is that chromatic lens aberrations do not deteriorate the energy resolution attainable. It is only necessary to stabilize the illumination conditions such as the focus condition and probe position. Figure 5 shows low-loss spectra from a Si crystal, taken by increasing the accelerating voltage (step size 0.1 eV). The spectra show the thickness-dependent plasmon-excitations.

The digital control system demonstrates a performance which is comparable with that of the internal electronics of the microscope. Due to the **freely** adjustable

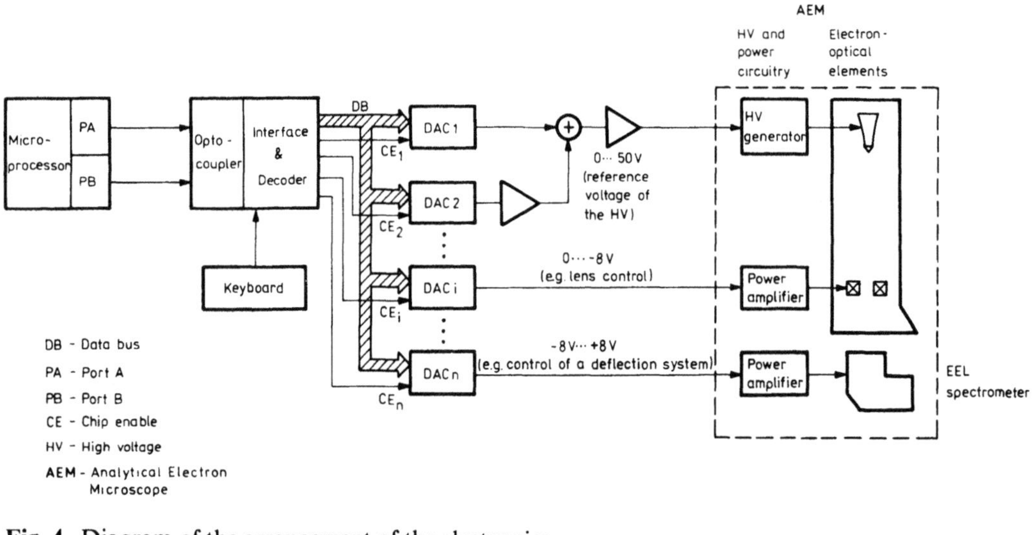

Fig. 4. Diagram of the arrangement of the electronics

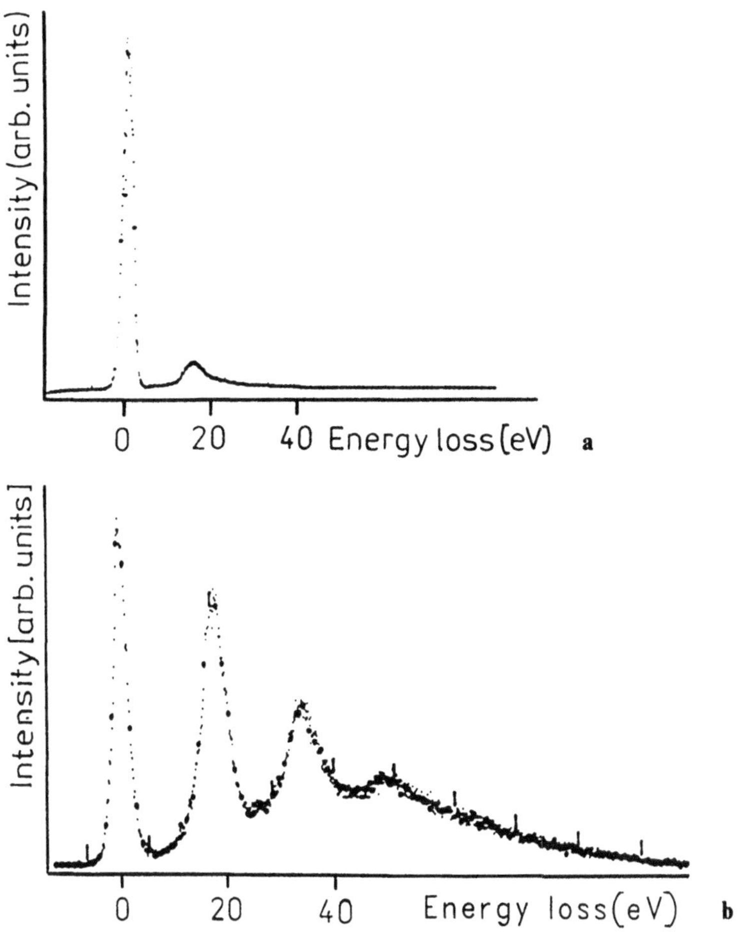

Fig. 5. Si energy low-loss spectra (60 kV accelerating voltage). **a** thin crystal region, **b** thick crystal region

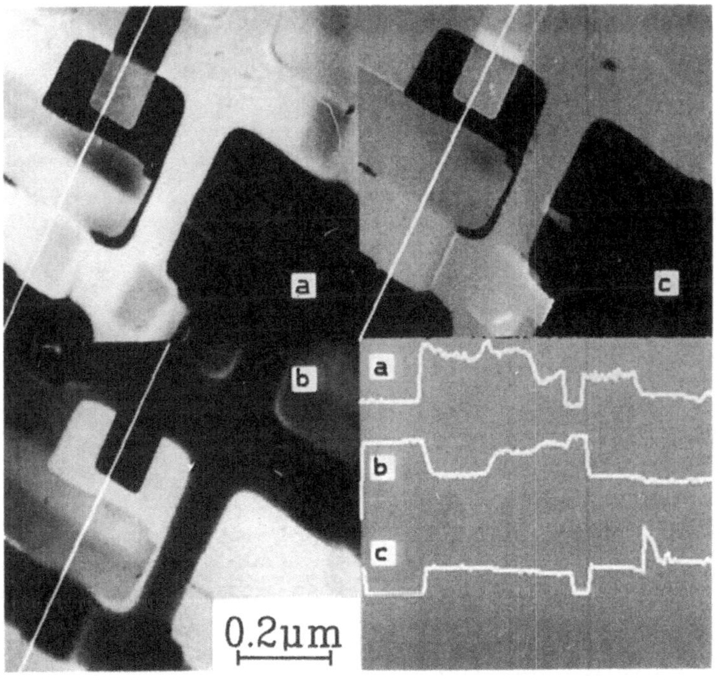

Fig. 6. Microelectronic test structure imaging and intensity profiles (60 kV accelerating voltage). **a** STEM bright field image, **b** secondary electron image, **c** 10–40 eV loss electron image

excitation of the microscope lenses the computer system is also useful for other analytical methods such as CBED.

To gather more detailed information about a specific specimen, the computer-controlled EELS is applied to the energy-filtered imaging in the STEM mode in combination with a system of digital image acquisition and processing [6]. Figure 6 shows the example of a microelectronic test structure with a chemically thinned Si structure, being imaged in the secondary electron (SE)-, the STEM- bright field and the energy-filtered STEM-mode.

Results and Discussion

Temperature treatment of Si-CZ crystals causes oxygen-precipitates, which play a role in device-processing. Spectra have been taken from inhomogeneities in the STEM bright-field image with a lateral resolution of 50 nm showing detection of oxygen (Fig. 7a). With increasing oxygen content, changes are indicated: a more pronounced oxygen K-shell ionization edge, a broadening and a shift of the plasmon

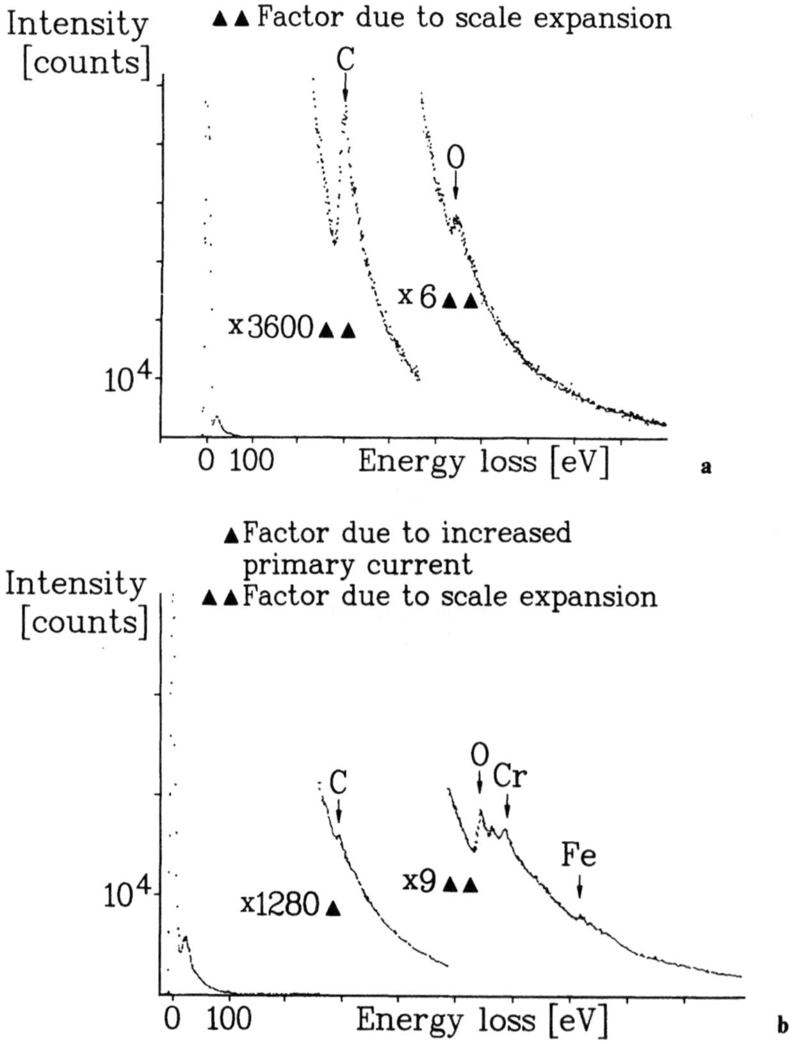

Fig. 7. Impurities in Si (60 kV accelerating voltage). **a** Oxygen, **b** Fe and Cr

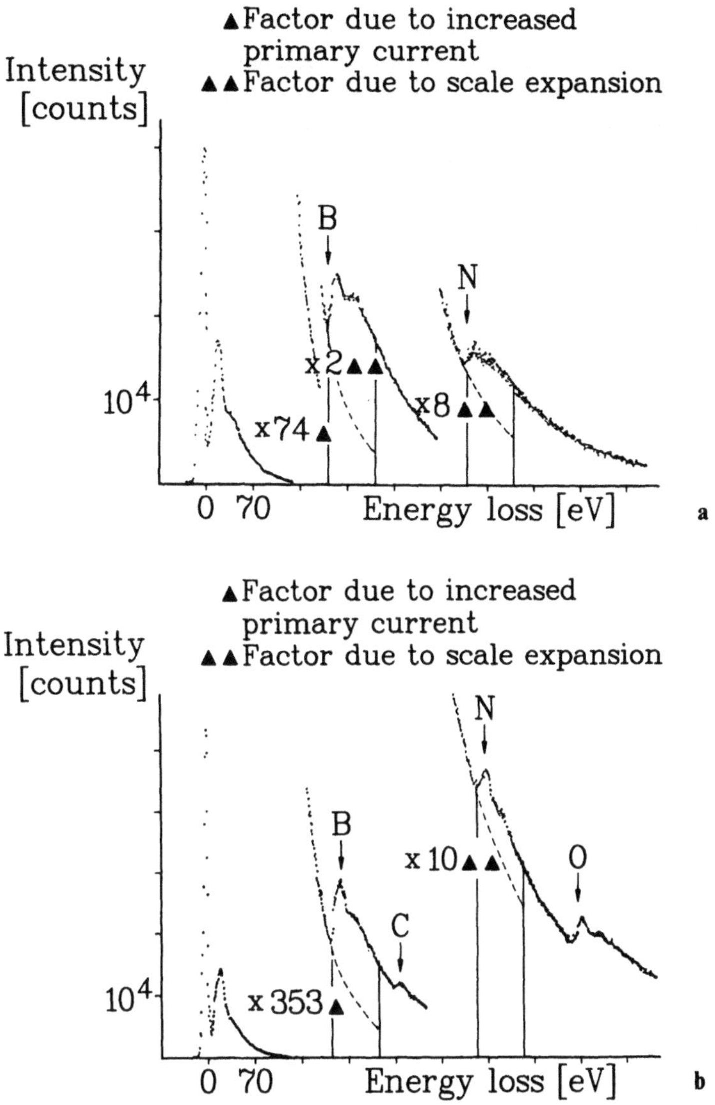

Fig. 8. EEL spectra of BN-films (60 kV accelerating voltage). **a** B, N; **b** B, N, C, O

peak from 17 to 22 eV. Impurities of Fe and Cr have been analysed in other crystal regions (Fig. 7b).

Figure 8a shows quantitative EELS for BN-films, prepared in different ways. The quantification was carried out by extrapolating the background using the least-squares method [7], and by calculating the partial cross-sections, according to the hydrogenic model [8]. The incident beam convergence was taken into consideration by correction of the partial cross-sections [9]. One film exhibits a B/N-ratio of 1.2. In the other film, prepared by ion plating, C and O were additionally detected with the B/N-ratio being about 1.9 (Fig. 8b).

References

[1] J. Heydenreichl, W. Rechner, *Mikrochim. Acta* [Wien] **1987**, *I*, 93.

[2] N. W. Parker, M. Utlaut, M. S. Isaacson, *Optik* **1978**, *51*, 333

[3] W. Rechner, *Proc. 10. Tagung Elektronenmikroskopie*, Leipzig, 1981, p. 254.
[4] W. Rechner, R. Schneider, *Proc. 11. Tagung Elektronenmikroskopie*, Dresden, 1984, p. 153.
[5] R. Schneider, W. Rechner, A. Pippel, J. Bischoff, *Measurement Sci. Technol.* **1990**, *1*, 887.
[6] M. Taege, W. Neumann, A. Pippel, *Proc. 12. Tagung Elektronenmikroskopie*, Dresden, 1988, p. 277.
[7] D. C. Joy, *Scanning Electron Microscopy* **1982**, *II*, 505.
[8] R. F. Egerton, *Ultramicroscopy* **1979**, *4*, 169.
[9] D. C. Joy, D. M. Maher, *J. Microsc.* **1981**, *124*, 37.

Mikrochim. Acta (1992) [Suppl.] 12: 205–212

Quantitative X-Ray Microanalysis of Bio-Organic Bulk Specimens

Abraham Boekestein

Technical and Physical Engineering Research Service-DLO, P.O. Box 356, NL-6700 AJ Wageningen, The Netherlands

Abstract. X-ray microanalysis carried out in the electron microscope on bio-organic bulk specimens, implies a number of specific problems in the case of quantitation. First it is essential that elemental loss and redistribution during preparation is prevented on the scale of the analysis. Further, the matrix is in most cases composed of very light elements (atomic number < 11), like C, N and O, which are normally not measured. Moreover most bio-organic specimens are non-conductors of electricity and this may affect the shape of the X-ray depth distribution curve. Finally, the conventional matrix correction programs supplied with the X-ray analysis equipment are in many cases 'tuned' for materials science analyses problems and this may have influenced the choice of specific correction formulae. This situation has resulted in special procedures to achieve reliable quantitation in X-ray microanalysis of bio-organic specimens. Most often the specimen composition in the microcompartment, which is to be analyzed, is simulated with 'ideal' standards, so that quantitation can be achieved by direct comparison of peak intensities or peak-to-background ratios neglecting the marginal matrix effects. In other cases an empirical matrix correction is achieved by a combination of ideal and non-ideal compound standards. In certain cases special matrix correction procedures have been developed for bio-organic specimens.

Key words: X-ray microanalysis, bio-organic, quantitation, standards

Biological bulk specimens, prepared for quantitative X-ray microanalysis, have specific characteristics, which may trouble the quantitation and the reliability of the results [1–3]. Many of the elements of interest are dissolved in water-containing compartments and may be lost easily during preparation. Therefore much attention is paid to the preparation procedures because most conventional techniques (like wet-chemical fixation and dehydration) cause leakage and redistribution of elements [4]. Further the matrix is in most cases organic and composed of different light (atomic number < 11) elements, which are normally not measured or quantified in biological X-ray microanalysis. As these elements are major constituents of the matrix, their effect on beam penetration and X-ray absorption is important.

Moreover, biological or bio-organic specimens are non-conductors of electrical current and this implies that bulk electrostatic charging takes place unless very rigid precautions have been carried out [5]. Bulk charging phenomena will seriously affect the shape of the excitation volume and hence the X-ray depth distribution curve.

Finally in many cases matrix correction software is used, which has been developed for materials science problems. This may imply that some of the expressions and constants in the different parts of the matrix correction are 'tuned' specifically for analysis of specimens with a relatively high mean atomic number and may perform poorly for light-element type specimens [6]. In order to overcome at least some of these problems, a number of specific procedures have been designed to increase the reliability of quantitation in biological X-ray microanalysis and to improve the performance of this analysis technique [1, 3, 7, 8].

Specimen Preparation Procedures

A number of specific preparation procedures have been developed to enable a reliable quantitative X-ray microanalysis of biological specimens. In Fig. 1 a schematic overview is presented of pathways leading to suitable specimens [9]. Much attention is given to the retainment of diffusible elements in their original ultrastructural locations in the specimen. In most cases the application of cryo-techniques give the best guarantee that this is achieved [4]. Most procedures start with a cryofixation which is in principle the fast cooling down of a specimen to temperatures of about 70 K. If the cooling rate is high enough (approx. 10000 K/s), a superficial layer in the biological specimen is frozen without ice-artefacts, i.e. the water is probably in a metastable amorphous state without crystals. If cooling rates are not so high and in deeper layers in the specimen, ice crystals can be observed as a freezing artefact (Fig. 2). For SEM-X-ray microanalysis this may not be so harmful as the excitation volume and hence the analytical resolution is on the order of micrometers.

Once a specimen is frozen, the next step will be to create a flat cross section through the specimen in order to enable an accurate X-ray microanalysis and localization of ultrastructural compartments. For this purpose cutting and milling

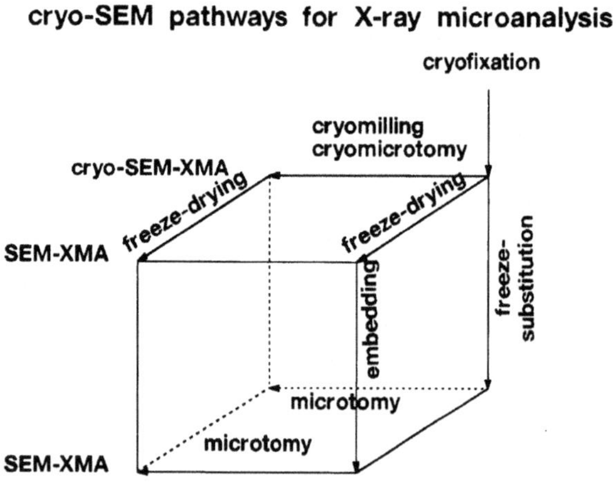

Fig. 1. Cryo-SEM pathways in order to perform X-ray microanalysis on biological specimens [9]. XMA = X-ray microanalysis. SEM = scanning electron microscopy

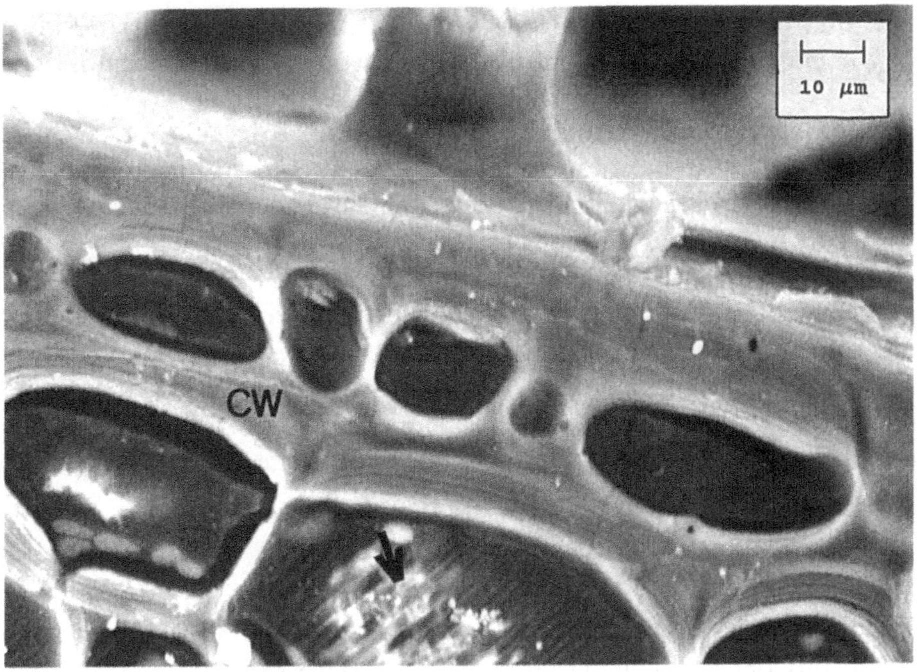

Fig. 2. Cryo-ultramilled cross-section through a *Vicia faba* pod showing cell walls (CW) and crystallized cell content (arrow) (courtesy of D. R. Verkerke, CABO-DLO)

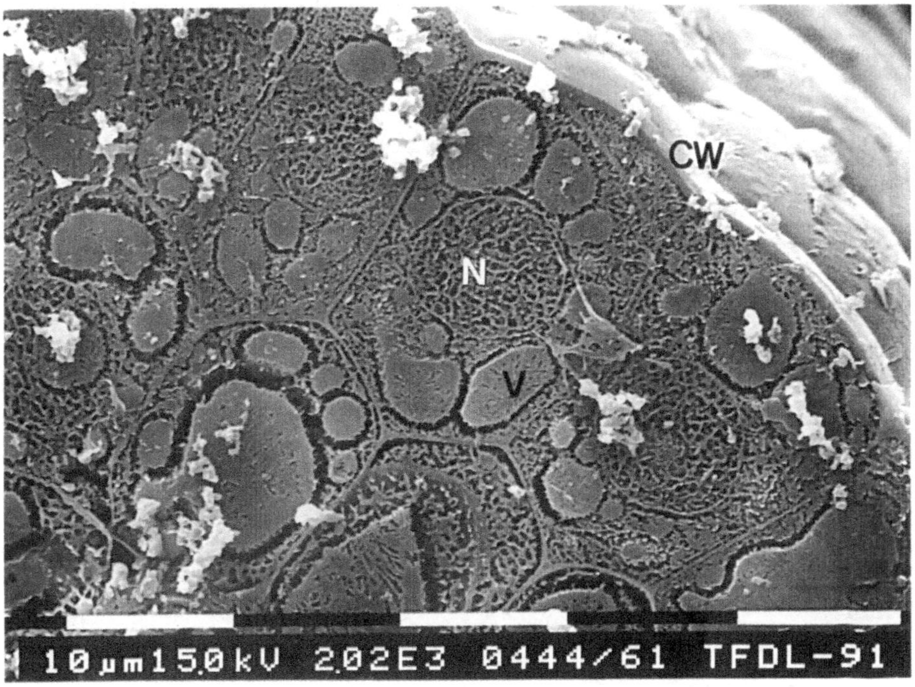

Fig. 3. Cross-section through a frozen-hydrated apex of potato. Bar = 10 μm. N = nucleus, CW = cell wall, V = vacuole (courtesy of C. van der Schoot, ATO-DLO)

devices have been developed in which the specimen is kept frozen. In Fig. 3 a cut cross section of frozen-hydrated potato apex is shown in which the main cell compartments are recognizable. The final procedure for frozen specimens is to coat them with a thin carbon layer in order to render the surface of the specimen conductive. The bulk charging phenomenon is in most cases not solved at all and ignored.

A second approach in bulk specimen preparation is to replace the frozen water with a polymer after freeze-substitution process [4]. Another variant of this approach is to freeze-dry the specimen first, after which it is embedded with a polymer [10]. The resulting specimen may be cross-sectioned with a glass or diamond knive at room temperature and carbon coated. The disadvantage of these plastic-embedded specimens is the danger of loosing diffusible elements like Na, K and Cl from the cellular compartments. The third approach in bulk specimen preparation is to freeze-dry the specimen only and offer it for subsequent analysis at room temperature. Apart from loosing diffusible elements there is also a disadvantage in the large excitation volume which is caused by the loose freeze-dried ultrastructure of the specimen.

Analysis and Analytical Conditions

For bulk specimen analysis it is important to have a good estimate of the penetration and beam broadening of the electron beam in the specimen, because these parameters directly influence the analytical resolution. In Fig. 4 a comparison has been made between the beam penetration and broadening in frozen-hydrated liver and biotite simulated by a Monte Carlo program for electron-matter interaction [11]. Specimens like liver, which are mainly composed of water, can in this case be

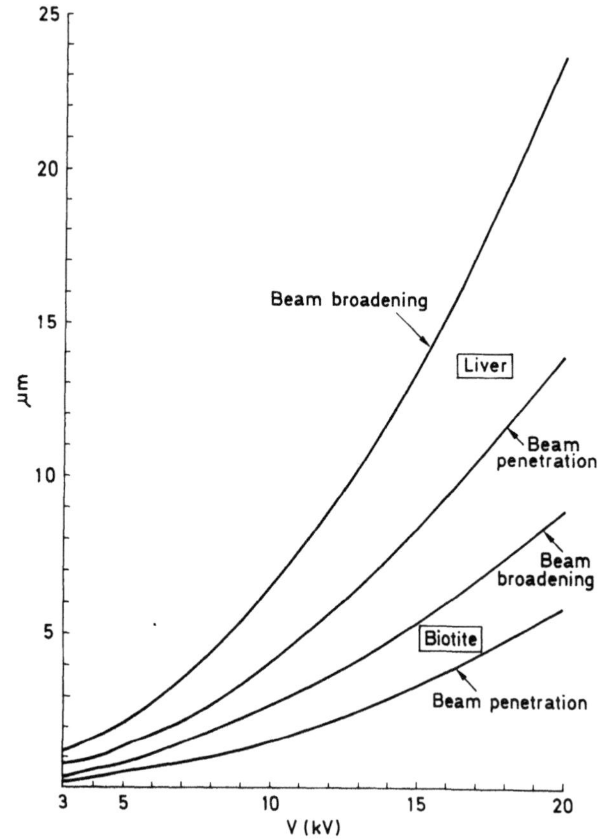

Fig. 4. Predicted beam broadening and penetration in frozen-hydrated liver and biotite, related to the accelerating voltage. Simulation was carried out with a Monte Carlo program for electron-matter interaction [11]

Table 1. Penetration depths of beam electrons and beam broadening in typical bio-organic specimens. Calculations based on Monte Carlo simulation of electron-matter interaction (see [11]). Conditions: 10 kV accelerating voltage, beam perpendicular to the specimen surface

Specimen	Penetration (μm)	Beam broadening (μm)
Frozen-hydrated liver	3.8	6.2
Frozen-dried liver	11.2	16.8
Epoxy resin	3.5	5.4

regarded as a model for frozen-hydrated biological specimens. It is clear that for plastic embedded biological specimens beam penetration and broadening will be even worse because the mean atomic number shifts in this case to a lower value (approx. 6) [1]. The curves in Fig. 4 indicate that for biological bulk specimens low accelerating voltages are to be preferred to attain good analytical resolution. In Table 1 data are given of calculated penetration depths for different kinds of biological specimens. These data indicate that for biological bulk specimens only major compartments like nuclei, large vacuoles, thick cell walls and whole cells, can be analyzed properly in the scanning electron microscope. Some improvement can be attained by embedding a freeze-dried specimen [10] and in some cases by tilting a specimen [12].

To avoid electrostatic bulk charging it is necessary to apply even lower beam energies in order to diminish the possible distortion of the excitation volume. The evaporation of a layer of carbon onto the specimen surface connects in principle only the surface of a specimen to ground potential but may not be sufficient to avoid all charging effects. According to calculations of Brombach for frozen-hydrated specimens [5], a flat pancake-like excitation volume may exist instead of the normal teardrop or half-sphere volume. The application of ideal standards in this respect gives at best a similar bulk charging in specimen and standard. If the standard is not completely ideal, serious errors may arise, due for instance to differences between standard and specimen in the effective mean beam energy at the point where the beam is entering the sample.

The specimen stage in the scanning electron microscope should be composed ideally of a cryo-stage option in order to keep the specimen frozen-hydrated, a Faraday cage to measure the beam current and separate sets of appropriate standards. In Fig. 5 a typical specimen stage is shown. Because in biological X-ray microanalysis concentrations are in many cases very low, it is also advisable to have a wavelength dispersive X-ray spectrometer attached to the microscope column to cover the concentration range down to approx. 100 μg/g. Wavelength dispersive spectrometry can also be useful in the deconvolution of energy dispersive X-ray peak overlap situations.

Quantitation

For quantitation in biological X-ray microanalysis in most cases the ideal standard approach is followed. This implies that a standard is developed specifically for an

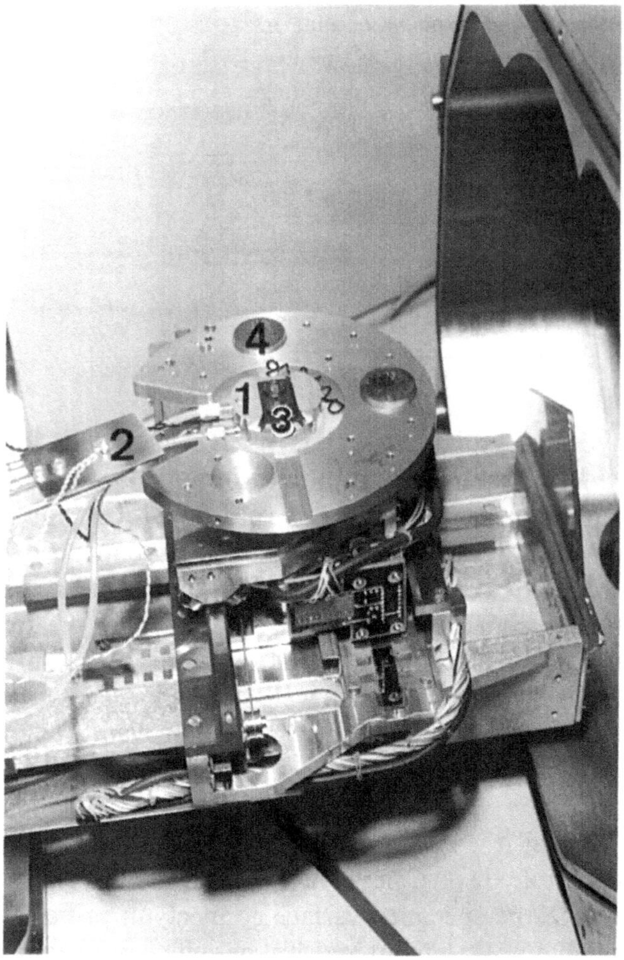

Fig. 5. Specimen stage in a Philips 535 scanning electron microscope showing cooled stage (1), cold trap (2), specimen (3) and standard sets (4)

analysis problem, which resembles the specimen both chemically and physically as much as possible [13, 14]. In most cases salts and proteins are used with good solubility in water, which implies that a wide range of matrices and element combinations can be simulated. In this case the matrix is almost identical which means that quantitation can be carried out by straightforward comparison of peak intensities. Important characteristics in this respect are the typical elemental combination and the homogeneity of the standard. For plastic embedded specimens organometallic compounds dissolved in epoxy resin have been used successfully [15].

The second best approach is quantification with secondary standards which is basically a calibration of the stable mineral compound standards with primary ideal standards [10]. In Fig. 6 such a calibration has been carried out showing straight lines for both types of standard. In this procedure the more stable secondary standard is used for daily analysis while the comparison with the primary standard incorporates the matrix correction.

If the specimen shows considerable surface microroughness, it appears that the peak-to-local background method yields more reliable results than the use of net peak intensities only [16–18]. The theory behind is the inherent correction of a different absorption path length from one place on the surface of the specimen to another, if the background intensity of the specimen is measured at the same

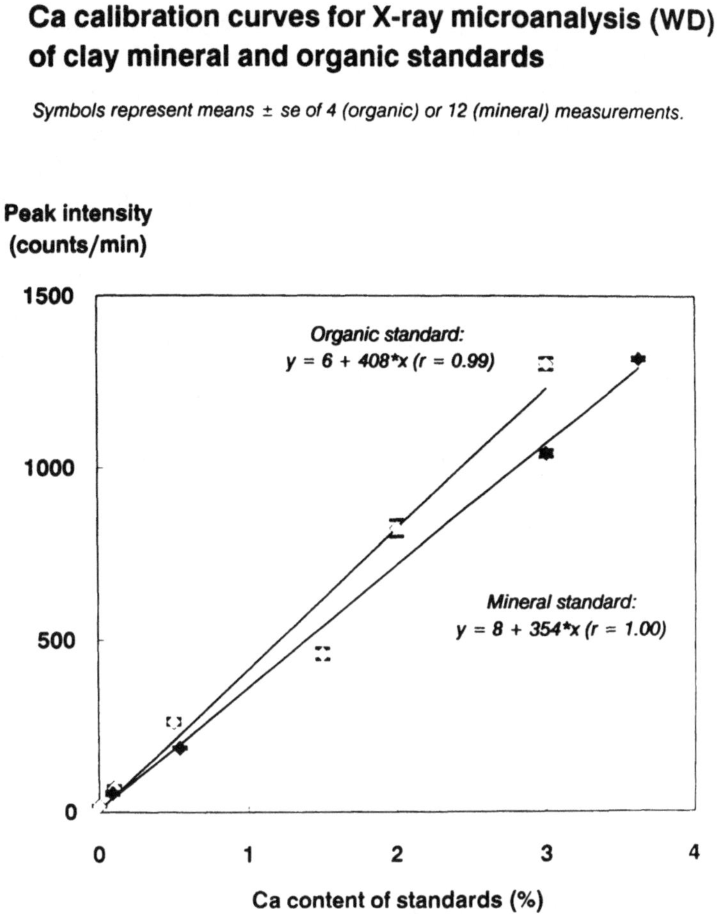

Fig. 6. Calibration curves for Ca obtained with mineral standards and organic standards (courtesy of D. R. Verkerke, CABO-DLO)

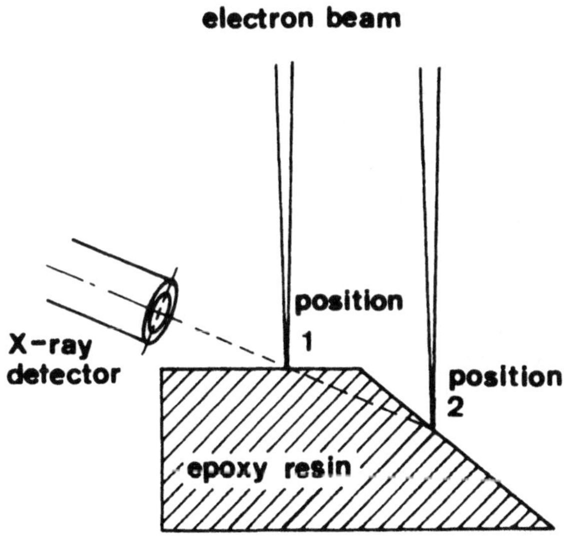

Fig. 7. Model situation for the effect of surface microroughness on the absorption path length differences between different positions on the surface

spectral energy as the peak (see Fig. 7) [19]. Recently other variants of the use of peak-to-background ratios in bulk specimen analysis have been published which adapt the continuum-normalisation method [20, 21] for the analysis of biological bulk specimens [22]. Zs.-Nagy et al. [23] used peak-to-background values to analyze water and dry mass content in the cell. Especially in these types of quantitation one has to pay special attention to the problem of extraneous background [24] because the true specimen background intensity is to be measured in these cases. Marshall and Condron [25] have followed an analogous approach but used the backscattered electron signal for normalization.

If there is no alternative for the relatively simple quantitation procedures described above, a matrix correction has to be applied. In this case several attempts have been made to design a ZAF-correction program specifically for biological specimens [1, 26]. In such programs it must be possible to specify the relative or absolute composition of the light-element matrix completely both for standard and specimen. Correction formulae be may optimally chosen for the light-element situation but this approach requires a microanalysis system that is 'open' for the user so that changes can be easily implemented in the software [1, 3, 6].

References

[1] A. Boekestein, A. L. H. Stols, A. M. Stadhouders, *Scanning Electron Microsc.* **1980**, *II*, 321.
[2] A. Boekestein, A. M. Stadhouders, A. L. H. Stols, G. M. Roomans, *Scanning Electron Microsc.* **1983**, *II*, 725.
[3] G. M. Roomans, *J. Electron Microsc. Techn.* **1988**, *9*, 3.
[4] R. A. Steinbrecht, K. Zierold (eds.), *Cryotechniques in Biological Electron Microscopy*, Springer, 1987, pp. 149, 240.
[5] J. D. Brombach, *J. Microsc. Biol. Cell.* **1975**, *22*, 233.
[6] A. Boekestein, A. M. Stadhouders, A. L. H. Stols, G. M. Roomans, *Ultramicrosc.* **1983**, *12*, 65.
[7] A. T. Marshall, *Scanning Electron Microsc.* **1982**, *I*, 243.
[8] K. Zierold, *J. Electron Microsc. Techn.* **1988**, *9*, 65.
[9] A. Boekestein, in: *Past, Present and Future of Electron Microscopy in Agriculture* (A. Boekestein, ed.), DLO, Wageningen 1991, 87.
[10] F. D. Ingram, M.-J. Ingram, *J. Microsc. Biol. Cell.* **1975**, *22*, 193.
[11] A. Boekestein, N. Amman, B. Otterloo, J. Suyker, P. Karduck, in: *Proc. 1st EMAS Workshop, European Microbeam Analysis Society*, Antwerpen, 1989, 186.
[12] I. Zs. Nagy, C. Pieri, C. Guili, *J. Ultrastruct. Res.* **1977**, *58*, 22.
[13] A. Warley, *J. Microsc.* **1990**, *157*, 135.
[14] L. E. Wyness, J. A. Morris, K. Oates, W. G. Staff, H. Huddart, *J. Pathol.* **1987**, *153*, 61.
[15] G. M. Roomans, H. L. M. van Gaal, *J. Microsc.* **1977**, *109*, 235.
[16] A. Boekestein, F. Thiel, A. L. H. Stols, E. M. Bouw, A. M. Stadhouders, *J. Microsc.* **1984**, *134*, 327.
[17] P. Echlin, S. E. Taylor, *J. Microsc.* **1986**, *141*, 329.
[18] P. Echlin, C. E. Lai, T. L. Hayes, *J. Microsc.* **1982**, *126*, 285.
[19] P. Statham, *Scanning Electron Microsc.* **1978**, *I*, 469.
[20] T. A. Hall, *J. Microsc.* **1986**, *141*, 319.
[21] K. Zierold, *Scanning Electron Microsc.* **1986**, *II*, 713.
[22] I. Zs.-Nagy, T. Casoli. *Scanning Microsc.* **1990**, *4*, 419.
[23] I. Zs.-Nagy, G. Lustyik, C. Bertoni-Freddari, *Tissue Cell* **1982**, *14*, 47.
[24] J. I. Goldstein, D. B. Williams, *Scanning Electron Microsc.* **1978**, *I*, 427.
[25] A. T. Marshall, R. J. Condron, *J. Microsc.* **1985**, *140*, 99.
[26] A. T. Marshall, R. J. Condron, *Micron Microsc. Acta* **1987**, *18*, 23.

Mikrochim. Acta (1992) [Suppl.] 12: 213–219

Quantitative Analysis of $(Y_2O_3)_x(ZrO_2)_{1-x}$ Films on Silicon by EPMA

Norbert Ammann[1,*], Andreas Lubig[2], and Peter Karduck[1]

[1] Gemeinschaftslabor für Elektronenmikroskopie der RWTH Aachen, D-W-5100 Aachen, Federal Republic of Germany

[2] ISI der K.f.A. Jülich, P.O. Box 1913, D-W-5170 Jülich, Federal Republic of Germany

Abstract. Approximately 70 nm thick films of non-conducting $(Y_2O_3)_x(ZrO_2)_{1-x}$ $(x \approx 0.1)$ were deposited on Si(100) substrates by electron beam evaporation. Subsequently, electron probe microanalysis (EPMA) was applied to determine the yttrium content. Experimental efforts were made to assess the accuracy of the EPMA results and to determine the detection limits of the technique.

The intensities of the characteristic X-ray lines Y Lα, O Kα, Zr Lα and Si Kα, emitted from the electron-beam excited samples, were measured with a Jeol JXA-50 A microanalyzer for beam energies of 4, 7, 10, 14 and 20 keV. Charging of the films did not occur. The intensities were calibrated with corresponding intensities of $Y_3Fe_5O_{12}$, Zr and Si standard samples. The number of Y, Zr and O atoms per cm^2 were determined independently for each beam energy by comparing the measured intensities with calculations of a Monte Carlo simulation program. The number of atoms per cm^2 were determined with an accuracy better than 5% and the uncertainties of the Y contents were below 0.25 atomic percent. The detection limits of the Jeol JXA-50 A for Y, Zr and O on a silicon substrate were 2.3, 4 and $13 \cdot 10^{14}$ atoms per cm^2, respectively, at 200 nA beam current and 1 min counting time.

Key words: thin film, electron probe microanalysis, quantitative analysis, oxide films, EPMA.

The present problem was to determine the Y content in approximately 70 nm thick non-conducting $(Y_2O_3)_x(ZrO_2)_{1-x}$ films on silicon with an accuracy of 1 atomic percent (x is the formula fraction of Y_2O_3 in the film). Since the lattice parameter of cubic $(Y_2O_3)_x(ZrO_2)_{1-x}$ can easily be varied by the Y_2O_3 content [1], this problem is important in the epitaxial growth of the films on Si(100) substrates.

Secondary ion mass spectroscopy (SIMS) could not be applied to the problem, because matrix effects make the absolute calibration of the data very difficult. Similarly, Rutherford backscattering spectroscopy (RBS) can be no adequate tool

* To whom correspondence should be addressed

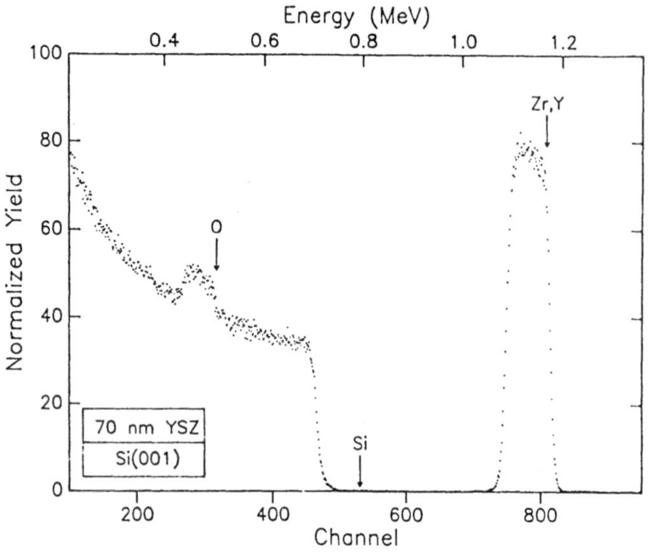

Fig. 1. RBS spectrum of 1.4 MeV He^+ ions incident on a 70 nm $(Y_2O_3)_x(ZrO_2)_{1-x}$ film on silicon. The vertical arrows denote the energy of He^+ ions backscattered from the surface atoms of the respective elements

in this case. Y and Zr are neighbouring elements so that their backscattering signals coincide in the RBS spectrum as shown in Fig. 1. Small variations of the Y content thus result in only marginal variations of the signal height, which can not be detected within measurement accuracy. Therefore electron probe X-ray microanalysis (EPMA) has been applied to the problem.

EPMA is well known in quantitative bulk analysis, but it is also used to analyze films with thicknesses far below the penetration depth of the beam electrons. Several authors have proposed parameter-models for the interpretation of the measured X-ray intensities of such films [2–5], but accuracy and reproducibility have not been assessed systematically. Therefore we have simulated the electron-sample interaction by a Monte Carlo (MC) program, which has shown close agreement with experimental data [6, 7]. Because the ionisation depth distributions are calculated on a strictly physical basis, results of comparable accuracy can be expected in the present study.

Problems could be expected because of the insulating property of the films, because insulating bulk materials are difficult to analyze even if these are coated with a conducting surface film [8]. However it has also been reported, that insulating surface films might not charge if the beam energy is high enough for the electrons to penetrate into the substrate [9].

Experimental

The 100 eV energy resolution of a Ge X-ray detector should be sufficient in energy dispersive X-ray spectroscopy (EDX) to distinguish between Si Kα and Y Lα, as well as between Y Lα and Zr Lα peaks. In the present case, however, the intensity differences are too large to detect the Y Lα peak in the EDX spectrum, as shown in Fig.2. The intensities of Si Kα, Y Lα and Zr Lα will be approximately 100 : 1 : 10. This is demonstrated by the corresponding wavelength dispersive X-ray spectrum (WDX) in Fig. 3.

At beam energies above 2.5 keV, X-ray intensities of any element in the $(Y_2O_3)_x(ZrO_2)_{1-x}$ films on silicon are measurable by WDX. The mass absorption coefficients of Y and Zr for O Kα radiation are large (15.1 and 14.8 cm^2/mg) [10], but at film thicknesses of approximately 40 $\mu g/cm^2$, these are not as significant as in bulk analyses. The mass absorption coefficients of the films for Si Kα, Y Lα and

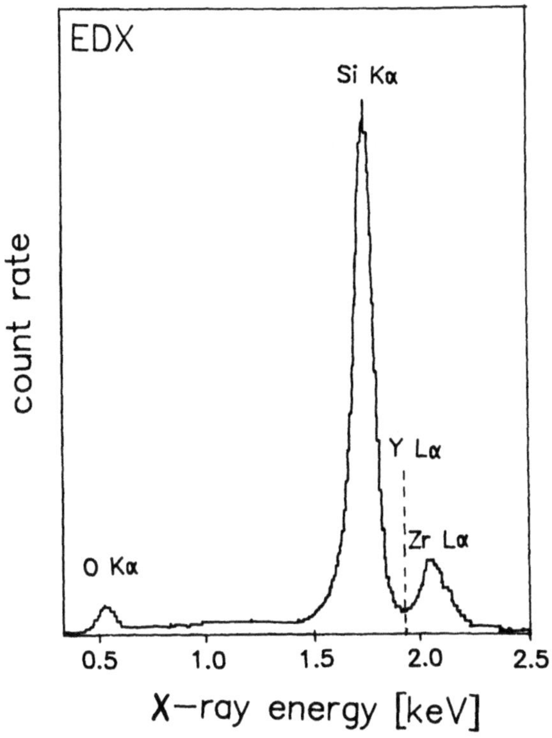

Fig. 2. EDX spectrum of 40 $\mu g/cm^2$ $(Y_2O_3)_{0.11}(ZrO_2)_{0.89}$ on silicon, measured with a Ge X-ray detector at 10 keV beam energy and normal incidence

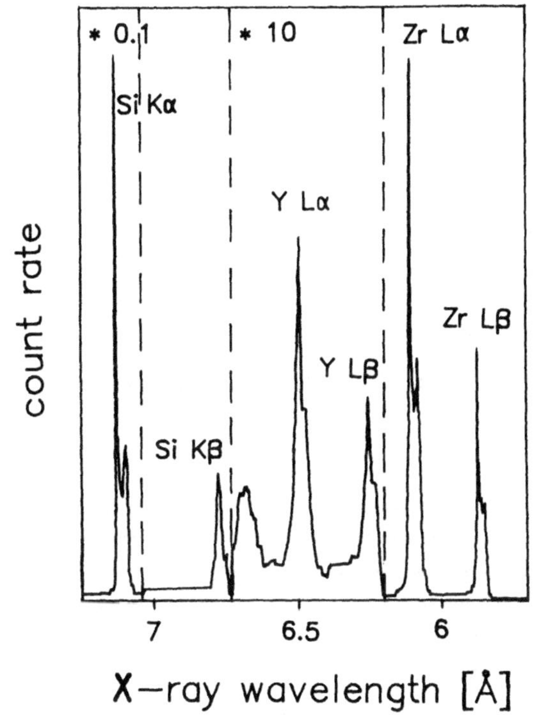

Fig. 3. WDX spectrum of 40 $\mu g/cm^2 (Y_2O_3)_{0.11} \cdot (ZrO_2)_{0.89}$ on silicon, measured with a PET crystal at 10 keV beam energy, normal incidence and 35 deg take off angle. Y lines magnified by x10, Si Kα x0.1

Zr Lα radiation (about 1 cm^2/mg) are small and therefore not critical. Calculations show that the fraction of secondary fluorescence radiation, generated by the continuum and the primary characteristic radiation, is below 1% of the primary radiation itself. This allows the interpretation of the emitted X-ray intensities neglecting fluorescence effects.

Before starting the actual measurement, the variations of the X-ray intensities at beam energies of 3 up to 30 keV are calculated by MC-simulation. As shown in Figs. 4 and 5, maximum X-ray intensities

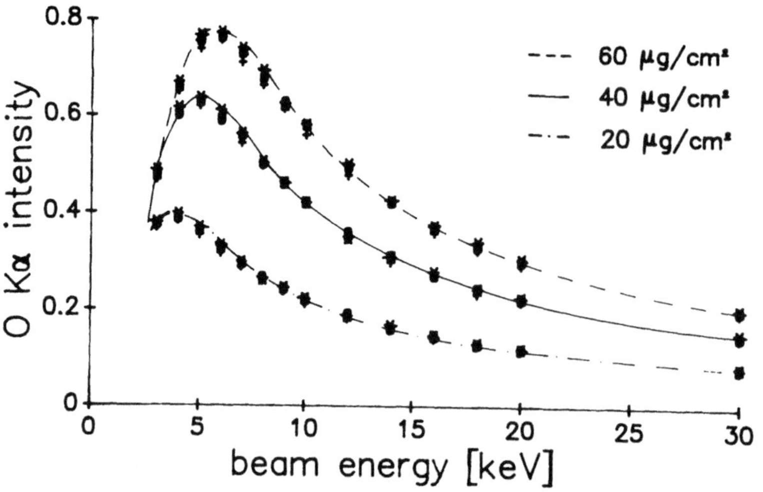

Fig. 4. Simulated O Kα intensities of 20, 40, 60 $\mu g/cm^2 (Y_2O_3)_x(ZrO_2)_{1-x}$ on silicon, at 35 deg take off angle and normal incidence ($\times : x = 0.1$, $\bullet : x = 0.15$, $+ : x = 0.2$)

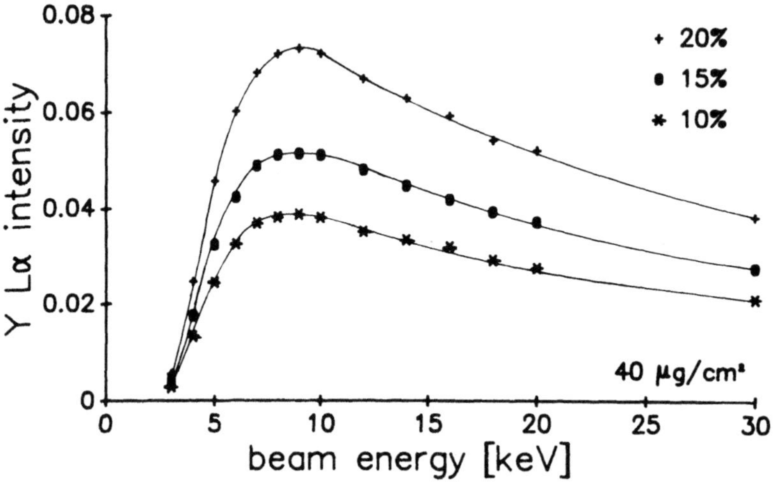

Fig. 5. Simulated Y Lα intensities of 40 $\mu g/cm^2 (Y_2O_3)_x(ZrO_2)_{1-x}$ on silicon, at 35 deg take off angle and normal incidence ($\times : x = 0.1$, $\bullet : x = 0.15$, $+ : x = 0.2$)

(counts/current/time) from the film-elements are observed between 5 and 12 keV, for thicknesses of 20, 40 and 60 $\mu g/cm^2$ and x-values of 0.1, 0.15 and 0.2. Therefore this should be an appropriate beam energy range for the analysis. The O Kα and Si Kα intensities are insensitive to the film composition, but depend strongly on the thickness of the films, whereas the Y Lα and Zr Lα intensities depend on both, thickness and composition of the films. This illustrates the possibility of determining film thickness and composition simultaneously from the X-ray intensities at a single beam energy.

The X-ray intensity measurements are performed with a Jeol JXA-50 A microanalyzer. A Pb-STE crystal is used for O Kα, PET for Y and Zr Lα, and RAP for the Si Kα line. To minimize statistical errors, each measurement is carried out more than five times at several points on each sample, with a minimum of 10,000 counts per intensity. To detect systematical errors, the measurements are carried out at beam energies of 4, 7, 10, 14 and 20 keV with counting times of 8, 8, 8, 15 and 40 s respectively and 200 nA beam current.

Each X-ray measurement is performed twice, with and without an additional 20-nm carbon film on the sample surfaces, although secondary electron images of the films show no charging effects.

Table 1. Measured k-values of sample hY/SI 6/1 and the average ratios $\langle k_{exp}/k_{MC}\rangle$ to the MC-simulated k-values of the determined sample composition in Table 2

	Experimental k-values				
E_0 [keV]	Y Lα	Zr Lα	O Kα	Si Kα	$\langle k_{exp}/k_{MC}\rangle$
4	0.338	0.548	0.828	0.046	0.98
7	0.189	0.314	0.528	0.393	1.01
10	0.101	0.169	0.374	0.635	1.01
14	0.053	0.081	0.282	0.766	1.01
20	0.031	0.044	0.270	0.908	0.99

Table 2. Composition of sample hY/SI 6/1, as determined at various beam energies (values in brackets are not considered in the statistical data evaluation)

	n_Z [10^{16} atoms/cm^2]					
E_0 [keV]	Y	Zr	O	Atomic% Y	Ratio O/O_{st}	Comp. x
4	[3.36]	[17.57]	[40.77]	5.447	1.01	0.087
7	3.67	19.63	46.71	5.237	1.04	0.085
10	3.68	19.93	46.24	5.264	1.02	0.084
14	3.53	18.53	42.33	5.483	1.00	0.086
20	3.60	18.17	43.38	5.528	1.04	0.090
Mean	3.62	19.07	44.66	5.392	1.02	0.087
s_{n-1}	±0.07	±0.85	±2.14	±0.132	±0.02	0.002
RSD_{n-1}	±1.9%	±4.5%	±4.8%	±2.45%	±1.8%	±2.7%

Differences between the results with and without carbon coating are negligible, confirming that charging does not occur. As an example, the measured k-values of one sample are listed in Table 1.

Results and Discussion

The number of atoms per cm^2, n_z, resulting in the minimum mean square deviation of simulated to measured intensities, is determined by iteration at each beam energy. The results for one sample are shown in Table 2. The atomic percent values are given by:

$$\text{atomic}\%_{(Z)} = 100 \cdot n_Z/(n_O + n_Y + n_{Zr}); \qquad Z \in \{O, Y, Zr\}.$$

The accuracy of the results can be estimated by the standard deviations s_{n-1}. In the calculation of the mean values and standard deviations for the n_Z's the values at 4 keV have not been considered. At 4 keV the excitation depth is only slightly larger than the film thickness, and this condition is too critical for thickness measurements. In the appropriate beam energy range ($E_0 \geq 7$ keV) thickness and composition of $(Y_2O_3)_x(ZrO_2)_{1-x}$ films on silicon can be determined within an accuracy better than 5%. The accuracy of the yttrium atomic fraction is far below the 1 at% originally

Table 3. Calculated detection limits of the Jeol JXA-50 A for Y, Zr and O on silicon at 200 nA, 1 min counting time and without specimen tilt

		Detection limits in 10^{14} atoms per cm^2				
Element	Crystal/line	4 keV	7 keV	10 keV	14 keV	20 keV
O	STE/Kα	13	13	15	26	39
Zr	PET/Lα	6.5	4.1	4	5	5.7
Y	PET/Lα	5	2.8	2.3	3.7	5

required. These uncertainties are of the same order than those known from bulk analyses by EPMA [11].

Assuming a film of stoichiometrical ZrO_2 and Y_2O_3, the relation

$$n_O(n_Y, n_{Zr}) = 1.5 \cdot n_Y + 2 \cdot n_{Zr}$$

should be obtained. Although the measured n_O's are slightly larger than those according to stoichiometry, non-stoichiometry of the oxides in the film cannot be concluded because the deviations are in the order of the uncertainty of the measured n_O-values. The detection limits, shown in Table 3 are calculated by the formula

$$n_{limit} = 3 \cdot n \cdot \sqrt{B/(P\text{-}B)},$$

with P and B, the number of peak- and background counts detected in 1 min at a beam current of 200 nA from a film with n atoms/cm^2. n_{limit} is the number of atoms/cm^2 resulting in a net peak height three times the standard deviation of the background, assuming proportionality of n and the X-ray intensities from the film [12].

The detection limits for O, Y and Zr on Si are calculated to be 2.3, 4 and $13 \cdot 10^{14}$ atoms per cm^2. This should allow the quantitative analysis of such films with a thickness down to 1 nm. Because detected count rates in modern equipment are approximately 10 times higher than in the JXA-50 A, detection limits could be approximately 3 times smaller.

Conclusion

EPMA has been shown to be an easy and widely applicable tool for thin film analysis. The yttrium content of a 70 nm thick $(Y_2O_3)_x(ZrO_2)_{1-x}$ film, could be determined with an accuracy far below 1 at%. From the results it is estimated that even a 1 nm thick film could be analyzed properly. The absence of matrix and mass resolution effects makes EPMA superior to SIMS or RBS in many cases.

References

[1] H. Fukumoto, T. Imura, Y. Osaka, *Jpn. J. Appl. Phys.* **1988**, *27*, L1404.
[2] J. L. Pouchou, F. Pichoir, *Rech. Aerosp.* **1984**, *5*, 47.
[3] H. J. Hunger, W. Baumann, S. Schulze, *Cryst. Res. Tech.* **1985**, *20*, 1427.
[4] P. Willich, D. Obertop, *Surf. Interface Anal.* **1988**, *13*, 20.

[5] G. F. Bastin, H. J. M. Heijligers, J. M. Dijkstra, *Microbeam Analysis 1990* (J. R. Michael, P. Ingram, eds.), San Francisco Press, San Francisco, 1990, p. 159.

[6] N. Ammann, P. Karduck, *Microbeam Analysis 1990* (J. R. Michael, P. Ingram, eds.), San Francisco Press, San Francisco, 1990, p. 150.

[7] P. Karduck, N. Ammann, H. G. Esser, J. Winter, *Fresenius Z. Anal. Chem.* **1991**, *341*, 315.

[8] G. F. Bastin, H. J. M. Heijligers, *Quantitative Electron Probe Microanalysis of Oxygen*, Eindhoven University of Technology, Eindhoven, 1989.

[9] S. J. Krause, J. Mohr, G. H. Bernstein, D. K. Ferry, D. C. Joy, *Microbeam Analysis 1989* (P. E. Russel, ed.), San Francisco Press, San Francisco, 1990, p. 459.

[10] B. L. Henke, P. Lee, T. J. Tanaka, R. L. Shimabukuro, B. F. Fujikawa, *Atomic Data and Nuclear Data Tables* **1982**, *27*, 1.

[11] H. Hantsche, *Scanning* **1989**, *11*, 257.

[12] S. Baumgartl, P. L. Ryder, H. E. Bühler, *Z. Metallk.* **1973**, *64*, 655.

Mikrochim. Acta (1992) [Suppl.] 12: 221–227

EPMA of Surface Oxide Films

Peter Willich* and Kirsten Schiffmann

Fraunhofer-Institut für Schicht- und Oberflächentechnik, P.O. Box 540645, D-W-2000 Hamburg 54, Federal Republic of Germany

Abstract. Electron probe microanalysis (EPMA) is used to study oxide film thickness (1-100 nm), oxide stoichiometry, impurities of oxygen (~1 wt%) in materials covered by a surface oxide film, and multilayer oxide formation on alloys. Excellent sensitivity for the determination of O $K\alpha$ intensities is provided by W/Si multilayer monochromators. Modelling of k-ratio versus E_0 (electron energy) is demonstrated as an accurate and versatile procedure of data reduction. The precision of the results with respect to oxide film thickness and oxide stoichiometry is in the range of 5–20%, depending on the complexity of the problem, the range of E_0, the experimental precision of the k-ratios, and the magnitude of the oxide film thickness.

Key words: electron probe microanalysis (EPMA), thin films, oxygen, oxide film thickness, non-destructive in-depth analysis.

Quantitative electron probe microanalysis (EPMA) of oxygen [1] and other ultralight elements (B, C, N) has been developed to a high degree of precision and accuracy [2]. The introduction of synthetic multilayers as monochromators in wavelength-dispersive X-ray spectrometry has considerably improved the sensitivity for the detection of these elements. Quantitative analysis of ultralight elements can be performed with an accuracy of less than 5%, provided that effects of chemical shift are taken into consideration and refined procedures of data reduction are used [2–4]. If the matrix correction programs are based on a realistic description of the $\Phi(\rho z)$ depth distribution function, the application to thin films, of which the thickness is a fraction of the maximum depth of X-ray emission, has been studied intensively [3, 5–10]. In view of the recent developments in the determination of thin films and ultralight elements including oxygen, EPMA has to be considered as a valuable technique to characterize surface oxide films with respect to film thickness and composition. The typical oxide film thickness is less than 100 nm, for instance in natural oxide films on metals and in the formation of oxide films on metals and alloys, caused by annealing. Some novel applications and also the limits of EPMA are discussed in comparison with the results of surface analytical techniques.

* To whom correspondence should be addressed

Experimental

Figure 1 shows O $K\alpha$ spectra recorded by W/Si multilayer monochromators attached to a CAMECA SX50 microprobe. The W/Si (2d = 6 nm) multilayer was used most frequently; because it appeared the best compromise with respect to intensity and peak/background ratio. Sufficiently conductive oxides proved to be suitable as standards for the calibration of oxygen [1, 11]. The relative intensities (k-ratios) of O $K\alpha$ presented in this paper refer to a single crystal of $Y_3Fe_5O_{12}$ (25.8 wt% O), containing impurities of Pb (0.7 wt%) and Si (0.2 wt%) [8, 11, 12]. $Y_3Fe_5O_{12}$ was also used as a standard for the determination of iron. Calibration by use of pure metals, typically covered by a natural oxide film, should be avoided. If there is no other choice, the determination of the oxide thickness by EPMA leads to correction factors [8], e.g., the intensity of Ta $M\alpha$ on Ta was multiplied by factors of 1.028 ($E_0 = 4$ keV) to 1.003 ($E_0 = 12$ keV) to take into account the influence of the surface oxide film ($\sim$4 nm) on a freshly scratched surface of Ta-metal.

Results and Discussion

Determination of Oxide Film Thickness

From the spectra given in Fig. 1 one can derive that oxide films even on the level of a monolayer can be detected by EPMA. Provided that the stoichiometry of the surface oxide can be assumed with a certain probability, a $\Phi(\rho z)$ thin film correction program may be used to calculate the correlation between k-ratio of O $K\alpha$ and oxide film thickness. The calibration functions of Fig. 2 are based on the PAP thin film program of Pouchou and Pichoir [5, 8], assuming densities (g/cm^3) of 8.2 (Ta_2O_5), 3.5 (Al_2O_3), and 5.2 (Fe_2O_3) to convert mass thicknesses into more convenient geometrical film thickness data. Fig. 2 also includes the experimental k-ratios of

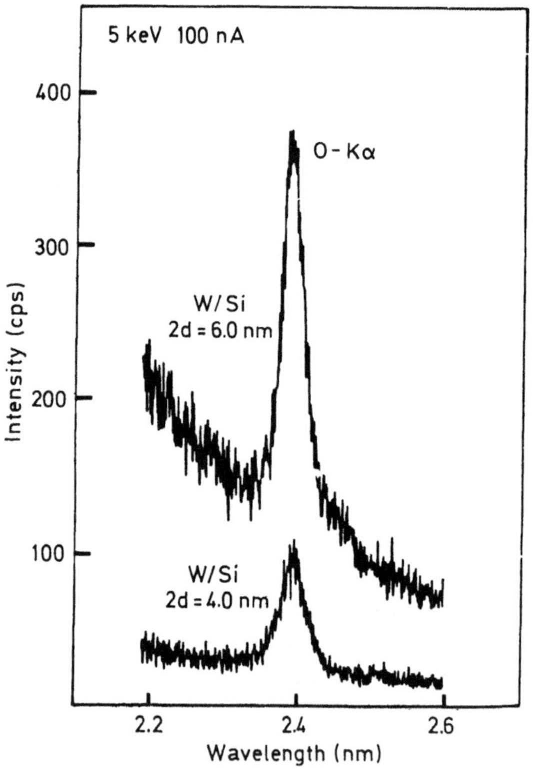

Fig. 1. O $K\alpha$ spectra of the natural Al_2O_3 film (4 nm) on Al recorded by use of W/Si monochromators. Recording time 512 s, PHA setting E = 1 V, $\Delta E = 1.5$ V

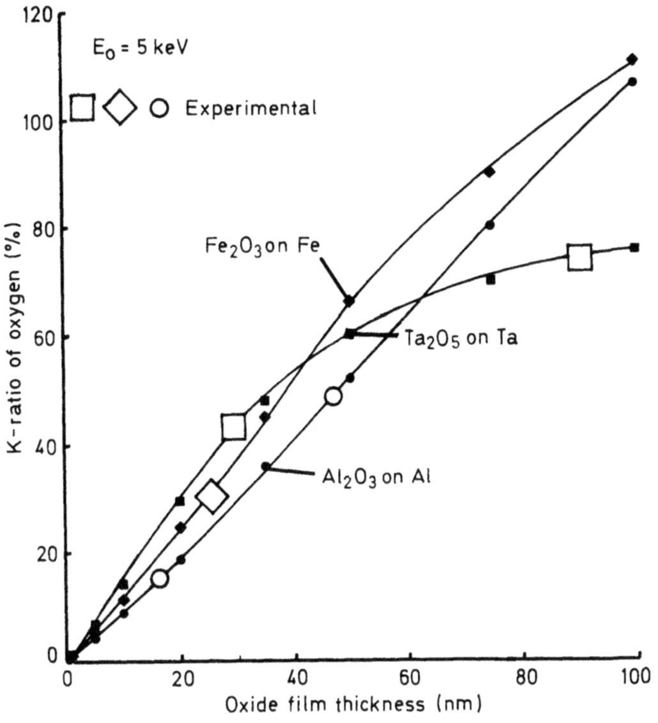

Fig. 2. Calibration functions calculated by the PAP thin film model to determine the oxide film thickness on metals. Experimental data refer to samples of known film thickness

O *Kα* from EPMA of Ta_2O_5 on Ta reference materials [13] and oxide films (Al_2O_3 on Al, Fe_2O_3 on Fe), the thickness of which was defined by SIMS depth profiles. The film thicknesses determined by EPMA show an average deviation of 3.5%, compared to the reference data. This deviation can be regarded as an estimation of the accuracy provided by the PAP thin film model. However, one should take into consideration the reliability of the reference data, which is in the order of 5–10 % in the case of the Ta_2O_5 on Ta.

Determination of Oxide Film Thickness and Composition

Simultaneous determination of oxide film thickness and oxide stoichiometry is based on experiments at various electron energies (E_0). As shown in Fig. 3 the results are represented as k-ratio versus E_0 plots. The PAP model [5, 8] enables one to compute the corresponding k-ratio versus E_0 plots for any configuration of oxide composition and film thickness. The model is modified until the calculated k-ratios are fitted to the experimental data (Fig. 3). In practice, for a given precision of the experimental k-ratios and a given range of E_0, the precision of the results with respect to oxide film thickness and oxide composition depends on the oxide film thickness (Fig. 4). An estimation of the analytical precision is obtained by comparing the k-ratios, computed for configurations close to the model giving the best fit, to the experimental k-ratios. For the example of Ta-oxide on Ta (Fig. 3), the best fit corresponds to a film thickness of 29.5 nm and 18.5 wt% O in the Ta-oxide. All configurations in the range of 26.5 nm/20.3 wt% O to 32.5 nm/16.7 wt% O give a fit to the experimental k-ratios of within ±3 %, the typical experimental precision of the k-ratio data. Therefore the precision of the final result (best fit) is limited to about ±10%; which means a configuration of 29.5 ± 3 nm/18.5 ± 1.8 wt% O. The

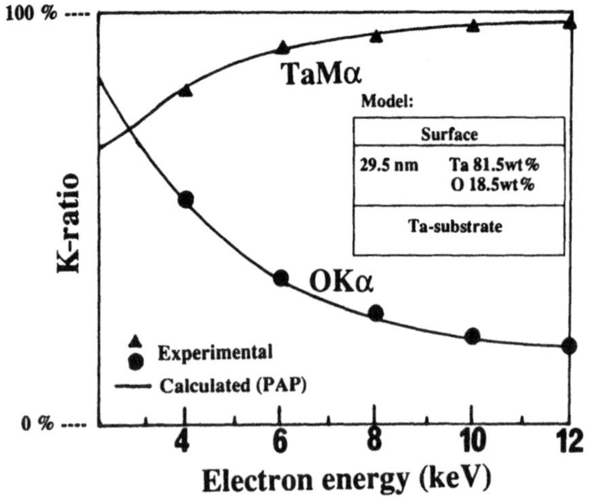

Fig. 3. K-ratios as a function of electron energy for Ta_2O_5 (18.1 wt% O, certified film thickness 28.5 ± 1.5 nm) on Ta. The model giving the best fit to the experimental k-ratio data is shown

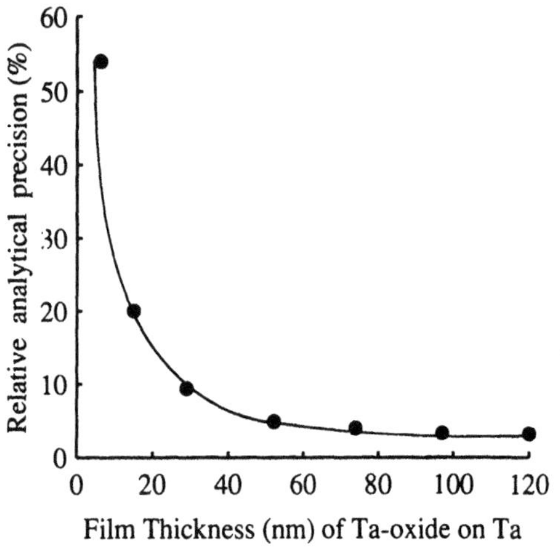

Fig. 4. Precision of the determination of oxide film thickness and composition. Calculations are based on assuming a precision of ±3% for the experimental k-ratios

certified reference data [13] are 28.5 ± 1.5 nm/18.1 wt% O (Ta_2O_5). For Ta_2O_5 with a film thickness of 96 nm (certified value), assuming again a precision of ±3% of the experimental k-ratios, the precision of the results derived from the best fit is about ±3.5% [Fig. 4]: which means a configuration of 97 ± 3 nm/17.9 ± 0.6 wt% O. The improvement of precision, as compared to the film of 29.5 nm, is due to the fact that for a sufficiently thick film the experiments at various E_0 give a better separation of oxide film thickness and composition: The k-ratios of the lowest electron energies become almost independent of the film thickness and enable a precise determination of the composition.

Oxygen in Materials Covered by a Surface Oxide Film

K-ratio versus E_0 modelling can also be applied to quantitative analysis of oxygen in materials covered by a surface oxide film, of which the thickness is determined simultaneously [8, 11]. Fig. 5 shows the example of oxygen in a "bulk" film (1.1 μm) of iron deposited on silicon. The lower limit of detection (LLD) for oxygen in a metal depends on the thickness of the surface oxide film. The estimation of the LLD as given in Fig. 6 for the example of Fig. 5, was obtained as follows: A first set of k-ratios

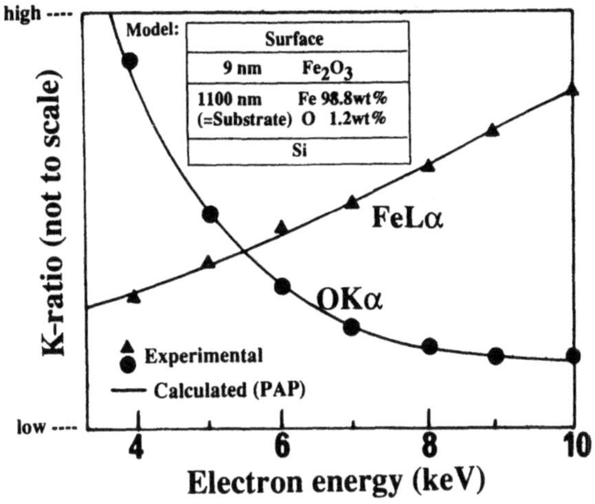

Fig. 5. K-ratios as a function of electron energy for Fe containing a low concentration of O and covered by Fe_2O_3. The model giving the best fit to the experimental k-ratio data is shown

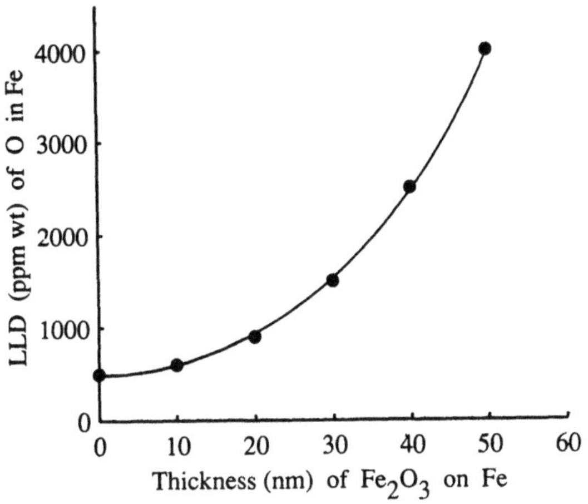

Fig. 6. Lower limit of detection for O in Fe covered by a surface oxide film. Calculations are based on assuming a precision of $\pm 3\%$ for the experimental k-ratios

is calculated assuming $c_O = 0$ (no oxygen in iron) and a certain oxide (Fe_2O_3) film thickness, e.g., $d_O = 10$ nm. Further sets of k-ratios are computed for increasing concentrations of oxygen and $d_O = 10$ nm. This leads to $c_O = 0.06$ wt% (600 ppm), to be regarded as LLD, for which the average shift of the k-ratios, compared to $c_O = 0$, exceeds 3%, which is the typical experimental precision of the k-ratio data. The precision for low concentrations of oxygen in a metal covered by an oxide film is about 5%: $c_O = 1.2 \pm 0.06$ wt% for $d_O \sim 10$ nm (Fig. 5). According to Fig. 6 the precision is limited to about $\pm 25\%$ ($c_O = 1.2 \pm 0.3$ wt%) for the determination of oxygen in iron covered by an oxide film of $d_O \sim 50$ nm.

Multilayer Oxide Formation on Alloys

In Fig. 7 an example is shown of EPMA applied to complex structures in the near surface region. The original sample was a magneto-optic alloy (film thickness $\sim$0.9 μm, Gd = 47.8 wt%, Fe = 52.2 wt%, O < 0.1 wt%, oxide thickness $\sim$10 nm), deposited on silicon. Oxide formation was studied after annealing for 350 h at 200°C in dry air. The k-ratios of Fig. 7 refer to a single crystal of $Gd_3Fe_5O_{12}$ as a standard. Carbon-coating on $Gd_3Fe_5O_{12}$ is required whereas correction factors are obtained by deriving the carbon thickness ($\sim$15 nm) from C *Kα* intensities [14]. The use

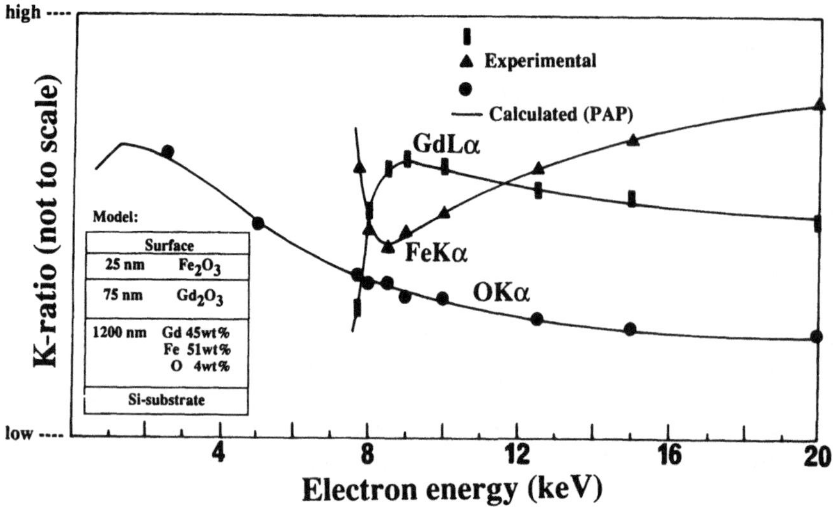

Fig. 7. Multilayer oxide formation on a Gd-Fe alloy caused by annealing. The model giving the best fit to the experimental k-ratio data is shown

of metal standards is critical with regard to experiments at very low overvoltages on materials which are extremely sensitive to surface oxidation. The statistical precision of the k-ratios given in Fig. 7 is at least 3%; this requires counting times in the order of 200–300 s for the determination of Gd *Lα* at $E_0 = 8$ keV ($E_0/E_c = 1.1$). From the view point of sensitivity it should be more advantageous to study the low-energy X-ray lines of Gd *Mα* and Fe *Lα*. However, additional problems arise due to the lack of accurate mass absorption coefficients and the influence of chemical shift. A simple qualitative interpretation of the experimental data (Fig. 7) leads to the conclusion that iron is probably enriched in the surface region and a high concentration of gadolinium is present in a region below. More quantitative results with respect to composition and film thickness are obtained by k-ratio versus E_0 modelling. The precision of the data corresponding to the model giving the best fit (Fig. 7), is in the order of 10–20%. This may be regarded as a rough estimation, however, the results of non-destructive in-depth analysis by EPMA are in principal in agreement with the model of oxide formation on Tb-Fe alloys established by Auger electron spectroscopy and composition depth profiling [15]. This leads to the conclusion that non-destructive EPMA is capable to provide usefull information about complicated oxide structures as a valuable complement to investigations based on surface analytical techniques.

References

[1] G. F. Bastin, H. J. M. Heijligers, *Microbeam Analysis* **1989**, 207.
[2] G. F. Bastin, H. J. M. Heijligers, *Scanning* **1990**, *12*, 225.
[3] J. L. Pouchou, F. Pichoir, *Microbeam Analysis* **1988**, 315.
[4] V. D. Scott, G. Love, *Scanning* **1990**, *12*, 193.
[5] J. L. Pouchou, F. Pichoir, *La Recherche Aérospatiale* **1984**, *5*, 47.
[6] R. H. Packwood, G. Rémond, J. D. Brown, *Proc. 11th ICXOM (London, Canada)*, 1987, p. 274.
[7] P. Willich, D. Obertop, *Surf. Interface Anal.* **1988**, *13*, 20.
[8] J. L. Pouchou, F. Pichoir, *Scanning* **1990**, *12*, 212.

[9] G. F. Bastin, H. J. M. Heijligers, J. M. Dijkstra, *Microbeam Analysis* **1990**, 159.
[10] N. Amman, P. Karduck, *Microbeam Analysis* **1990**, 150.
[11] P. Willich, D. Obertop, *Mikrochimica Acta [Wien]* **1985**, *Suppl. 11*, 299.
[12] P. Willich, D. Obertop, J. P. Krumme, *Microbeam Analysis* **1988**, 307.
[13] M. Seah, *Depth Profiling Reference Materials Ta_2O_5 on Ta*. Community Bureau of Reference (CBR), Brussels, Luxembourg, 1983.
[14] P. Willich, K. Schiffmann, *Microbeam Analysis* **1990**, 177.
[15] R. B. van Dover, E. M. Gyorgy, R. P. Frankenthal, M. Hong, D. J. Siconolfy, *J. Appl. Phys.* **1986**, *59*, 1291.

Mikrochim. Acta (1992) [Suppl.] 12: 229–233

Non-Destructive Determination of Ion-Implanted Impurity Distribution in Silicon by EPMA

Andrey P. Alexeyev

Research Centre for Surface and Vacuum Investigations, 2 Andreyevskaya nab.
Moscow 117334, Russia

Abstract. An urgent need exists for non-destructive methods of analysis which can provide information on the quantitative depth-distribution of impurities in semiconductors, including the ion-implanted impurity redistribution after the annealing processes. A proposal for using electron probe X-ray microanalysis is presented in this work. The advantage of this approach is demonstrated by analysis of Si-specimens with ion-implanted P and As. A part of the specimens has been annealed. The method is based on measurements of X-ray intensities of impurities at three values of primary electron energy E_0. Gaussian distributions were used as an approximation for ion-implanted profiles. A two-parameter fitting procedure has been developed for the determination of Gaussian parameters before and after annealing, which give minimal differences between the experimental and the calculated X-ray intensities at different E_0. The results, obtained, show good agreement with depth profiles from Auger electron spectroscopy and secondary ion mass-spectrometry. The proposed approach may be suitable for the quick control locally of thermal processes in semiconductor technology and for the calibration of ion-implanted reference materials.

Key words: Electron probe microanalysis, ion-implanted impurities.

Ion implantation has emerged in recent years as a common technique to dope semiconductors for integrated-circuits production. New applications of ion implantation, such as drain and source of MOS, emitter and base doping, pre-deposition for diffusion, require a dose in the 10^{15}–10^{16} cm^{-2} range, which means high dose implantation. The key problem is the annealing process. The damage createdby the ions must be removed and the implanted species must be electrically activated with minimal broadening of their depth profiles [1]. The different depth profiling techniques are used for the analysis of impurity distribution both before and after the annealing processes.

High-dose implanted impurities like P and As, may be easily detected by the method of electron probe X-ray microanalysis (EPMA) in the wavelength dispersive mode. In this case depth profiles of implanted impurity are determined by the measurements of their characteristic X-ray intensities at different energies of primary electrons. The depth distributions of impurities were determined by the assumption that they have a Gaussian shape which is described by two parameters Rp and ΔRp. The experimental results are intended to demonstrate the suitability of the proposed approach for the quick estimation of impurity distribution both before and after the annealing processes.

Theory

The characteristic X-ray intensity of the impurity atoms measured from the specimen, compared to the intensity measured of the standard (k-ratio) may be expressed as [2]:

$$k(E_0) = \frac{\{\int_0^\infty W_i(z)\rho(z)\Phi(z, E_0)\exp(-\mu\rho z \operatorname{cosec}\psi)\,dz\}_{\text{specimen}}}{W\{\rho\int_0^\infty \Phi(z, E_0)\exp(-\mu\rho z \operatorname{cosec}\psi)\,dz\}_{\text{standard}}}, \tag{1}$$

where $W_i(z)$ is the weight fraction of the impurity, which is a function of the depth z; W is the weight fraction of impurity atoms in a standard; ρ is the density; $\Phi(z, E_0)$ is the depth distribution of X-ray production for the characteristic X-rays of impurity atoms at energy of primary electrons E_0; μ is the mass absorption coefficient for the characteristic X-ray line from the impurity atoms; ψ is the X-ray take-off angle.

Considering Vegard's law, the expression for the weight fraction of the impurity atoms as a function of the depth becomes:

$$W_i(z) = n_i(z)A_i/[n_i(z)A_i + NA] = n_i(z)A_i/\rho(z)N_a, \tag{2}$$

where $n_i(z)$ is the concentration [cm^{-3}] of impurity atoms at a depth z and A and N are the atomic mass and concentration of matrix atoms (Si) respectively; N_a is Avogadro's number; A_i is atomic mass of impurity atoms.

The depth profile of an implanted impurity with Gaussian distribution can be described by the expression [3]:

$$n(z) = Df(z) = \frac{2D}{\sqrt{2\pi}\Delta\text{Rp}[1 + \operatorname{erfc}(\text{Rp}/\sqrt{2}\Delta\text{Rp})]}\exp[-(z - \text{Rp})^2/2\Delta\text{Rp}^2)], \tag{3}$$

where D is the dose of implanted impurity; $f(z)$ is the range distribution of implanted impurity, and $\int_0^\infty f(z) = 1$.

Hence the expression for D is:

$$D(E_0) = \frac{k(E_0)N_a W\{\rho\int_0^\infty \Phi(z, E_0)\exp(-\mu\rho z \operatorname{cosec}\psi)\,dz\}_{\text{standard}}}{A_i\{\int_0^\infty f(z)\Phi(z, E_0)\exp(-\mu\rho z \operatorname{cosec}\psi)\,dz\}_{\text{specimen}}}. \tag{4}$$

The value of $D(E_0)$ at different E_0 will remain constant if the function $f(z)$ is close to reality. If $f(z)$ is unknown one can try to determine $f(z)$ from the $k(E_0)$ data. The Gaussian distribution with two parameters Rp and ΔRp was used as a first approximation for $f(z)$. This approximation allows to estimate the depth distribution of impurity atoms. It is the aim of the optimization procedure to find a set of Rp and ΔRp parameter values which yield the best agreement between experimental and calculated X-ray intensities of impurity atoms at different E_0. In order to evaluate this agreement, the following criterion was used:

$$B(\mathrm{Rp}, \Delta\mathrm{Rp}) = \sum_{i=1}^{n} \{D(E_0, \mathrm{max.}) - D(E_0, i)\}^2, \tag{5}$$

where $i = 1, \ldots, n$ is the number of accelerating voltages at which $k(E_0)$ was measured; E_0, max. is the maximum value of E_0. In this case the depth of $\Phi(z, E_0)$ exceeds the impurity depth profile and the value $D(E_0, \mathrm{max.})$ is less sensitive to deviations of Rp and ΔRp in eq. (4) from the true values.

The minimum of $B(\mathrm{Rp}, \Delta\mathrm{Rp})$ is regarded as the solution of the problem. The gradient algorithm of Davidon-Fletcher-Powell [4] was incorporated in the computer program in order to find this minimum.

Results and Discussion

The specimen, implanted with 50 keV As-ions was annealed in a furnace at 1000°C during 20 min and the specimen, implanted with 60 keV P ions was annealed by 70-ns laser impulse at energy density 1.3 J/cm^2. The measurements were carried out in a CAMEBAX X-ray microanalyser. The intensities of $P - K_\alpha$ ($E_c = 2.143$ keV) and $As - L_\alpha$ ($E_c = 1.323$ keV) were measured at different E_0. Binary stoichiometric compounds like GaP and GaAs, were used as standard specimens. The background level was measured on a Si specimen containing no impurity. The measured X-ray intensities (after background subtraction) are presented in Tables 1 and 2. The statistical precision of measured intensities is within 1–3 %.

A two-parameter fitting has been developed to determine the minimum of $B(\mathrm{Rp}, \Delta\mathrm{Rp})$ for each specimen. Gaussian expressions for $\Phi(z, E_0)$ has been used [5, 6]:

$$\Phi(z, E_0) = \{\gamma - [\gamma - \Phi(0)] \exp(-\beta\rho z)\} \exp[-\alpha^2(\rho z)^2]$$

It was assumed in the calculations that $\Phi(z, E_0)$ is independent from the impurity concentration. The best results were achieved if β and γ-parameters were calculated according to [5] at $E_0 > 3E_c$ and according to [6] at $E_0 < 3E_c$. When the minimum of $B(\mathrm{Rp}, \Delta\mathrm{Rp})$ was found the standard deviation of the mean $\sum_{i=1}^{n} D(E_0, i)/n$ was smaller than 1 %.

Table 1. $P - K_\alpha$ intensities (pulse counts) of the implanted specimens and standard (after background subtraction) at different energies of primary electrons*

	Energy of primary electrons (keV)		
Specimen	3.5	4.0	10.0
Initial	149	240	449
Laser annealing	125	189	450
GaP	8216	12580	95460
Background	28	35	91

* 100 nA beam current, 20 s counting time

Table 2. As – L_{α} intensities (pulse counts) of the implanted specimens and standard (after background subtraction) at different energies of primary electrons*

	Energy of primary electrons (keV)		
Specimen	2.5	3.0	10.0
Initial	303	554	744
Furnace annealing	205	332	800
GaAs	2919	4838	30085
Background	54	65	246

* 100 nA beam current, 20 s counting time

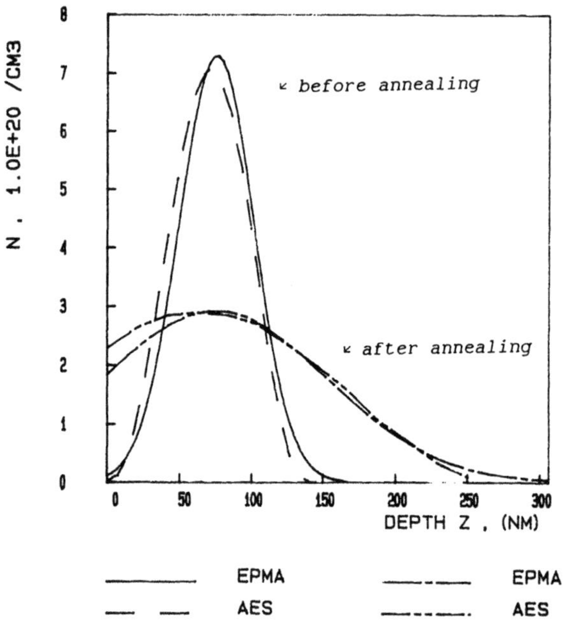

Fig. 1. Distribution of P (60 keV implantation energy) in Si before and after annealing.
Implanted dose = 4.75E + 15/cm^2.
The results of two-parameter fitting:
before annealing – Rp = 75.5 nm, ΔRp = 26.0 nm;
after annealing – Rp = 74.7 nm, ΔRp = 77.9 nm

Implanted impurity distributions were derived according to Eq. (3) from the results of Rp and ΔRp fitting. These profiles are presented on Fig. 1 and 2. The results of depth profiling of these specimens by methods of Auger electron spectroscopy (for P) and secondary ion mass-spectrometry (for As), are presented in these figures too. Concentration scaling of these profiles were made according to the sensitivity factors which have been determined from EPMA data [7]. A good correlation of the results was observed.

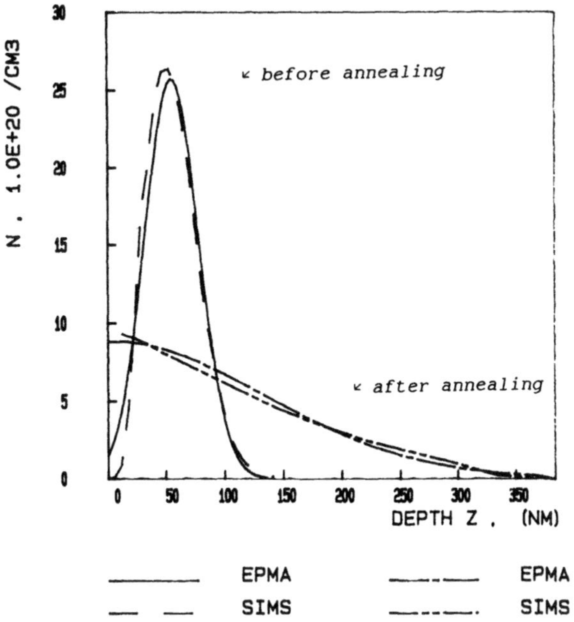

Fig. 2. Distribution of As (50 keV implantation energy) in Si before and after annealing.
Implanted dose = 1.48E + 16/cm^2.
The results of two-parameter fitting:
before annealing – Rp = 55.0 nm, ΔRp = 23.0 nm;
after annealing – Rp = 1.3 nm, ΔRp = 131.8 nm

The proposed method can be suitable for the local quick estimation of impurity distribution before and after annealing processes in IC-fabrication. It is enough to measure $k(E_0)$ at three values of E_0. In this work $k(E_0)$ has been measured at 4 and 5 values of E_0 also and the results were close to the results presented in this paper.

The approach presented may be used also for the calibration of ion-implanted reference materials [8]. In general, further experiments and calculations for the determination of $\Phi(z, E_0)$-functions at low E_0 values are needed.

References

[1] P. N. Butcher, N. H. March, M. P. Tosi (eds.), *Crystalline Semiconducting Materials and Devices*, Plenum, New York, 1986, p. 591.
[2] J. D. Brown, *Spectrochim. Acta* **1983**, *38B*, 1411.
[3] H. Russel, H. Glawischnig (eds.), *Ion Implantation Techniques*, Springer, Berlin, Heidelberg, New York, 1982, p. 177.
[4] B. D. Bunday, *Basic Optimization Methods*, Edward Arnold, London, 1984, chapter 4.
[5] G. F. Bastin, H. J. M. Heijligers, F. J. van Loo, *Scanning* **1986**, *8*, 45.
[6] G. F. Bastin, F. J. J. van Loo, H. J. M. Heijligers, *X-Ray Spectrom.* **1984**, *13*, 91.
[7] A. P. Alexeyev, V. I. Zaporozchenko, *Vacuum* **1990**, *41*, 1725.
[8] W. H. Gries, W. Koschig, *Surface Interf. Anal.* **1990**, *16*, 321.

Mikrochim. Acta (1992) [Suppl.] 12: 235–239

An Electron Spectroscopy Study of a-SiN_x Films

Alexander G. Fitzgerald*, Heather L. L. Watton, and Mervin J. Rose

Department of Applied Physics and Electronic and Manufacturing Engineering, University of Dundee, Dundee DD1 4HN, Scotland U.K.

Abstract. The variation of composition and chemical bonding in the range of a series of amorphous silicon nitride (a-SiN_x : H) films deposited by glow discharge has been investigated. A comparison has been made between quantitative X-ray photoelectron spectroscopy (XPS) and compositions deduced from electron energy loss spectroscopy (EELS). Core-level and valence-band XPS and extended electron loss fine structure (EXELFS) in EELS were used to investigate the chemical bonding and structure present in these films.

Key words: PECVD a-SiN_x : H, XPS, EELS, EXELFS.

Amorphous silicon nitride (a-SiN_x) is widely used in microelectronics, being an excellent dielectric layer and barrier against oxidation. It is mainly used as an insulator in thin film transistors and in MNOS FET memory devices. A knowledge of the manner of variation of the chemical bonding with a-SiN_x film composition may give information on the microscopic nature of trap states in these devices. In this investigation the variation of a-SiN_x film composition with NH_3/SiH_4 gas volume ratio obtained by Fitzgerald et al. [1] has been confirmed and the study of these films has been extended to an investigation of the variation of chemical bonding with film composition by EXELFS and core and valence band XPS.

Experimental

XPS spectra were collected with a VG HB 100 Multilab UHV-SEM fitted with a CLAM 100 concentric hemispherical analyser. The spectrometer control unit was controlled by an OPUS PC II IBM—compatible microcomputer running VGS2000 software supplied by VG Scientific. The XPS spectra were recorded with the spectrometer operating in the constant analyser-energy mode at a pass-energy of 20 eV. Peak intensities were measured after background subtraction, and integration to obtain the peak area. SIMS spectra were acquired with a VG AG61 primary ion source and an SQ300 mass spectrometer. To remove surface contamination, specimen were etched by argon ion beam from an Ion Tech B11 fine-beam saddle-field cold-cathode ion gun with a maximum current of 20 μA. EELS spectra were collected with a JEOL 100C TEM fitted with a Gatan 666 PEELS™ spectrometer,

* To whom correspondence should be addressed

with a typical electron dose of 2.38 nA/μm^2. This spectrometer is controlled by a photodiode interface unit and voltage scan module, together with an Apple Macintosh microcomputer.

The films of a-SiN$_x$: H were deposited onto a number of different substrates by radio frequency glow discharge decomposition of $NH_3 + SiH_4$ mixtures. Deposition conditions for all films were: power = 5 W, temperature = 330 °C, total pressure = 33.3 Pa and flow-rate = 8 standard cm^3/min. The NH_3/SiH_4 gas volume ratio, R, was varied between 1.0 and 12.0. Films prepared for XPS were deposited onto aluminium foil substrates; those used for transmission electron microscopy were deposited onto carbon-coated mica sheets to a thickness of approximately 15 nm, and those used for SIMS were deposited onto glass. The films for TEM work were floated off the carbon-coated mica sheets in distilled water and mounted onto TEM copper grids. The gold-flashed films were prepared by evaporation of gold onto the a-SiN$_x$: H films on aluminium foil substrates under vacuum. The Si_3N_4 standard used for analysis of the XPS results was obtained from Goodfellow Metals Ltd. The composition of the film deposited at a gas volume ratio of R = 8 and determined by XPS was used as a standard for EELS measurements. This secondary standard was necessitated by the difficulty in obtaining a Si_3N_4 standard of similar thickness to the deposited a-SiN$_x$: H films, and the large separation between the N K-edge (at 401 eV) and the Si K-edge (at 1839 eV) meant, that it was likely that the collection efficiency over this range was not uniform.

Results and Discussion

Figure 1 and Table 1 summarise the compositions of a series of a-SiN$_x$ films, determined by XPS and EELS. Compositions were calculated from the XPS results by using software developed by Moir et al. [2] and from the EELS results by using software supplied by Gatan [3]. For the compositions calculated by XPS the error is $\pm 5\%$, and for those calculated by EELS $\pm 15\%$. SIMS spectra collected from the range of films verified that hydrogen was present in the films and therefore Si and N atomic concentrations are expressed as percentages of (N + Si), eliminating the need to consider the approximate 15 atom% H content. EELS was applied to films produced in a different deposition run from those analysed by XPS. This was because of the difference in film thickness required for the two analytical techniques.

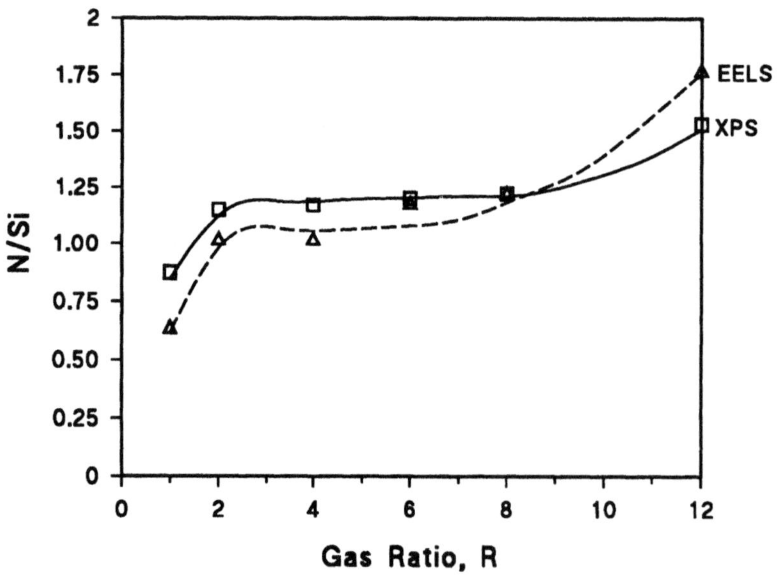

Fig. 1. N/Si Ratio as obtained by XPS and EELS, plotted against gas volume ratio $R(=NH_3/SiH_4)$

Table 1. XPS and EELS analyses of a-SiN_x : H films. EELS results are calculated relative to the XPS value at $R = 8$

	Composition atom %			
	XPS		EELS	
R	Si	N	Si	N
1.0	53.4	46.6	61.4	38.6
2.0	46.6	53.4	49.8	50.2
4.0	46.0	54.0	48.8	51.2
6.0	45.4	54.6	46.1	53.9
8.0	45.1	54.9	45.1	54.9
12.0	39.5	60.5	36.2	63.8

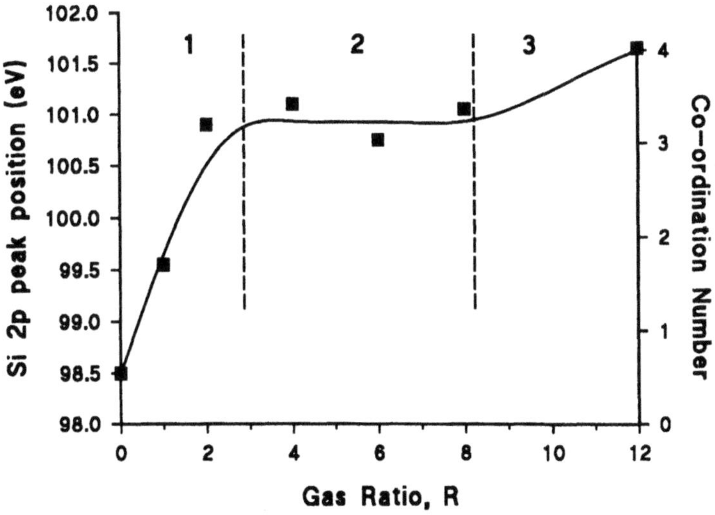

Fig. 2. Si 2p peak position (eV), obtained by XPS, plotted against gas volume ratio $R(=NH_3/SiH_4)$

The disagreement between the two sets of compositions can most likely be explained by an alteration in deposition parameters between the two runs. Since the films used for EELS are required to be much thinner, this may also explain the differences in composition between the two sets.

To obtain information on the chemical bonding in the films the Si 2p peak positions were measured by XPS. This was done by gold-flashing the films and calculating the actual peak positions from the shift of the Au $4f_{5/2}$ peak position. The results are shown in Fig. 2 and Table 2. In Table 2 the shifted peak positions are the peak positions measured from the XPS spectra and the actual peak positions are the shifted peak positions minus the shift in eV of the Au $4f_{5/2}$ peak. This is carried out because with each film different amounts of charging will occur when XPS is performed, while the position of the Au $4f_{5/2}$ peak is known to be 83.8 eV, hence the shift in the position of this peak due to charging can be calculated and subsequently removed from all peaks in the spectrum. Three regions can be distin-

Table 2. Si $2p$ peak positions in a-SiN_x : H films

R	Shifted peak centre (eV)	Actual peak centre (eV)
1.0	103.9	99.6
2.0	107.2	101.5
4.0	107.6	101.1
6.0	107.0	100.8
8.0	106.6	101.1
12.0	107.1	101.7

guished in Fig. 2, with the binding energy increasing with gas ratio from 98.5 eV, the value for an a-Si : H film, to approximately 102 eV. Nitrogen is more electronegative than silicon and hence the charge transfer from silicon to nitrogen accompanying the replacement of Si–Si bonds by Si–N bonds explains this shift in peak position to higher binding energies, as the Si atoms will be left with a positive charge. In a previous XPS study of a-SiN_x : H film [4] it was proposed that the observed Si $2p$ peak is actually comprised of five superimposed peaks. These five peaks correspond to the possible bonding configurations where zero, one, two, three or all four of the Si–Si bonds in a-Si are replaced by Si–N bonds. This gives a good approximation for calculating the co-ordination number of nitrogen atoms around each silicon atom. In region 1 of Fig. 2, this co-ordination number rises to approximately 3, in region 2 it stays at 3, whilst in region 3 it rises to 4, as in stoichiometric Si_3N_4.

The valence-band spectra contained two peaks at 11 and 24 eV. The 11 eV peak can be attributed to Si $3s$, Si $3p$ and N $2p$ electrons, and the 24 eV peak is attributed to the N $2s$ peak. The ratio of the intensities of these two peaks varies with R and indicates a change in bonding over the range of film compositions, Si–Si bonds being replaced by Si–N bonds as the N content of the films increases.

The extended fine structure modulations after an ionisation edge in EELS can be used to obtain nearest neighbour distances. This was done by using software supplied by Gatan [3]. The Si K-edge was analysed to obtain this information for Si in this range of films. The N K-edge was not used, as it was found that the first maximum of the EXELFS modulation was obscured by the inflection of this edge. Table 3 summarises the information obtained. In stoichiometric Si_3N_4 the Si–N

Table 3. Nearest neighbour distances (nnd) for silicon atoms in a-SiN_x : H films

R	nnd (Å)
1.0	2.20
2.0	2.10
4.0	1.96
6.0	1.92
8.0	1.86
12.0	1.84

bond length is 1.73–1.75 Å [4] with the silicon atoms each tetrahedrally surrounded by 4 nitrogen atoms and each nitrogen atom occupying a planar, triply co-ordinated site with Si atoms as nearest neighbours. The Si–Si bond length in a-Si is 2.35 Å [5]. As the nitrogen content of the films increases, the average of the distance between a Si atom and its first nearest neighbours decreases from just below the value for a-Si to that for Si_3N_4. This would imply that as the nitrogen content of the films increases, Si–Si bonds are being replaced by Si–N bonds.

Acknowledgements. One of us (HLLW) wishes to thank the Carnegie Trust for the Universities of Scotland, the C. R. Barber Trust, the President's Fund and EMAS for help with expenses in attending EMAS '91. Thanks are also due to the Science and Engineering Research Council for their support (grant reference number GR1F92978).

References

[1] A. G. Fitzgerald, P. A. Moir, C. P. McHardy, *Inst. Phys. Conf. Ser. No. 93, Vol. 1*, 1988, chapter 5.
[2] P. A. Moir, A. G. Fitzgerald, B. E. Storey, *Surf. Interface Anal.* **1989**, *14*, 295.
[3] M. Kundmann, K. Chabert, K. Truong, M. Leber, O. Krivanek, *Gatan EL/P Software Version 2.0.5.*, Gatan, 1990.
[4] R. Karcher, L. Ley, R. L. Johnson, *Phys. Rev. B* **1984**, *30*, 1896.
[5] D. R. Lide (ed.), *Handbook of Chemistry and Physics*, CRC, Boston, 1990, p. 9–11.

Mikrochim. Acta (1992) [Suppl.] 12: 241–245

Electron Probe Microanalysis of Glass Fiber Optics

Marija Kern[1,*], Eva Perman[1], and Peter Pavli[2]

[1] Inštitut za elektroniko in vakuumsko tehniko Teslova 30, YU-61000 Ljubljana, Slovenija
[2] Hidrometeorološki zavod Slovenije Vojkova 1b, YU-61000 Ljubljana, Slovenija

Abstract. With respect to the composition of optoelectronic glasses careful research using spectrochemical analytical procedures has been carried out to prevent interelement effects. Flame atomic absorption and emission spectrometry, as compared to wet chemical analysis and XRF shows satisfactory agreement. The results of EPMA bulk analysis obtained using oxides or proper compounds of the analytes as standards, and applying a ZAF correction, show good agreement with other analytical techniques. The method has been improved using glass standards of similar composition, that were analyzed in different laboratories. EPMA appears to be a suitable method for the investigation of glassy optoelectronic components.

Key words: EPMA, glass material, fiber optics, alternative bulk analyses.

The applicability of fiber optic components consisting of several kinds of glasses, has been increasing rapidly in the research of optoelectronics. The composition of these glasses is rather unique in order to achieve the required physical properties as refractivity, absorption and reflectivity [1, 2]. Because of the small dimensions electron probe microanalysis (EPMA) is applied for the characterization of single glass phases. In this respect it is important that the accuracy of EPMA results is comparable to bulk analysis results in order to enable a reliable determination of eventual compositional changes in the bordering areas. For the investigation of optoelectronic components manufactured from different sorts of glasses, as well as for example fiber optics or microchannel plates, electron microscopy and microanalysis are, because of the small glass phase dimensions of a few μm, very suitable methods [3, 4, 5].

Experimental

For bulk analysis of glasses flame atomic emission spectrometry (FAES), flame atomic absorption spectrometry (FAAS) and chemical analyses were applied. Bulk analyses were performed also with

---

* To whom correspondence should be addressed

EPMA using oxides or standard glasses as reference materials. The sample structure was investigated in the JEOL JSM 35 scanning electron microscope. Different glasses in fiber optic structures were analyzed with the TRACOR TN-2000 EDX microanalyzer and the JEOL FCS-35 WDX microanalyzer. The TRACOR super ML computer program including also the ZAF correction procedure (version 11) for calculation of element concentration was used in spectrum processing. The same ZAF correction procedure was used for calculations of the wavelength dispersive EPMA results.

The investigated analytical procedures can be applied in the field of fiber optic components production and application. In the order to identify a single glass phase the accuracy of the EPMA should be comparable to the accuracy of the bulk analysis, assuming that samples were sufficiently homogeneous. To carry out a comparison, bulk samples of glasses were analyzed by flame atomic absorption (FAAS) and emission spectrometry (FAES), EPMA and by other laboratories applying X-ray fluorescence (XRF), FAAS, inductively coupled plasma atomic emission spectrometry (ICP-AES), gravimetry and spectrophotometry.

Bulk samples of lanthanum and absorption glasses for atomic absorption or emission analysis were crushed and ground to 100 mesh. For the determination of most constituents, except boron oxide, the sample was dissolved in a mixture of HF, HNO_3 and H_3BO_3. A special procedure was applied to determine boron oxide in order to prevent losses of volatile boron fluoride. The sample was dissolved in HF and HNO_3 at 100°C and the excess of HF was then neutralized by the addition of silica to form fluorosilicic acid [6]. Careful matching of sample and standard solutions was required for calibration because matrix and interelement effects can be expected. Wherever possible, the standard addition method was applied. Instrumental setting and precision of determination have been investigated for all analytes. Methods were assessed for the determination of BaO, La_2O_3, B_2O_3, SiO_2, K_2O, Na_2O, Al_2O_3, Fe_2O_3 and CoO [5].

The bulk samples for EPMA were prepared by casting the glass sample in a resin. The samples were polished and coated with a conductive carbon film in vacuum. Standard oxides or other compounds of the elements to be determined were cast together with the samples. The standards were preliminarily tested for homogeneity and volatility under the electron beam and found suitable as standards for EPMA [7, 8]. We have prepared ZrO_2, Ta_2O_5, TiO_2 and Nb_2O_5 by sintering pure powdered oxides in vacuum. Because of hygroscopy some standards, for instance La_2O_3 were soon decomposed. In this case borides, chlorides and sulfates, like LaB_3, $BaSO_4$ and KCl were used as standards.

Special care was taken when preparing the fiber optic samples that were first cut in resin and polished perpendicularly to the fibers. To illustrate dimensional problems of fiber optic investigation the cross sectional EM micrograph of a fiber optic sample is shown in Fig. 1. Lanthanum glass is used

5 μm

Fig. 1. Cross sectional EM micrograph of fiber optics; light areas from lanthanum glass and dark areas from the absorption glass

for optical fibers due to its refractivity (light areas). The absorption glass forms a very thin dark film around the optical fibers or is put into the fiber optic structure as thin fibers (dark areas).

Quantitative EPMA was performed under at an accelerating voltage of 15 to 20 kV, a take off angle of 35° and a specimen tilting angle of 0°. Both the energy dispersive and the wavelength dispersive EPMA modes were applied. For qualitative determination of the elements present, and for quantitative analysis of K_2O and Na_2O the energy dispersive X-ray analyzer was used. A low accelerating voltage and a beam current below $1 \cdot 10^{-10}$ A, enabled insignificant migration of K and Na ions under the electron beam. Contents of other oxides were mostly determined by the wavelength dispersive X-ray analyzer [9] because large overlapping of peaks in the X-ray spectra recorded with the energy dispersive X-ray analyzer were observed.

Results and Discussion

Bulk analysis results and results obtained on fiber optics of lanthanum and absorption glass samples are shown in Tables 1 and 2 respectively. Taking into consideration the accuracy $\pm 5\%$ of our EPMA results, the results in Table 1 demonstrate good agreement between EPMA bulk analysis, spectrochemical results (FAAS, FAES, ICP-AES, XRF), spectrophotometry and gravimetry. To determine SiO_2 by EPMA in the element combination shown in Table 1, the ZAF correction applied gives too high results with respect to other methods. Difficulties were encountered when determining BaO by FAES or FAAS. Therefore gravimetric analysis was required to obtain a reliable concentration value for comparison of results.

The EPMA analysis results of the dark absorption glass are reported in Table 2. As before, there was good agreement between various techniques of bulk analysis. The EPMA results for Na_2O and K_2O were surprisingly close to the results obtained by FAES or FAAS, knowing that Na and K may migrate under electron

Table 1. Quantitative analysis of lanthanum glass (% by weight)

	Laboratory					
oxide	A	RSD(%)	B	C	D	E
BaO	17.6	1.6	19.1 FAES	16.9 FAAS, XRF	16.9	18.2
La_2O_3	28.9	1.3	25.8 FAES	26.2 FAAS	—	28.4
ZrO_2	7.5	2.0	—	7.3 XRF	7.2	7.4
TiO_2	3.2	5.2	—	3.1 XRF	—	3.1
B_2O_3	—	—	10.0 FAAS	10.0 FAAS, ICP-AES	—	—
Nb_2O_5	5.5	2.1	—	4.9 XRF	5.6	5.2
Ta_2O_5	16.5	1.9	—	16.0 XRF	—	14.7
SiO_2	19.0	1.5	16.0 FAAS	15.9 FAAS	15.9	20.0

A—EPMA bulk analysis of La-glass
RSD—relative standard deviation (%); (five measurements)
B—FAAS and FAES analysis
C—Different analysis methods
D—Gravimetric analysis
E—EPMA analysis of La-glass in fiber optics

Table 2. Quantitative analysis of lanthanum glass (% by weight)

	Laboratory				
Oxide	A	RSD(%)	B	C/D	E
K_2O	9.4	2.1	11.1 FAAS	10.3	0.4
Al_2O_3	2.4	4.9	1.4 FAAS	1.7	2.6
B_2O_3	—	—	11.0 FAAS	11.0	—
Fe_2O_3	6.3	3.1	6.1 FAAS	6.6	4.6
CoO	4.1	3.5	4.3 FAAS	—	2.6
TiO_2	2.9	5.1		—	1.6
SiO_2	60.7	0.9	62.0 FAAS	64.0	52.9
Na_2O	1.4	7.1	1.3 FAES	1.3	2.9
La_2O_3	—	—	—	—	6.7
ZrO_2	—	—	—	—	0.8
BaO	—	—	—	—	2.3
Nb_2O_5	—	—	—	—	0.9

A—EPMA bulk analysis of absorption glass
RSD—relative standard deviation (%); five measurements
B—FAAS and FAES analysis
C,D—Different analysis techniques
E—EPMA analysis of absorption glass in fiber optics

bombardment [10, 11]. This agreement may be explained by the use of a low beam current in order to prevent the migration of Na and K. Higher concentrations obtained for Al_2O_3 by EPMA are a consequence of the correction procedure applied, which is known to have a poor performance for elements with a low atomic number [8].

EPMA results obtained on fiber optic structures are acceptable for the optical fibers. There are, however, too large differences for the dark glass area between bulk and fiber optic analysis. Despite the small analytical volume (1 μm^3) the dimensions of the structures have an essential role in determining the composition. The EM micrograph in Fig. 1 shows that the glass phase surrounding the fiber measures only 1 to 2 μm in width. It is practically impossible to analyse this phase without exciting also a part of the bordering fiber area. While the EPMA results of the fiber area show agreement with bulk analysis of the same glass, the composition of the small glass phase probably includes also the constituents of the neighbouring fiber glass phase, as shown in Table 2. Hence only a partial identification of the small glass phase composition is possible. Furthermore, it is possible to identify larger diffusion phenomena between the phases but due to possible interferences between the neighbouring phases, EPMA results for phases smaller than 1.5 μm in width must be carefully examined.

References

[1] G. W. McLellan, E. B. Shand, *Glass Engineering Handbook*, McGraw-Hill, New York, 1986.
[2] L. A. Ketron, *Am. Ceram. Soc. Bull.*, **1987**, *66*, 1571.

[3] P. Pavli, E. Kansky, E. Perman, R. Zavašnik, IEVT, Ljubljana 1982, 1986.
[4] E. Perman, R. Zavašnik, P. Pavli, E. Kansky, *Zbornik referatov*, MIDEM SD 81, 1981, 79.
[5] M. Kern, E. Perman, P. Pavli, *Zbornik referatov*, MIDEM SD 90, 1990, 343.
[6] B. E. Andrew, *Am. Ceram. Soc. Bull.*, **1976**, *55*, 583.
[7] S. J. B. Reed, *Electron Microprobe Analysis*, Cambridge University Press, Cambridge, 1975.
[8] J. I. Goldstein, H. Jakowitz, *Practical Scanning Microscopy*, Plenum, New York, 1975.
[9] A. L. Albee, L. Ray, *Anal. Chem.* **1970**, *42*, 1408.
[10] D. J. Bloomfeld, G. Love, V. D. Scott, X-Ray Spectrom. **1986**, *14*, 139.
[11] W. Weisweiler, R. Neff, *Glastechn. Ber.* **1986**, *53*, 282.

Mikrochim. Acta (1992) [Suppl.] 12: 247–254

Quantitative Microanalysis of Low Concentrations of Carbon in Steels

Jacky Ruste

Electricité de France, Direction des Etudes et Recherches, EMA P.O. Box 1,
F-77250 Moret sur Loing—France

Abstract. Conditions of quantitative X-ray microanalysis of carbon in industrial steels are given. The different problems are described and a solution is proposed for each of them. Examples of applications are given.

Key words: X-ray microanalysis, quantitative analysis, carbon, steels.

Carbon plays a essential role in the chemical and mechanical behaviour of steels which is due to the possibility of forming, e.g. carbides, nitrocarbides and borocarbides. It is very important for the metallurgist to know the average bulk carbon concentration in steel which can be obtained easily by chemical analysis. In many cases, however he also needs to know the local composition in a special micro-area (segregated zone, welded joint). Electron probe microanalysis (EPMA) is one of the most suitable techniques to solve this problem. However, quantitative microanalysis of the very light elements ($Z = 5 - 8$) presents some specific difficulties; most of them are well-known for more than fifteen years [1–3]. In industrial steels, especially the simultaneous localisation of carbon in very small carbides and in solid solution makes the average local analysis difficult.

In this paper, after a brief recall of the main analysis difficulties (with suitable solutions in the case of low carbon concentration in steels), we propose a method to analyse carbon in such specimens.

Specific Experimental Problems

There are the different problems which occur when we want to determine very light elements, and a number of these are essential for quantitative analysis of carbòn especially in the case of low concentrations:

Specimen Preparation

Embedding, polishing and cleaning of the specimen must be done very carefully. If possible, epoxy resins must be avoided and replaced by a low fusion-temperature metallic alloy. Alumina polishing is preferable to polishing with diamond and the final cleaning must be carried out with freon in an ultrasonic tank followed by a high-pressure gas drying.

Contamination

During analysis, a thin carbon layer is contaminating the specimen surface which is caused by the preparation (whatever the quality of the latter) and to the cracking by the electron beam of organic molecules (primary and secondary pump oil, joint grease, etc.). Several devices can be used to reduce this contamination (Fig. 1): a cooled baffle under the diffusion pump, a cooled trap on the third condenser pole piece (near the specimen) and a adjustable air jet. The best results are obtained with both of these devices. We can observe (Fig. 2) during about one or two minutes, a decrease in carbon intensity, which corresponds to a local decontamination; then the contamination is stabilized during several minutes, which allows analysis. At last, contamination is increasing again because of specimen cooling down.

Peak Overlap

Line interferences of higher orders of diffraction are very frequent, but can generally be reduced by pulse height discrimination. However, overlapping of carbon K line with L spectra of chromium (2nd order) and nickel (3rd order) has an important

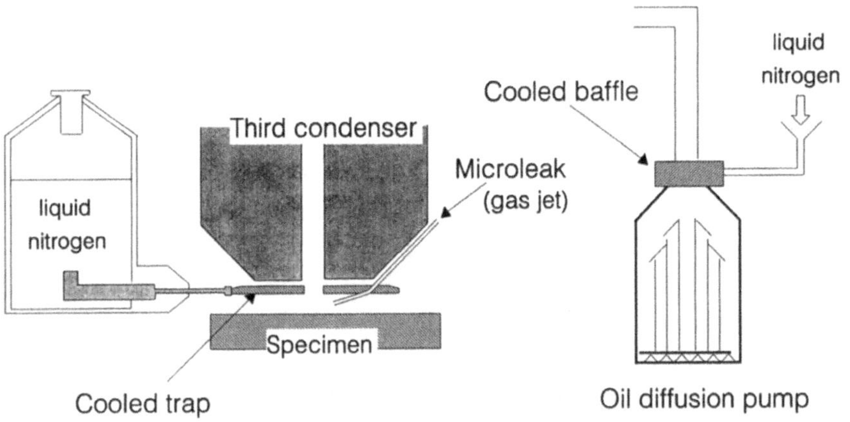

Fig. 1. Anticontamination devices used on a CAMECA "Camebax" microprobe

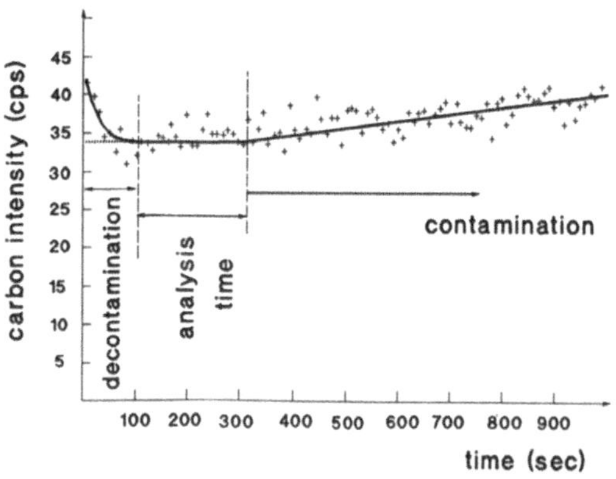

Fig. 2. Variation of carbon emission on pure iron vs time ($E_0 = 10$ kV, $I = 60$ nA, focused probe)

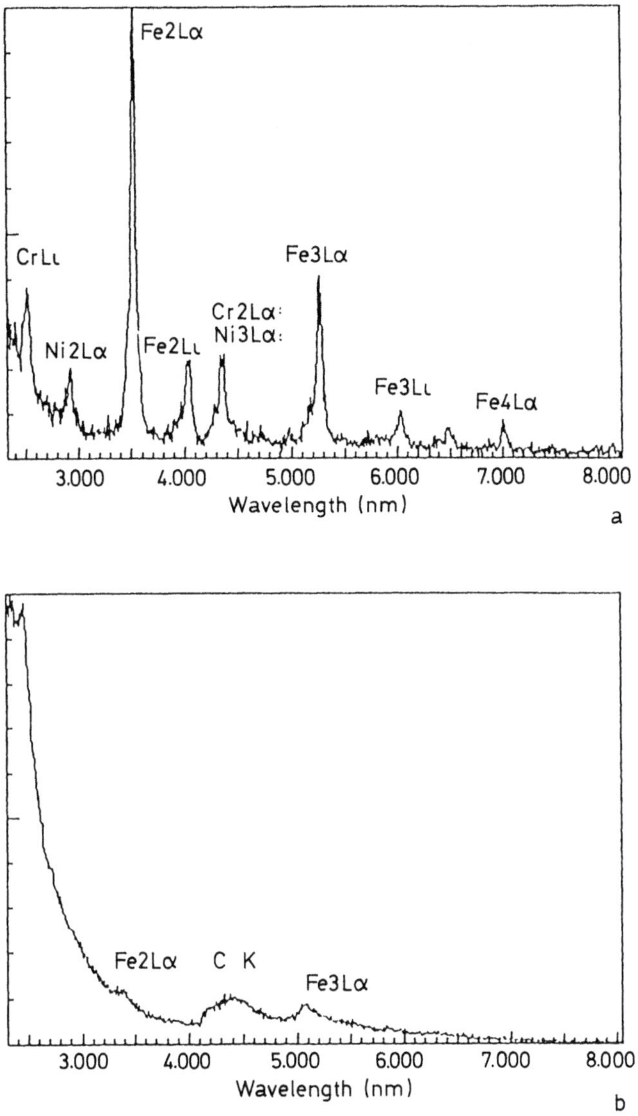

Fig. 3. Wavelength dispersive X-ray spectra of an austenitic steel, obtained with a lead stearate crystal (**a**) and a Ni/C multilayer crystal. (E_0 = 10 kV, I = 100 nA)

effect on background measurements and detection limit of carbon in austenitic steels. The use of multilayer crystals (W/Si with 2d = 5.98 nm or Ni/C with 2d = 9.5 nm), instead of lead stearate, can reduce these overlap situations (Fig. 3).

Detection

The major experimental difficulties for carbon analysis are the low count rates and the pulse distribution at the amplifier output. The former can be reduced by the use of multilayers, while a satisfactory pulse distribution can be obtained by optimizing serious electronic parameters like counter voltage and the low and upper levels before each analysis.

Choice of a Standard

A good standard must be homogeneous, stoichiometric or with a known composition, easy to obtain and a reliable reference for every laboratory. Furthermore, a general standard is needed which can be used for all specimens. Different standards have been proposed like steels, carbides, diamond, glassy carbon. We have chosen for diamond in which case we must take into account the chemical shift which appears between carbon line in diamond and in the steel carbide.

Background Measurements

The actual background intensity for the carbon line is composed of contributions of "Bremsstrahlung", carbon contamination and, occasionally overlapping lines (e.g., Cr and Ni lines in austenitic steels). Under these conditions, it is impossible to obtain a rigorous determination of the background intensity which is measured on each side of the carbon peak, followed by a linear or logarithmic interpolation. Background must be measured on a specimen which is identical to the analysed one, but without carbon (or with a very low concentration, less than the detection limit). This specimen must be prepared under the same conditions (and if possible, in the same mount), in order to have the same surface contamination. For low carbon steel analysis, iron with 26 ppm of carbon is used, and for austenitic steels, a "316" steel (Ni 14%–Cr 16.3%–Mo 1.44%) with 73 ppm of carbon. It is also possible to use the same specimen as background reference, if a local zone with a known concentration can be found. In this case, the measured intensity is the sum of two contributions: the steel carbon emission, which can be calculated with the correction method and the contamination emission, which can be obtained by subtraction.

Correction Methods

As conventional ZAF methods are not suitable for the quantification of very light element analysis, new correction procedures have been proposed (modified ZAF [2], $\Phi(\rho z)$ curves [4]) and generally give good results. The only problem is the choice of the mass absorption coefficients (m.a.c.'s). The coefficients we used in this work are given in Table 1.

For a voltage of 10 kV and a take-off angle of 40°, the carbon correction factor relative to a diamond standard is 2.17, for low carbon or austenitic steel.

Specimen Structure

Finally, one of the most important difficulties for carbon analysis in industrial steels lies in the metallographic structure. Carbon is usually contained in two phases: the matrix, in solid state solutions, and in some carbides. The content in the matrix is

Table 1. Mass absorption coefficients (m.a.c.'s) of carbon K line in several elements

Element	Ti	Cr	Mn	Fe	Ni	Mo
m.a.c.'s (cm^2/g)	8093	10480	11840	13300	16580	15500

often very low. In low-carbon steel, ferrite has a solubility limit less than 200 ppm; the other phase is pearlite, which is composed of a succession of several thin layers of ferrite and carbide Fe_3C (the average carbon content of this phase is about 0.77%).

Under these conditions, it is impossible to obtain accurate value of the average carbon concentration, even in a local area. It is not possible to use the rigorous procedure and to wait one or two minutes for decontamination of the surface, followed by analysis for several minutes to have a good statistical precision. The total counting time would be prohibitive.

We propose to reduce the counting time to a very small value (between 1 and 10 seconds), to decrease the waiting time and to increase the number of analysis points (from several hundred to several thousand), in order to obtain a correct average carbon content with a good statistical precision.

Applications

Low Carbon Steels

The background intensity has been measured on an iron specimen which contains about 26 ppm of carbon. Before each analysis, this intensity is obtained by means of several series of 150 counts of 10 seconds.

Figure 4 shows an intensity profile obtained on a 0.31% steel. The background intensity is indicated. Figure 5 shows the same profile, after background removal and application of a correction procedure. The average carbon content is about 0.34% after correction, which is close to the chemical analysis.

Chromium-Molybdenum Steels

A chromium-molybdenum steel (Cr 2.25%–Mo 1%) with about 800 ppm of carbon has been analysed. Results of different analyses are given in Table 2. This example shows analysis difficulties with heterogeneous alloys: we can observe a large varia-

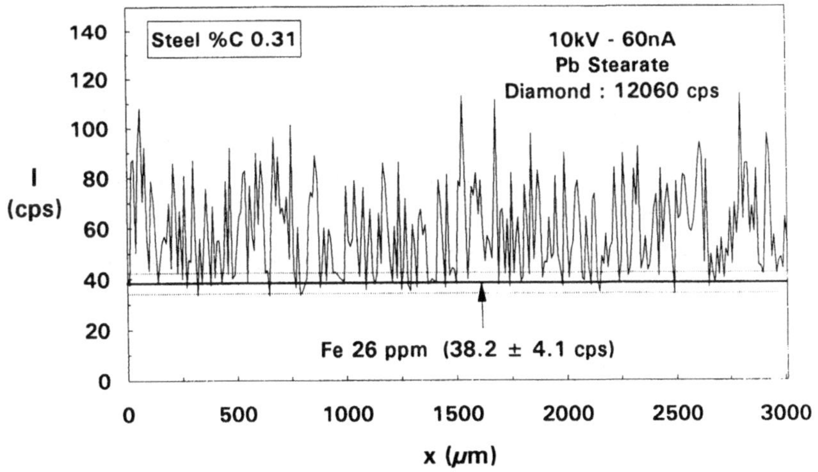

Fig. 4. Intensity profile of carbon on a low-carbon steel (carbon content: 0.31%) 301 analysis points of 10 seconds, measured on distances of each 10 microns

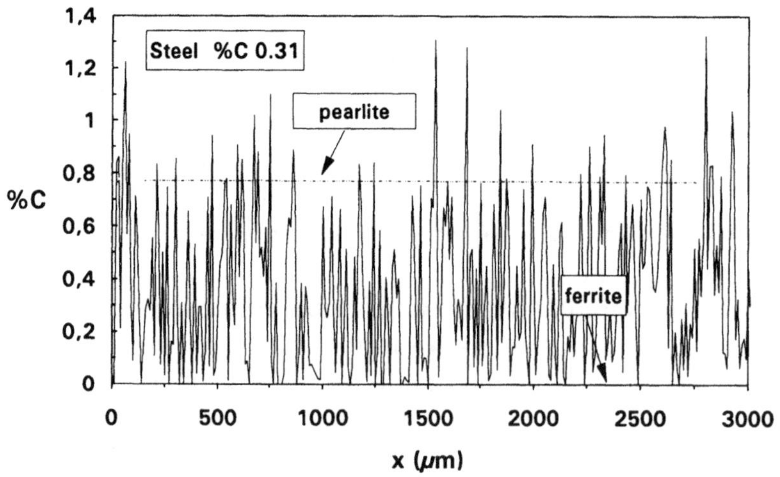

Fig. 5. Carbon concentration profile on a low carbon steel (chemical content: 0.31%) after background subtraction and correction procedure

Table 2. Carbon analysis of a chromium-molybdenum steel, with a carbon content of 800 ppm. ($E_0 = 10$ kV, I = 60 nA)

Area	# of analyses	Counting time (s)	C-content (ppm)
1st area	151		300
2nd area	151	10	890
average	302		600
3rd area	75	100	980

tion in the results due to the localisation of the analysed areas. On this case, a large number of analyses is necessary to obtain a good average measure.

Segregated Zone

In some low-alloy steels, segregated zones exist, which can induce cracks under certain conditions. These segregation zones can also facilitate the embrittlement of the steel. For nuclear applications, it is very important to know the segregation level for all the elements, carbon included.

Figure 6 shows a carbon concentration profile across a segregation zone which presents a crack. The carbon concentration on a distance from this zone is in good agreement with the chemical value (0.16%). In the segregated zone, the average carbon content is about 0.26%, which represents a segregation rate of 62%.

Welded Joint

Nuclear reactor vessels, nozzles and tube plates are made in low-alloy steel with an austenitic steel cladding, deposited by fusion. This process induces a diffusion of

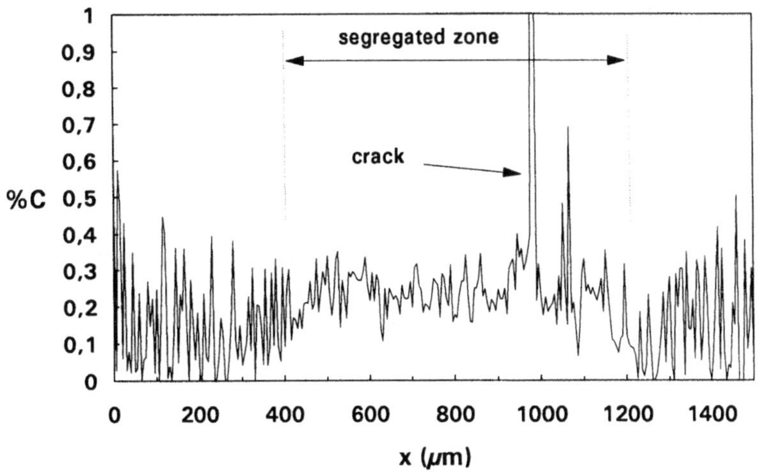

Fig. 6. Carbon concentration profile across a segregated zone, in a low-alloy steel. The background was measured on a 26 ppm steel

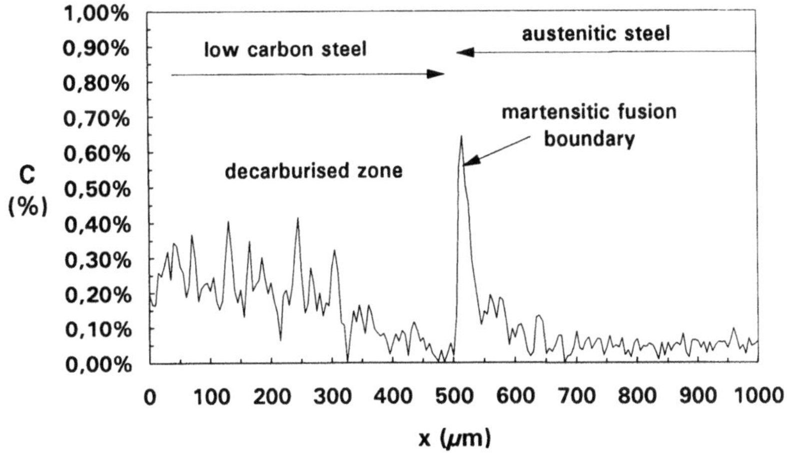

Fig. 7. Carbon concentration profile across a welded joint between a low carbon steel (C = 0.20%) and an austenitic steel

carbon across the welded joint, which causes a decarburisation of the low-alloy steel, an important carbon increase at the boundary between the two steels (martensitic fusion boundary) and a local carburisation of the austenitic steel.

Figure 7 shows a carbon profile across a welded joint, after background removal and correction procedure. We can see that the decarburisation is very important and extends to about 300 microns. The carbon concentration is 0.65% in the martensitic fusion boundary and about 0.15% in the austenitic steel.

Conclusions

Despite of specific difficulties, quantitative microanalysis of carbon in industrial steels is possible, with a satisfactory precision (better than 10%). We must carefully check specimen preparation, counting electronics adjustments and the anticontamination process. It is also very important to select a good specimen for background measurement. The number of analysis points must be determined as a function of specimen heterogeneity.

References

[1] J. Ruste, *J. Microsc. Spectrosc. Electron.* **1979**, *4*, 123.
[2] F. Coppola, F. Maurice, J. Ruste, *VIIIth International Congress on X-Ray Optics and Microanalysis* (D. R. Beaman, R. E. Ogilvie, D. B. Wittry, eds.) Pendell, Midland, U.S.A., 1980, p. 851.
[3] G. F. Bastin, H. J. M. Heijligers, *J. Microsc. Spectrosc. Electron.* **1986**, *11*, 215.
[4] J. L. Pouchou, F. Pichoir, *J. Microsc. Spectrosc. Electron.* **1986**, *11*, 229.

Mikrochim. Acta (1992) [Suppl.] 12: 255–259

Electron Configuration of the Valence-Conduction Band of the Mineral Wustite

Djordje M. Timotijevic[1,*] and Miodrag K. Pavicevic[2]

[1] RTB Bor, Copper Institute, YU-19210 Bor, Yugoslavia

[2] University of Belgrade, Faculty of Mining and Geology, University Laboratory for Electron Microanalysis, Djusina 7, YU-11000 Belgrade, Yugoslavia

Abstract. Soft Fe $L_{III,II}$ X-ray emission spectra of wustite have been recorded by EPMA in the 690–730 eV energy range. Using these spectra for different electron beam excitation voltages, the self-absorption Fe L_{III}-spectrum was formed. By comparing Fe $L_{III,II}$ X-ray emission and Fe L_{III}-self-absorption spectra of wustite, relative energies of the valence-conduction band orbitals could be determined as well as the splitting of these orbitals in the crystal field and the size of their spin splitting.

Key words: EPMA, Fe $L_{III,II}$, X-ray, self-absorption, wustite.

The complex energy levels configuration of the valence-conduction band orbitals of the mineral wustite, results from the interaction between the electrons of the 3*d* and 4*s*-orbitals of the central metal ion, Fe^{2+}, and the electrons of 2*s* and 2*p* orbitals of the oxygen ligand ions as well as from the 3*d* electrons unpaired spins of the Fe^{2+} ions. The central Fe^{2+} ion in wustite is surrounded by oxygen ligand ions in octahedral coordination. Such a complex electron configuration of the valence-conduction band of wustite is reflected in the shape of its Fe $L_{III,II}$ emission and Fe $L_{III,II}$ absorption spectra, which are not represented as simple peaks but as a complex band structure.

The relative energy levels of valence-conduction band as well as other relevant parameters were determined to take over interpretation of both types spectra with MOT (molecular orbital theory [1]), on the base the so-called SCF Xα-Scattered Wave Cluster Method [2].

If the Fe $L_{III,II}$ emission band of wustite is recorded and the Fe L_{III} self-absorption spectra are constructed by the method of differential self-absorption [3], it can be shown that information on both occupancy and vacancies of the valence-conduction band is obtained as well as information concerning the position of the absorption edge, splitting of 3*d* orbitals under the influence of the crystal field and the spin splitting of the same orbitals when $2t_{2g}$ and $3eg$ energy stages are concerned.

Recording of Fe $L_{III,II}$ emission band within 690–730 eV energy range was performed with the ARL SEMQ EPMA microanalyser. The unit is completely

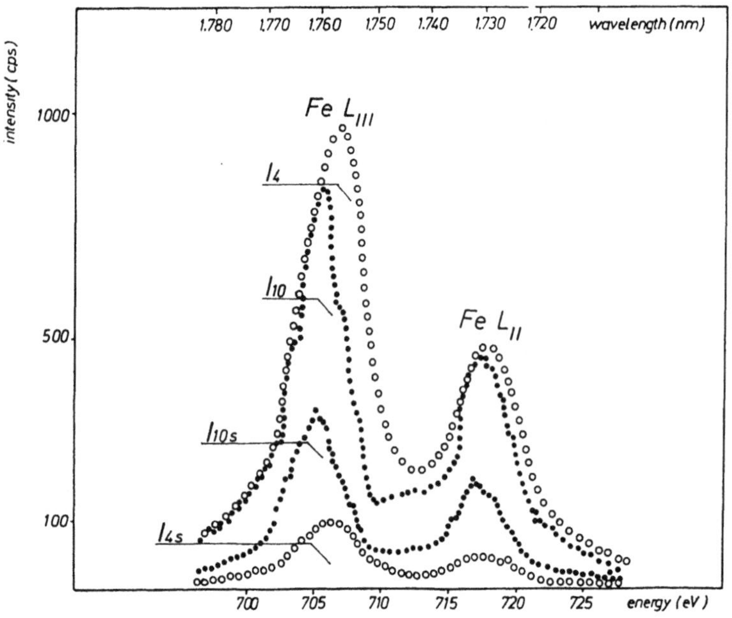

Fig. 1. X-ray emission Fe $L_{III,II}$ spectra of wustite. I_{4s} and I_{10s}, normalized spectrum I_4 and coincided spectrum I_{10}. see page 2

automated and equipped with a full-focusing X-ray spectrometer with a TAP analyzing crystal and proportional flow counter. The recording was done at 4 kV and 10 kV voltage excitation in steps of 0.12 eV and a recording time of 4 s. After subtracting background counts, these spectra are presented in Fig. 1 in legend, 4 kV is marked by open rings (I_{4s}) and 10 kV by black dots (I_{10s}).

The formation of self-absorption spectra of the same bands involved a two-step procedure: First normalization and then generation of the self-absorption spectrum according to the method described in [3].

The spectra were normalized by first arbitrarily setting the highest I_{4s} peak equal to 1000 (Fig. 1, I_4) and then multiplying the intensity of the I_{10s} spectrum by a constant factor (Fig. 1, I_{10}), so that it coincides with the adjusted I_4 spectrum on the low energy side of the Fe L_{III} emission peak.

In comparing these spectra, the reduction of Fe L_{III}-maximum intensity at the high energy side is evident as well as the shift of the peak towards lower energies with increase in voltage excitation. This phenomenon is also noticed in other iron compounds [4–7].

The iron self-absorption spectrum Fe $L_{IIIsabs}$ and Fe L_{IIsabs} (marked by squares in Fig. 2) was generated by plotting the intensity ratio I_4/I_{10} of the normalized spectra.

Results and Discussion

The determination of defined peak energies in Fe $L_{III,II}$ emission and in Fe $L_{IIIsabs}$ self-absorption spectra, was carried out by establishing the positions of the maxima. In case of insufficiently defined structures ("humps" and "shoulders") energies were estimated after approximation by corresponding Gaussian distribution curves [8].

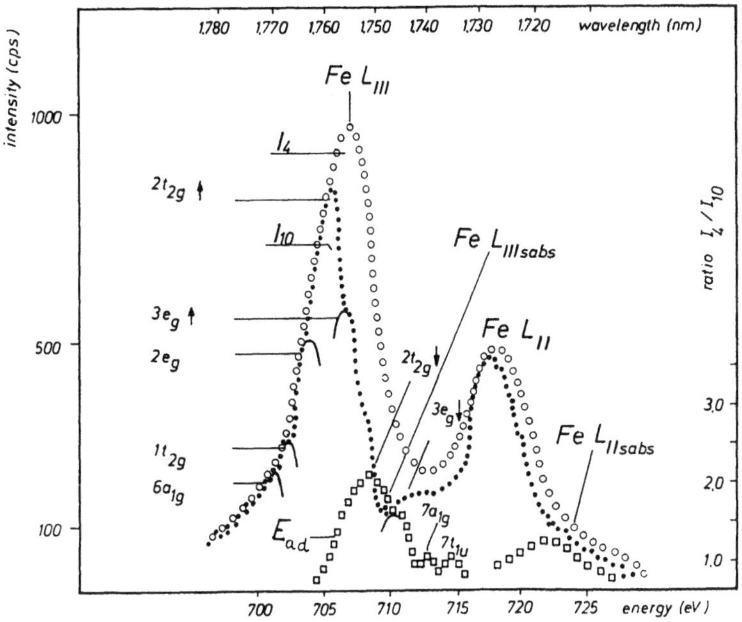

Fig. 2. Construction of an X-ray self-absorption spectrum Fe $L_{IIIsabs}$ of wustite. The energy of the absorption edge E_{ad}

Table 1. Results for the dominant peaks and structures occurring within the X-ray Fe $L_{III,II}$, emission and Fe L_{III} self-absorption spectra of wustite. Literature data are given for ferrosilicate

Spectrum structure description	Electron transition	Relative (eV) energy levels	Literature data [13]
Shoúlders and humps on the low energy side	$6a_{1g} \rightarrow$ Fe $2p^{3/2}$	701.5	
	$1t_{2g} \rightarrow$ Fe $2p^{3/2}$	702.5	
	$2e_g \rightarrow$ Fe $2p^{3/2}$	703.5	
Fe L_{III}-peak	$2t_{2g}\uparrow \rightarrow$ Fe $2p^{3/2}$	705.0	705.0
Weak peak on the high energy side	$3e_g\uparrow \rightarrow$ Fe $2p^{3/2}$	707.0	706.0
Absorption edge		707.5	706.1
Fe $L_{IIIsabs}$ · peak	Fe $2p \overset{3/2}{\rightarrow} 2t_{2g}\downarrow$	709.0	707.0
Weak hump on the high energy side	Fe $2p \overset{3/2}{\rightarrow} 3e_g\downarrow$	710.3	709.1
Weak peaks on the high energy side	Fe $2p \overset{3/2}{\rightarrow} 7a_{1g}$	713.0	
	Fe $1t_{2g} \rightarrow 7t_{1u}$	714.0	713.9
Fe L_{II}-peak	$2t_{2g}\uparrow \rightarrow$ Fe $2p^{1/2}$	718.8	717.4

The energy of the experimental absorption edge (E_{ad}) is estimated, from the self-absorption spectrum, according to the Bonnell criterion [9]. The interpretation of energy-peaks thus determined was carried out on the basis of the Tossell energy diagram of molecular orbitals of the valence-conduction band calculated for the cluster FeO_6^{10-} unit by SCF Xα method [10] and according to [8, 11, 12]. A review of identified peaks, i.e. the energies of molecular orbitals of the valence-conduction band of wustite together with a description of the electron transition and literature data for ferrosilicate [13] is given in Table 1.

The origin of dominant peaks in X-ray emission spectra, identified as Fe $L_{III,II}$ peaks has mostly been attributed to transitions from mixed Fe $3d$, $4s$ molecular orbitals of valence-conduction band to $2p^{3/2}$ and $2p^{1/2}$ sublevels of the L iron shell. According to our interpretation these peaks are formed during transition from antibonding, occupied $2t_{2g\uparrow}$ orbitals of the crystal field that contains three electrons and is characterized by an energy of 705.0 eV. The small hump at the low energy side of the Fe L_{III} peak, at energies of 703–704 eV corresponds to electron transition from bonding occupied $2e_g$ orbital. At even lower energies of 702.5 to 701.5 eV, a weak shoulder corresponds to transitions from the bonding, occupied $1t_{2g}$ and $6a_{1g}$ orbitals. A weak hump at the side of high energy of the Fe L_{III} electron transition from the antibonding, occupied $3e_{g\uparrow}$ orbital of the crystal field that contains two electrons. The weak intensity of this hump can be explained by self-absorption caused by the vicinity of $2t_{2g\uparrow}$ orbital [13].

The Fe L_{III} self-absorption spectrum results from electron transition from Fe$2p^{3/2}$ sublevel of the iron shell to bonding, partially occupied $2t_{2g\downarrow}$ and vacant $3e_{g\downarrow}$ orbitals of the crystal field as well as to vacant orbital $7a_{1g}$ and $7t_{1u}$ of the conduction band and thus the peaks characterized for the transitions mentioned, are evident. The dominant peak in this spectrum identified as Fe $L_{IIIsabs}$, corresponds to the transition to the bonding $2t_{2g\downarrow}$ orbitals of the crystal field that contains one electron and two vacancies and is characterized by an energy of 709.0 eV. The position of the absorption edge is estimated to the half way at the low energy side of this peak and corresponds to 707.5 eV. The weak hump at the high energy side results from electron transition to the highest, antibonding $3e_{g\downarrow}$ vacant orbital containing two vacancies. Two peaks at energy of 713 and 714 eV appearing in the part of the self-absorption spectrum that corresponds to the minimum between the Fe L_{III} and Fe L_{II} emission peaks. The former corresponds to the electron transition to bonding, vacant, diffusion $7a_{1g}$ orbital of the conduction band that belongs to $4s$ iron level which is, according to the selection rules, allowed. The latter corresponds to electron transition to the $7t_{1u}$ orbital but, belongs to $4p$ iron level. Although this transition is not allowed, it cannot be completely excluded [10, 13]. An important

Table 2. Evaluation of the accuracy of the measurements of the relative energies of wustite molecular orbitals and evaluation of the crystal field splitting size for $2t_{2g\uparrow}$ and $3e_{g\uparrow}$ orbitals and determination of the spin splitting size for the $2t_{2g}$ and $3e_g$ stages

Wustite, $Fe_{1-x}O$	This study	Tossell SCF X [10]	Burns opt. spec. [10]	Faye opt. spec. [10]
Estimation of crystal field orbitals splitting (eV) $2t_{2g}\uparrow \rightarrow 3e_g\uparrow$	1.5	1.7	1.27	1.18
Determination of spin $2t_{2g}$ stage splitting (eV) $2t_{2g}\uparrow \rightarrow 2t_{2g}\downarrow$	4.0	3.2	2.83	
Determination of spin $3e_g$ stage splitting (eV) $3e_g\uparrow \rightarrow 3e_g\downarrow$	3.8	2.9	2.72	

property in the self-absorption spectrum of wustite is a discontinuity the 716–718 eV energy range, which is untypical for iron oxides [13]. As the Fe L_{II} self-absorption spectrum, regarding low intensities of Fe L_{II} peaks in I_4 and I_{10} spectra, was obtained with a somewhat less precision, the interpretation of the same has not been given.

The energies of the crystal field orbitals, obtained in the manner described, are used while estimating the splitting size of $2t_{2g\uparrow}$ and $3e_{g\uparrow}$ orbitals by a crystal field and during determination of the spin splitting of $2t_{2g}$ and $3e_g$ stages. These results together with the corresponding literature data [13] are given in Table 2.

The results of Table 1 are either in good agreement with literature data or are somewhat higher ranging from 0.1 to 2.0 eV. The same holds for Table 2, in which our results are 0.23 to 1.17 eV higher. One of the explanations concerning these discrepancies might be the imperfectness of the procedure of normalization and coincidence. Second, technical restrictions do not allow a better measuring statistics. Overall, this method has proved to be complemetary while microexcitation has an advantage in the case of dispersed systems and boundary phases.

References

[1] J. A. Tossell, G. V. Gibbs, *Phys. Chem. Minerals* **1977**, *2*, 21.
[2] D. J. Vaughan, J. A. Tossell, K. H. Johanson, *Geochim. Cosmochim. Acta* **1974**, *38*, 939.
[3] R. J. Liefeld, *Soft X-ray Band Spectra*, Academic Press, New York, 1968, p. 133.
[4] M. K. Pavicevic, Dissertation, Univ. Heidelberg, 1971.
[5] M. K. Pavicevic, P. Ramdor, A. El Goresy, *Geochim. Cosmochim. Acta*, 1972, 3, [Suppl.] 295.
[6] M. K. Pavicevic, in: *Analyse extraterrestischen Materials* (W. Kiesl, H. Malissa jun., eds.), Springer, Berlin, Heidelberg, New York, 1974, p. 289.
[7] M. K. Pavicevic, Predavanja odrzana u JAZU, Svezak 42, Zagreb, 1974, p. 1 (in Serbo-Croatian).
[8] D. W. Fisher, *J. Phys. Chem. Solids* **1971**, *32*, 2455.
[9] C. Bonnell, *Physical Methods in Advanced Inorganic Chemistry*, Interscience, New York, 1968, p. 45.
[10] J. A. Tossell, D. J. Vaughan, K. H. Johnson, *Am. Mineral.* **1974**, *59*, 319.
[11] R. K. O'Nions, D, G. W. Smith, *Am. Mineral* **1971**, *56*, 1452.
[12] M. Romand, J. S. Solomon, W. L. Brun, *J. Phys. Chem. Solids* **1973**, *34*, 1765.
[13] C. G. Dodd, P. H. Ribe, *Phys. Chem. Minerals* **1978**, *3*, 145.

Mikrochim. Acta (1992) [Suppl.] 12: 261–268

Structural Analysis of Silver Halide Cubic Microcrystals with Epitaxial or Conversion Growths by STEM-EDX

Shijian Wu[1,*], André Van Daele[2], Willem Jacob[2], Renaat Gijbels[1], Ann Verbeeck, and René De Keyzer[3]

[1] Department of Chemistry, University of Antwerp (UIA), B-2610 Wilrijk, Belgium
[2] Department of Medicine, University of Antwerp (UIA), B-2610 Wilrijk, Belgium
[3] Agfa-Gevaert N. V., B-2640 Mortsel, Belgium

Abstract. Elemental distributions and contents of silver halide cubic microcrystals with epitaxial or conversion growths were analyzed by backscattered electron imaging, X-ray mapping and X-ray analysis in the scanning transmission electron microscope (STEM) combined with energy dispersive X-ray analysis (EDX). A liquid nitrogen cryostage was used to minimize sample damage and drift during electron bombardment. The quality of backscattered electron imaging was improved by selecting the optimal parameters of the image processing system and by off-line image processing. The backscattered electron images and X-ray maps show the elemental distribution in individual microcrystals. X-ray analyses in the spot mode yield semiquantitative results. Although the cubic shape of the microcrystals and their distorted appearance after tilting the sample make them difficult to be analyzed, this work indicates that the combination of these methods can be used to determine the elemental distribution and content in cubic microcrystals and that the analyses provide a valuable advice for the choice of the preparation conditions.

Key words: Silver halide photographic microcrystals, backscattered electron imaging, X-ray mapping, X-ray microanalysis.

There is a considerable interest for studying the chemical composition and elemental distribution in silver halide photographic microcrystals, particularly in individual microcrystals. Several techniques have been applied for that purpose, such as X-ray photoelectron spectroscopy [1], analytical electron microscopy [2], secondary ion mass spectrometry [3], and the scanning ion microprobe [4]. Gao et al. used backscattered electron imaging and X-ray analysis in the spot mode to analyze the iodide distributions and contents in single tabular silver iodobromide micro-

* To whom correspondence should be addressed

crystals [5]. Wu et al. studied the distributions and contents of Br, Cl, I and other elements in various types of silver halide microcrystals (tabular grains and tabular grains with AgCl or AgCNS epitaxial growths) using X-ray mapping and X-ray spot microanalysis (STEM-EDX) [6]. Because of the shape of the silver halide tabular microcrystals and their limited thickness, microanalysis in single microcrystals can be carried out conveniently relative to grains with other shapes.

In this work we studied the halide distribution in cubic silver halide microcrystals with silver bromide epitaxial growths and with bromide conversion at the corners, using scanning transmission electron microscopy and energy dispersive X-ray spectrometry (STEM-EDX).

Experimental

The cubic Ag(Cl, Br) microcrystals with AgBr epitaxial growths or bromide conversions were prepared in different ways. After removing the gelatin, repeated centrifugation and washing in distilled water, the grains were resuspended in water and dispersed onto carbon coated 100 mesh copper grids for electron microscopic analysis. All preparation processes were carried out under dark room conditions.

The analyses were carried out in a JEOL 1200 EX TEMSCAN electron microscope equipped with an ASID10 scanning device, an annular semiconductor backscattered electron (BSE) detector, a 30-mm^2 active area Si(Li) energy dispersive X-ray detector and a multichannel analyzer (Tracor Northern 5500). A Kontron SEM-IPS image processing system with an Intel 80286 based host control processor, was used for acquiring and processing secondary electron images, backscattered electron images and elemental X-ray maps. The role of the image processing computer system was twofold: it allows optimization of the acquisition conditions and subsequent processing of the images. For different scanning modes (SEM, BSE, STEM, X-ray mapping), different programs were used. After specifying the optimum acquisition parameters such as the dwell time per pixel, the number of pixel integrations and frame loops, the total number of pixels and the windows of analysis, the computer hardware takes over the scanning of the microscope. This makes it easy to acquire images with high signal to noise ratio. In addition, the images can be stored in the processing system and processed off line, which obviously enables the analysis time to be decreased. Image processing can be performed so as to enhance the contrast of the image and see more details. Background subtraction of X-ray maps can also be performed by arithmetic operations.

A liquid nitrogen cryostage (Gatan Model 636 double tilt cooling holder) was used to minimize sample damage and drift during electron bombardment. In normal operation, the cooling process took about 1 h from adding liquid nitrogen to the beginning of the analysis. The temperature of the specimen was kept at 106 K. In these conditions, no radiation damage was observed after the beam was placed on the same location for 500 s. The sample sometimes drifts when the analysis begins which necessitates to correct the sample position. After several minutes, no further sample drift is usually observed during a measurement of 100 s live time. X-ray counts measured repeatedly from the same location using a finely focused electron beam ("spot mode") can be considered as constant. This conclusion was reached by comparing the counts of the first 100 s dwell time and those of the tenth 100 s dwell time.

The electron microscope was operated at 80 kV accelerating voltage, and at a typical magnification of 150000 for the secondary electron (SE) and BSE imaging mode and 80000 magnification for the X-ray microanalysis. The latter corresponds to a probe size of about 2.5 nm. The X-ray counts were measured during 100 s live time. Monte Carlo simulation of the electron distribution in a cubic silver chloride microcrystal (see Fig. 1) shows the primary electron beam broadening and also allows to estimate the spatial resolution in X-ray analysis of cubic microcrystals. It appears that X-rays are produced within a much larger and less well defined volume than is indicated by the electron probe

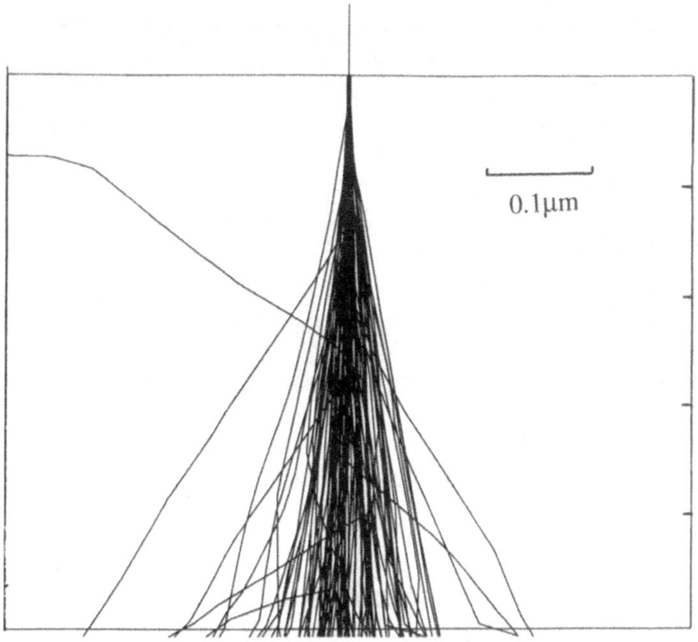

Fig. 1. Monte Carlo simulation of 100 trajectories in a silver chloride cubic microcrystal with 0.5 μm thickness using 80 kV accelerating voltage and a 2.5 nm electron probe size

size. The lateral diffusion of secondary X-rays increases also with the thickness of the silver halide microcrystal. A sample tilting of 35° and 40° respectively is used for X-ray mapping and for local X-ray microanalysis. Due to this tilt angle, the image of the microcrystals appears distorted. The selection of the microcrystals and the analysis spot must satisfy some special demands according to the characteristics of the cubic microcrystals with epitaxial or conversion growths because of the penetration and broadening of the electron beam and the diffusion of the X-rays. The analyses remain semiquantitative only.

Results and Discussions

In order to improve the quality of BSE imaging, three steps were taken: the image processing system IPS was applied to record the BSE image, optimal parameters were selected for obtaining a high signal to noise ratio in the image and the image was processed off-line to enhance the contrast. Fig. 2a shows a BSE micrograph taken directly from an AgCl cubic microcrystal converted by bromide (sample #4). Fig. 2b is a BSE image acquired by selecting optimal IPS acquisition parameters. Not only can the image be observed immediately using the IPS which is advantageous in selecting spots for X-ray analysis, but there is also more detail and high image quality is obtained (Fig. 2b). In Fig. 2a, two bright regions are found in the microcrystal, while Fig. 2b clearly shows four bright corners. According to the Everhardt [7] theory, the electron backscattering coefficient $\eta(\chi)$ is a function of the average atomic number ($\bar{Z}$), the average atomic mass ($\bar{A}$), the density (ρ), the specimen thickness (χ) and the initial electron energy (E_0). The initial electron energy (E_0) is constant under given experimental conditions, a cubic silver halide microcrystal has a certain thickness (χ) and we can calculate that the average atomic number $\bar{Z}_{(AgBr)}$ of AgBr is 41 and $\bar{Z}_{(AgCl)}$ is 32 for AgCl. Hence, the brighter parts in

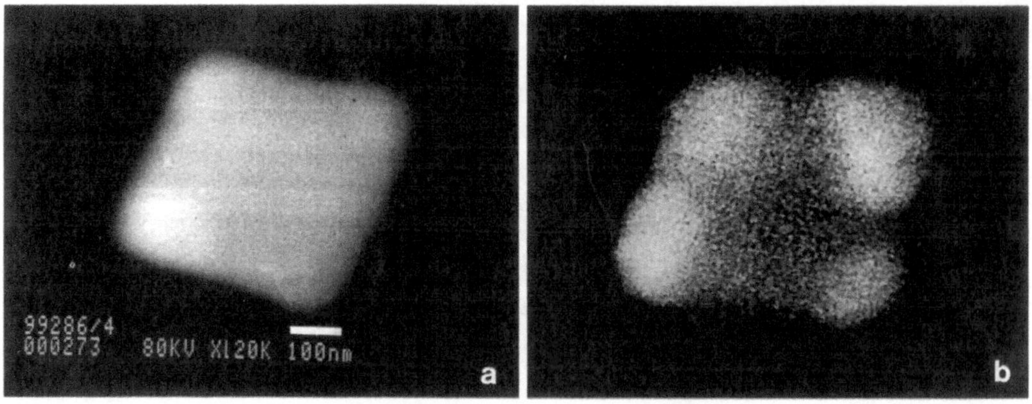

Fig. 2. Electron micrographs of an AgCl cubic microcrystal converted by bromide (sample #4). Accelerating voltage 80 kV. **a** BSE micrograph taken directly from the STEM screen. **b** BSE image acquired by the image processing system IPS

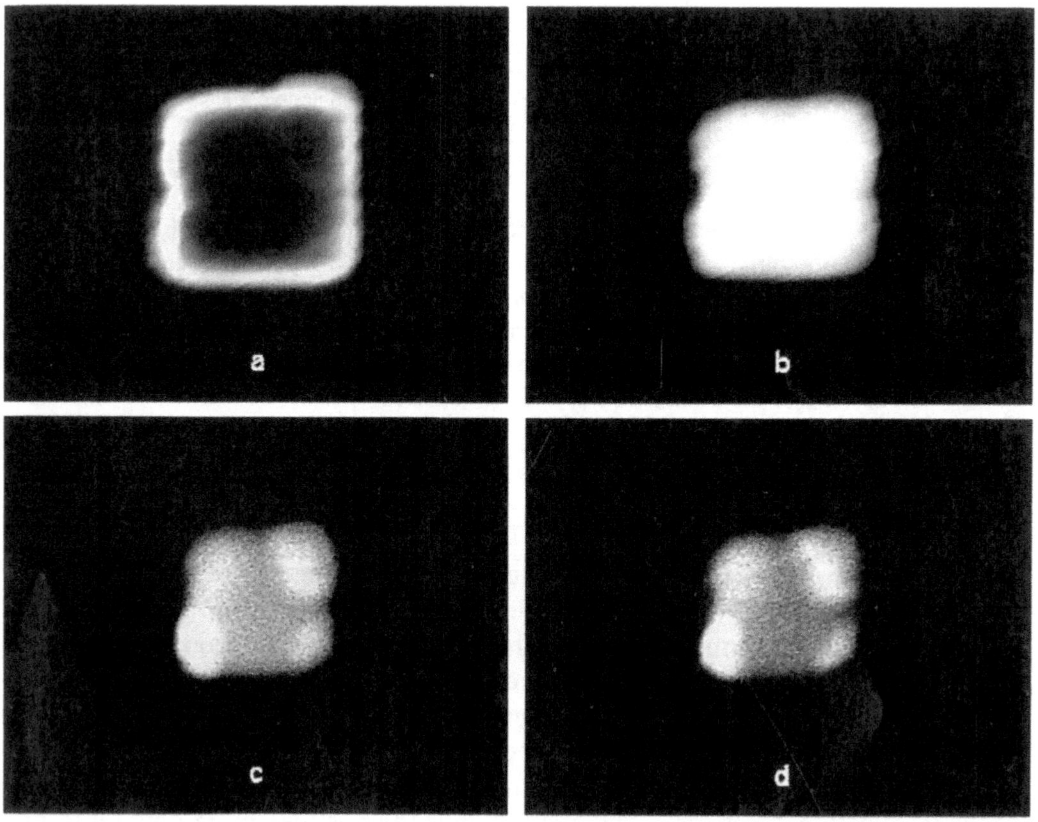

Fig. 3. Influence of the selection of BSE acquisition parameters in IPS (sample #4). **a** secondary electron image; **b** BSE image with 5 time integrations, 2 frame averaged loop times (5 × 2); **c** BSE image with 5 × 4; **d** BSE image with 5 × 6. The latter condition yields a better contrast.

the BSE image indicate a high bromide concentration in the Ag(Br, Cl) microcrystal, in this case this is in the corners of the cube.

Figure 3a shows the secondary electron image of the microcrystal in sample #4. Fig. 3b, 3c and 3d show the influence of the selection of the BSE acquisition parameters. In BSE acquisition operation, the parameters including the operation of the electron microscope and of the IPS, the dwell time per pixel, the integration times, the number of frame loops and the total number of pixels, determine whether the compositional distribution can be shown clearly. It appears that even under the same analytical conditions, a difference in the number of frame averaging loops (the images are averaged in the IPS memory) results in a different outcome. Obviously, Fig. 3d shows more clearly the bromide distribution in the converted cubic microcrystal. However, in a practical acquisition process, one has not always enough time to select the optimum operation parameters; even though the operation conditions are kept completely constant, it appears that changing the specimen, which often implies a different composition and structure, will also lead to a discrepancy in BSE imaging. A convenient way is to save these parameters in the IPS program. In every operation only the most important parameter influencing the BSE image directly would be changed a little, according to the structural outline of the specimen. Image processing was carried out off line to improve the image contrast. This step decreased the operation time and correspondingly decreased also the specimen damage which would result from a long analysis time. Figure 4a shows the BSE image of an AgCl microcrystal (sample #1) with AgBr epitaxial growths at the corners, acquired directly from the electron microscope screen after selecting the parameters of IPS operation. Fig. 4b shows the BSE image processed off line by IPS. The contrast of the BSE image of the microcrystal appears to have been increased after digital image processing. Figure 4b indicates more clearly a high bromide concentration at the four corners and shows that the growth at the corners occurred at the inside of the cubic microcrystals.

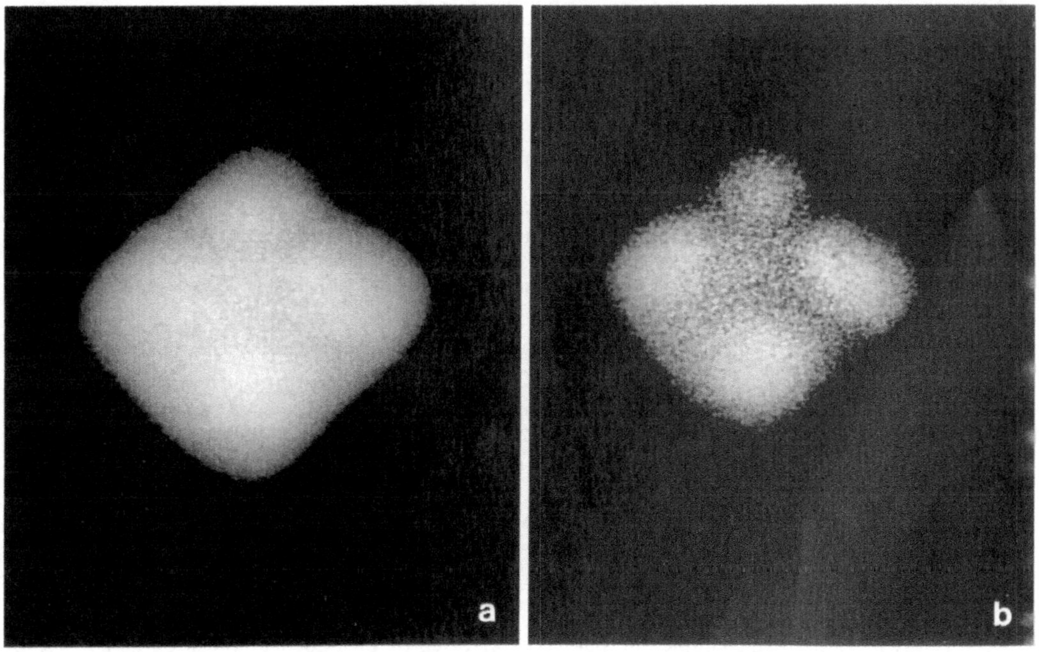

Fig. 4. Backscattered electron image of an Ag(Cl, Br) cubic microcrystal with epitaxial growths inside the corner (sample #1). **a** before image processing; **b** after image processing to enhance contrast

X-ray maps were acquired from single microcrystals. Considering that X-ray mapping needs a long acquisition time (about 10–15 min, i.e. 5–10 times more than BSE imaging) the specimen might drift a little during the measurement. A long acquisition time is necessary, in order to obtain enough sensitivity. The dwell time per pixel determines the limit of detection for the elements; therefore a long dwell time per pixel was selected with several integrations per pixel, but only one time frame loop was used. X-ray maps can be acquired for four elements simultaneously. After tilting over an angle of 35°, at least one side face of the cube can be seen and the image of the cubic microcrystal appears distorted. The X-ray map shows also a distorted brightness which is different in the middle and at the side of the crystal, but we can still get an outline of a different distribution of Br and Cl. Besides sample #3 whose epitaxial growths at the outside of the corner can be seen clearly, we have also imaged Br and Cl distributions of other samples, Fig. 5 shows X-ray maps of a microcrystal from sample #6. The BSE image did not show any brighter region which could indicate a high bromide concentration since the conversion growth only took place at the outside boundary of the cubic corners and the growth parts were very small. This is because the BSE intensity is directly proportional to the thickness of the microcrystal and the thin conversion parts with higher bromide content could not show a brighter BSE signal than the thick chloride parts. Com-

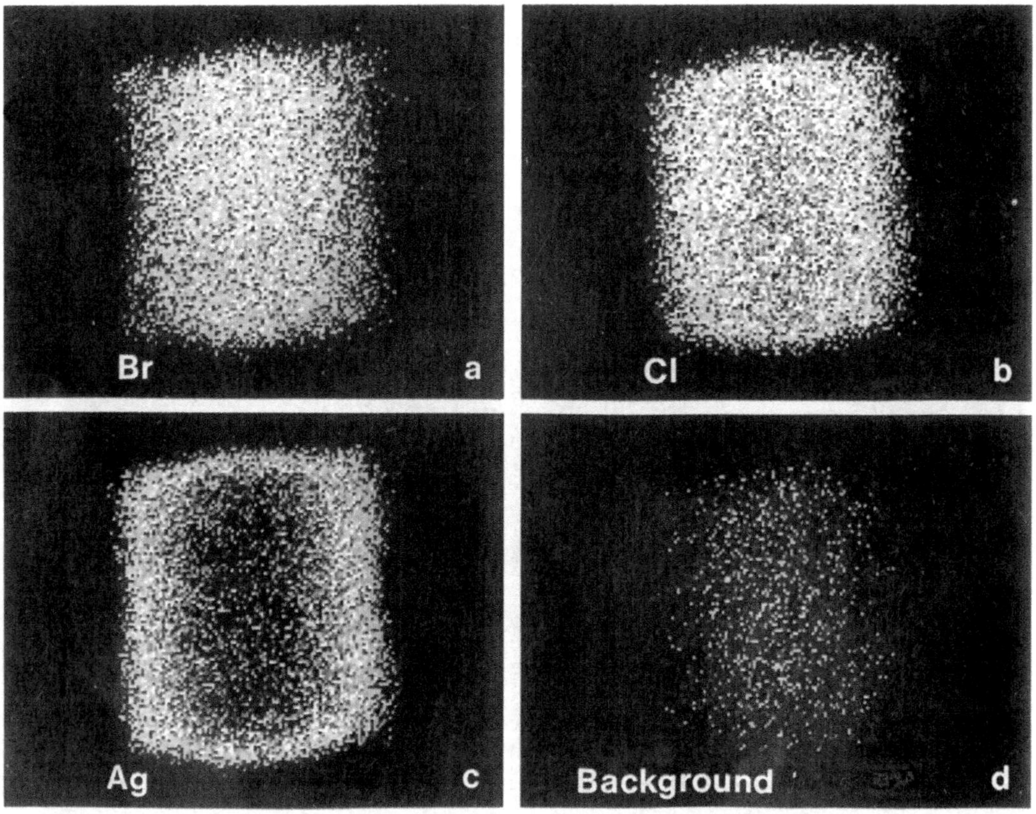

Fig. 5. X-ray map of an AgCl cubic microcrystal converted by bromide (sample #6). Upper left: Br; upper right: Cl; bottom left: Ag; bottom right: Background. X-ray map of Br Lα shows Br in the four corners (the height of the crystal is 0.5 μm)

paring the distributions of bromide and chloride in the X-ray maps (Fig. 5), we observed a higher Br/Cl ratio at the four corners.

Some quantification is possible by X-ray microanalysis. For the X-ray measurement of bromide, unfortunately, the interference of Al Kα on Br Lα from the sample holder can be very strong depending on the location on the grid and the orientation of the holder. This is caused by X-rays excited by electron scattering in the specimen support. Therefore it is necessary to use the less intense Br Kα radiation at 11.907 keV. By establishing a correlation between the X-ray intensity ratio and bromide concentration ratio, semiquantitative results could be obtained. The results of X-ray spot analysis are shown in Table 1. All samples are Ag(Cl, Br) emulsions containing 2% bromide. There are different bromide concentrations at the centres, the sides and the corners in the four samples. When conversion growths appear at the outside of the corners of the crystals, the concentration ratios of Br to Ag in these sections are about 0.8 (sample #3). The result indicates that the conversion growths did not take place into the microcrystals and that the bromide precipitated

Table 1. X-ray spot analysis results and Br/Ag concentration ratios

Sample #		1	3	4	6
Crystal scheme	Spot				
# 1	1	0.007	0.002	0.000	0.012
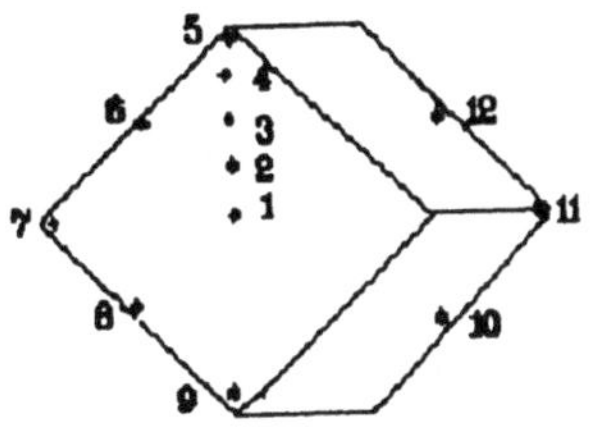	2	0.005	0.028	0.000	0.034
	3	0.007	0.713	0.035	0.345
	4	0.032	0.000	0.000	0.079
# 3	5	0.100	0.825	0.244	0.406
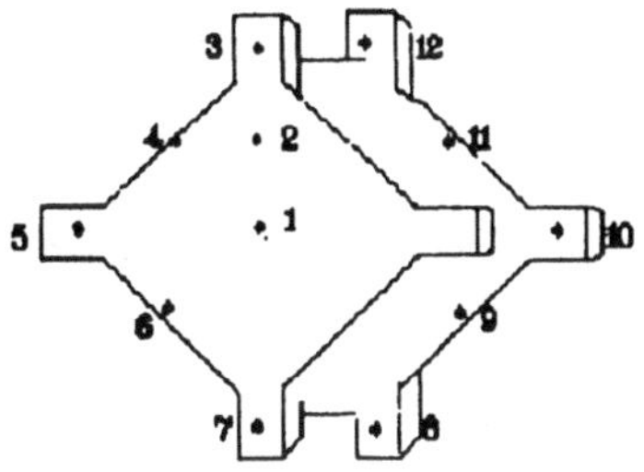	6	0.063	0.000	0.000	0.094
	7	0.221	0.605	0.297	0.296
	8	0.051	0.096	0.062	0.415
# 4 & 6	9	0.236	0.006	0.000	0.070
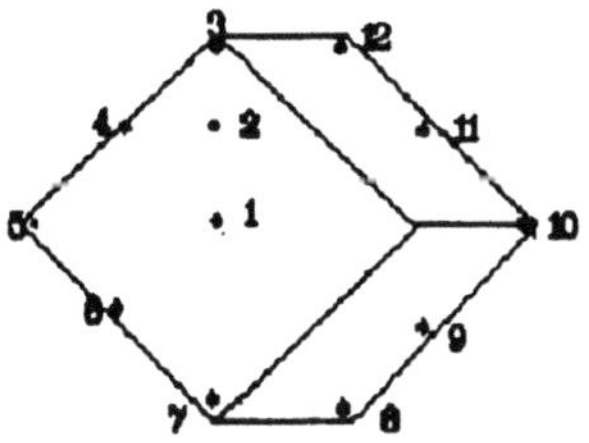	10	0.063	0.817	0.070	0.194
	11	0.245	0.004	0.000	0.081
	12	0.081	0.923	0.312	0.184

only at the outer corners of the microcrystals. For the microcrystals, converted inside the cubes, the concentration ratios of Br to Ag can also reach high values at corner sections of the microcrystals (e.g. 0.2 for sample #1, about 0.3 for sample #4). However, for sample #6, whose BSE image could not show a high bromide content at the corners, the X-ray analysis indicates higher bromide concentrations at the corners (Br/Ag concentration ratio about 0.5) and at the centre than those of sample #4. The data make clear that the structure of the sample is situated between sample #3 and sample #4. X-ray microanalysis allows confirmation of the element-non-specific information of the backscattered electron images and provides semiquantitative results.

Conclusion

Elemental distributions and contents of several kinds of cubic Ag(Cl, Br) microcrystals were analyzed by backscattered electron imaging, X-ray mapping and X-ray analysis in the spot mode. The acquisition of backscattered electron images was improved by careful selection of IPS parameters. The compositional contrast of processed BSE images showed the elemental distribution in some samples. X-ray maps also provided some indications of elemental distributions. X-ray analysis in the spot mode yielded semiquantitative results. The analytical results indicate whether different epitaxial or conversion growths occur in different microcrystals, at the inside of the corners or at the outside. The analyses yield bromide concentrations in the centre, sides and corners and indicate which sample is composed of ideal Ag(Cl, Br) cubic microcrystals with AgBr epitaxial or bromide conversion growths at the inside of the corners.

Although the cubic shape of the microcrystals and the distorted images after tilting the samples make them difficult to be analyzed, this work indicates that the combination of the different operating modes can be used to determine the elemental distribution and content in silver halide cubic microcrystals with epitaxial or conversion growths. The analyses also provide a valuable advice to the choice of the preparation conditions.

References

[1] T. M. Kelly, M. G. Mason, *J. Appl. Phys.* **1976**, *47*, 4721.

[2] M. A. King, M. H. Loretto, T. J. Maternaghan, F. J. Berry, in: *Progress in Basic Principles of Image Systems* (F. Granzer, E. Moisar, eds.), Vieweg, Braunschweig, 1987, p. 79.

[3] R. Gijbels, J. Van Puymbroeck, L. W. Ketellapper, in: *Progress in Basic Principles of Image Systems* (F. Granzer, E. Moisar, eds.), Vieweg, Braunschweig, 1987, p. 99.

[4] T. J. Maternaghan, C. J. Falder, R. Levi-Setti, J. M. Chabala, *SPE 42nd Annual Conf.*, Boston, 1989, p. 27.

[5] X. L. Gao, R. Gijbels, B. Nys, W. Jacob, Y. Gilliams, *J. Imaging Sci.*, **1989**, *33*, 87.

[6] Shijian Wu, A. Van Daele, I. Geuens, W. Jacob, R. Gijbels, A. Verbeeck, R. De Keyzer, in: *The Advancement of Imaging Science and Technology* (B. Peng, Y. Huang, S. Wang, Z. Qi, eds.), International Academic Publ., Beijing, 1990, p. 65.

[7] H. Niedrig, *Scanning Electron Microsc.* **1981**, *1*, 29.

Mikrochim. Acta (1992) [Suppl.] 12: 269–273

Characterization of the Bony Matrix of the Otic Capsule in Human Fetuses by EPMA

Silvia S. Montoro[1,*,**], Frank F. Declau[2], and Pierre J. Van Espen[1]

Departments of [1]Chemistry and [2]E.N.T., University of Antwerp, U.I.A., B-2610 Wilrijk, Belgium

Abstract. The fetal ossification pattern of the human otic capsule is studied using electron probe microanalysis (EPMA). By scanning electron microscopy the differences in geometry and orientation in the specimen, as well as the non-uniform matrix density and thickness, are clearly observed. The "peak-to-background" method was applied in order to compensate for the variation in density and surface topography. Standards were used as reference samples and the reproducibility, precision and accuracy of the analysis were studied for the three layers composing the otic capsule.

Key words: Peak-to-background method, Ca/P ratio, mineralization, otic capsule, bony layers.

Up to now, the normal degree of mineralization in the three separate layers, present in the otic capsule, is unknown. The determination of the Ca/P ratio at the microscopic level is required in the diagnostic evaluation of ear pathology and to establish the physiological role of the otic capsule, beyond its passive protective function.

The present study is an approach to the analytical characterization of the human fetal ear in its primary calcification front [1]. The otic capsule consists of three layers called endosteum, endochondrium and periosteum. In this study these will be simply called inner, middle and outer layer, respectively. The inner layer presents a very pronounced tilted surface which is capsule shaped. In the middle layer the surface is rough and the tissue presents a porous structure with differences in matrix density. The outer layer shows a surface which is compact and quite smooth. The micrograph in Fig. 1 shows the surface of the inner and middle layers as well as a section through the outer layer of a fetal sample in its earliest formation.

The main goal is to study the degree of mineralization in the bony matrix by determining the concentration ratio of the elements calcium and phosphorus in each of the layers. The latest stage of mineralization expected in a sample corresponds,

* To whom correspondence should be addressed

** On leave from Centro Regional de Investigación y Desarrollo de Santa Fe, Argentina

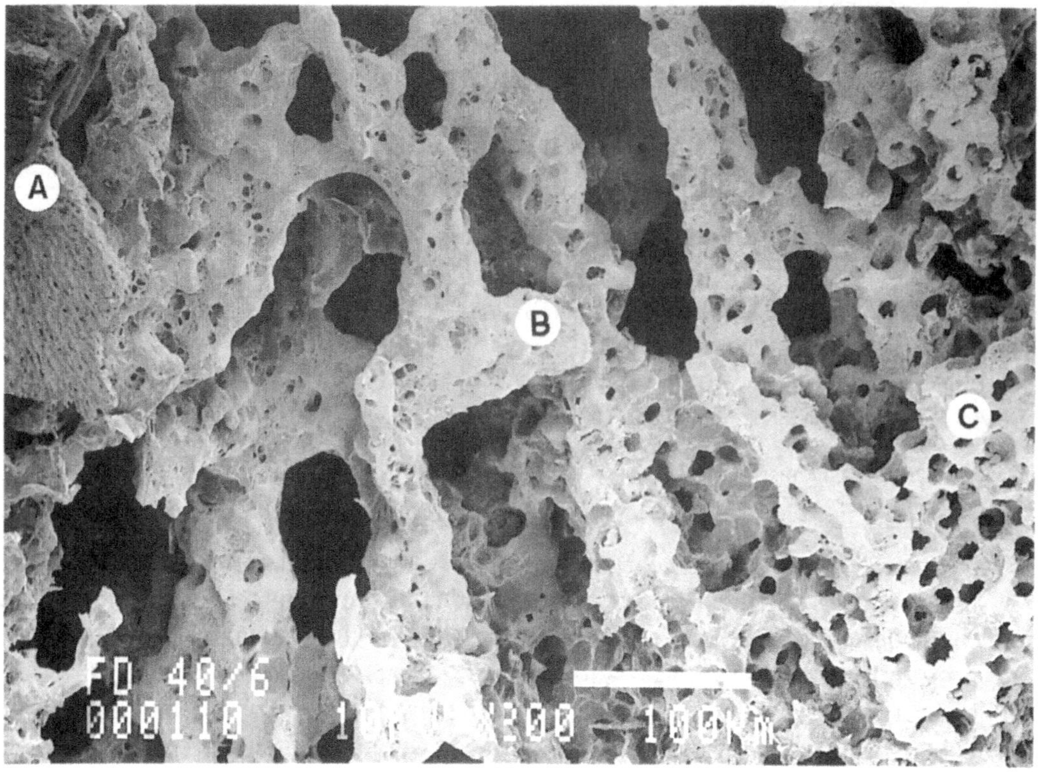

Fig 1. Surface micrograph of a fetal otic capsule sample.
A: Surface of the inner layer (endosteum)
B: Surface of the middle layer (endochondrium)
C: Section through outer layer (periosteum)
(bar is 100 μm)

to a Ca/P molar ratio of 1.67 which is similar to the ratio in an adult bone and in apatite.

In a previous study, a semiquantitative analysis by EPMA of these bony samples, after embedding in poly-methylmetacrylate and polishing, was carried out by comparing the net characteristic intensities of the peaks with intensities of pure standards [1]. By critically reviewing the results, some inconsistencies in the Ca/P ratios due to the presence of embedding material were found. Therefore it became necessary to analyze the samples in their natural state.

Edie and Glick applied ZAF corrections to organic compounds containing calcium and several calcium phosphates with a relative error of 5% for most analysis. In all cases the samples considered in their work had flat or polished surfaces [2]. Several correction procedures specifically designed for biological material analysis are reported in the literature [3].

A complex geometry of the samples makes the net peak intensity very much depending on the roughness of the surface [4]. In this work, net peak intensities could not be used because the matrix density and local tilt conditions are unknown and changing from site to site. On this basis, conventional ZAF corrections can not be carried out.

Therefore the "peak-to-background" (P/B) method was applied in order to compensate for the variation in density and topography of the different layers.

According to Roomans [5] the P/B ratio is less sensitive to the local variations of tilt and take-off angles than absolute intensities are when analysing rough surfaces because the background radiation under or very near the peaks is affected in the same way as the characteristic radiation.

Experimental

Six otic capsules from 16 to 39 weeks old fetuses and six adult capsules were analyzed. The fetal head was obtained fresh at post mortem and immediately chemically fixed with glutaraldehyde 2.5% in sodium cacodylate buffer (pH 7.4). Macrosections were obtained by embedding the heads in polymethylmetacrylate (Technovit 3040; Kulzer & Co GmbH, Wehrheim, Germany) and sectioning them into slices with a semi-automatic cutting system (Exakt; Walter Messner, Oststeinbek, Germany). All specimens were treated by critical point drying procedure in CO_2 after dehydration in ethanol followed by immersion in amylacetate [1].

As standards, the following materials were used: fluorapatite (crystal) with a certified molar Ca/P ratio of 1.67; dentine (hard tissue) extracted from adult tooth with a molar Ca/P ratio equal to 1.52, assessed by atomic absorption spectrometry (AAS); calcium phosphate analytical grade pelletized powder with stoichiometric Ca/P ratio of 1.49; outer layers from two adult bones with a Ca/P molar ratio of 1.76 and 1.67 respectively, also determined by AAS.

Several authors consider the first three specimens as adequate standards [2, 3, 6, 7, 8]. Among dentine, apatite and fully mineralized bone small differences in density are expected.

No other treatment was given to the surface of the bony samples before the analysis. The apatite and dentine were cut and polished down to 1200 mesh abrasive paper. All the samples and standards were mounted onto the sample holder by means of copper tape, graphite or silver paint and coated with approximately a 20 nm thick carbon layer in a Balzers Union coating device.

The measurements were carried out in a JEOL 733 Superprobe equipped with an Tracor TN 2000 energy dispersive Si(Li) detector.

The analyzing conditions adopted were 25 kV accelerating voltage, a beam current of 1 nA with a beam spot size of 0.1 μm, a magnification of X3000 (corresponding to an analytical area of 20 × 25 μm) and a counting time of 60 seconds. Twenty spectra were obtained for each layer in each sample and also for each standard. The analysis spots were selected at random.

The AXIL software was employed to evaluate the spectra using a non-linear least squares fitting procedure [9].

Results and Discussion

Apart from Ca and P, no other characteristic peaks were present in the spectra with rare exceptions of sulphur which was found in areas with rests of cartilage. The Ca and P net peak intensities, the Ca/P peak intensity ratios, the peak-to-background ratios for each element separate and the Ca/P peak-to-background ratios measured on the standards were used to construct calibration curves with the latter showing the best linear regression statistics (data not shown). Therefore equation 1 was used to relate the Ca/P peak-to-background ratios with the concentration ratio in the sample:

$$\frac{I_{Ca}/B_{Ca}}{I_P/B_P} - a + b\frac{C_{Ca}}{C_P}\,. \tag{1}$$

Where I_{Ca} and I_P are the net intensities of the Ca and P lines, B_{Ca} and B_P the correspondent background intensities and C_{Ca} and C_P the molar concentrations. Constants a and b are obtained from the calibration curve [7, 10, 11].

Table 1. Selected background regions, in keV

Element	Left-to-peak	Under-the-peak	Right-to-peak
P	1.41–1.69	1.81–2.21	3.09–3.41
Ca	3.09–3.41	3.45–3.89	4.29–4.61

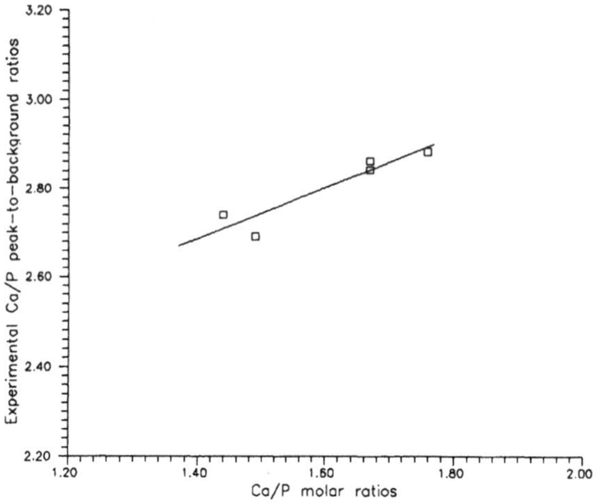

Fig. 2. Calibration curve for the standard samples, the background region on left side of the peak is considered. Correlation coefficient = 0.929

Different background areas were considered for the equation: one under the peak and two located at both sides. The background regions in keV are listed in Table 1. Calibration curves were established using the three different background estimates. The background on the left side of the characteristic peaks gave the best fitting results. The following correlation coefficients were obtained using equation 1: left-to-peak region 0.929, under-the-peak region −0.523, right-to-the peak region 0.288. In Fig. 2 the calibration curve corresponding to the left region is plotted.

Although absorption edges for the peaks are considered in the spectrum evaluation, there are some uncertainties in fitting the continuum radiation under the peak in the phosphorous region. The standard deviation of the mean for 20 measurements are in the range 0.01 – 0.02 for the five standard samples.

The molar Ca/P ratios for each layer in each fetal sample are shown in Table 2; the order of listing of the samples is not in age sequence. A final ratio of 1.64 and a standard deviation of 0.11 was obtained for the six adult otic capsules.

The reproducibility of the method was verified by measuring two samples twice, in which the ratios and the standard deviations obtained were found identical.

By comparing the estimated standard deviations for the three layers from all the samples, it is clear that the outer layer presents the lowest variation; which is probably related to the characteristics of the specimen surface and matrix structure. The standard deviation obtained for the Ca/P ratio for each layer in each sample was between 0.06 and 0.18.

Through analysis of variance (ANOVA), it was found that there is no statistical significant difference between the Ca/P ratio in the three layers, the three layers agree within an estimated error of 3%. This means, in contrary to what was expected, that the mineralization process is similar for all the layers and independent of the calcified matrix structure, although their organization is different [12, 13, 14].

Table 2. Molar Ca/P ratios in otic capsules of fetuses

Samples		Ca/P Molar Ratios Inner	Middle	Outer	Mean	Std Dev.
	49	1.42	1.39	1.40	1.40	0.01
	50	1.33	1.26	1.32	1.30	0.04
	51	1.39	1.49	1.40	1.43	0.05
Fetuses	54	1.35	1.44	1.37	1.39	0.05
	56	1.52	1.51	1.33	1.45	0.11
	57	1.21	1.46	1.28	1.32	0.13
	Mean	1.37	1.42	1.35	1.38	
	Std Dev.	0.10	0.09	0.05	0.06	

The accuracy of the analysis was tested by analyzing two fetal samples also by AAS. The Ca/P ratios obtained varied within a range of 1.50 to 1.69, except for one of the outer layers showing a Ca/P ratio of 2.1. The main disagreement in the results between both techniques may be due to uncertainties obtained in the analysis of P by AAS that affects drastically the final Ca/P ratio.

From the clinical point of view, by comparing the results from otic capsules of fetuses and adults, it can be concluded that the mineralization process has not yet come up to the hydroxyapatite stage, and that some amorphous calcium phosphate still remains. That makes the otic capsule bones relatively different from any other bone, in the sense that its mineralization process develops at a much slower rate. This could also explain the lack of correlation between the obtained Ca/P and the age of the fetus [14].

Acknowledgements. The authors thank W. Van Mol for carrying out the AAS analysis. S. M. would like to thank specially the University of Antwerp for the scientific and financial support given to her.

References

[1] F. Declau, W. Jacob, W. Dorrine, B. Appel, J. Marquet, *J. Laringol. & Otology* **1989**, *103*, 113.
[2] J. W. Edie, P. L. Glick, *J. Microsc.* **1979**, *117*, 285.
[3] A. Boekestein. A. L. H. Stols, A. M. Stadhouders, *Scanning Electron Microsc.* **1980**, *II*, 321.
[4] K. Kiss, *Appl. Spectr.* **1983**, *37* (1), 19.
[5] G. M. Roomans, *Scanning Electron Microsc.* **1981**, *II*, 344.
[6] W. J. Landis, *Scanning Electron Microsc.* **1979**, *II*, 555.
[7] T. A. Hall, *J. Microsc. Biol. Cell* **1975**, *22*, 271.
[8] H. J. Höhling, W. A. P. Nicholson, *J. Microsc. Biol. Cell.* **1975**, *22*, 185.
[9] P. Van Espen, K. Janssens, J. Nobels, *Chemon. Instell. Lab. Systems* **1986**, *1*, 109.
[10] T. A. Hall, *Scanning Microsc.* **1989**, *3*, 461.
[11] I. Zs.-Nagy, T. Casoli, *Scanning Microsc.* **1990**, *4*, 419.
[12] F. Declau, W. Jacob, W. Dorrine, J. Marquet, *Acta Otolaryngol* [*Stockholm*] **1990**, *470* (Suppl.) 56.
[13] F. Declau, W. Jacob, S. Montoro, J. Marquet, *Int. J. Pediatric Otorrinolar.* **1991**, *21*, 21.
[14] F. Declau, *Ph. D. Thesis*, University of Antwerp, 1991.

Mikrochim. Acta (1992) [Suppl.] 12: 275—278

Overview

On consideration of all the work published in this issue, we can sum up as follows:

— Progress has been made in analytical procedures for all electron-beam micromethods and microscopy.
— Real problems have been brought out, the solution of which leads to the improvement and applicability of specific methods.
— The currently most important application areas have been set out.

Conclusions

1. EPMA with WDS or EDS in widely used, being comparable in applicability with TEM, EELS and AES. Multi-element qualitative analysis of all kinds of samples (massive, thin-film, monolayer, multi-layer and thin-section) of organic or inorganic origin is possible for all elements from boron onwards. It must, however, be emphasised that the X-ray emission spectra of ultra-light elements show significant differences in detectability and interpretation from those of elements of higher atomic number.

2. Quantitative EPMA for ultra-light elements shows significant differences in procedure, reproducibility and accuracy for different sample types. The total error arises from the four most important steps in quantitative procedure.

a. The nature and treatment of the sample.
b. Spectroscopic excitation and detection, and recording and processing of signals.
c. The selection of suitable standards.
d. The use of computer programs for quantitation.

From these observations the following conclusions can be drawn.

2.1 The high reproducibility in quantitative analysis with EPMA often gives relative standard deviations less than 1% for massive and thin-film samples. The agreement with the true value principally depends on the choice of standards for massivc samplcs and on the software used for quantitation. In both cases, improvements in quantitation can be achieved only with the "modular" approach, which uses theoretical and empirical parameters and experimental data. In this way one gains an exact description of the function $\Phi(\rho, z)$ with parameters such as surface ionization, maxima and position of the $\Phi(\rho, z)$ curve and the energy range of X-ray. Theoretical treatment by the Monte Carlo method always assumes experimental checks by means of the "sandwich" technique, which is applicable to all kinds of samples. This makes possible the determination of "depth profiles", i. e. of the thickness and composition of, for instance, ten layers whose total thickness is no more than 1—2 µm, by means of changes in excitation potential.

2.2 The limitations on quantitative analysis of the ultra-light elements ($Z=5$ to $Z=8$) lie in the properties inherent in their spectra. The main differences between these spectra ,and those of heavier elements are the small peak to background ratio, the complex structure of the emission bands and the strong dependence of the spectral properties on the nature of the chemical bonding of the element concerned. This appears as wavelength shifts, peak asymmetry, the origination of new lines and disappear the exciting line. For quantitative analysis of the ultra-light elements there is as yet no universal correction program like, for instance, ZAF-programs. In certain cases, however, one can obtain results quantitative within 2—5%, if the appropriate matrix-correction program is chosen in combination with consistent mass-absorption coefficients.

2.3 Quantitative analysis of organic thin sections, with the use of "simulated" standards, uses the same correction procedures as for inorganic substances. Correct interpretation of the results requires stability of the micro-structure and constancy of the elemental speciation in the sample. In order to achieve these ends, sample preparation involves physical fixation by means of rapid deep-freezing. The main problems for complete and accurate quantitative analysis are: electric charging on surfaces and surface contamination (especially arising from light and very light elements, including hydrogen).

3. HR TEM with high-energy incident radiation (0.2 to 1 MeV or more) can reveal ultramicrostructure at the atomic level. Simultaneous investigations with HR TEM and other analytical techniques such as EDS, EELS and ED provides additional information on chemical composition, the nature of the chemical bonding and the crystal structure, with detection of all elements from $Z=3$. If one further considers that the resolution of this method is in the nm-region (i. e. smaller by a factor of 10^3 or more compared with EPMA), then it is unparalleled in materials science for the investigation of phase boundaries and highly dispersed multi-phase systems. The limitation of this method lies principally in the need to prepare ultra-thin samples that are transparent to electron beams and in the damage caused by high-energy electron radiation.

4. AES and EPMA are complementary and together are very useful, especially in demanding situations near the detection limit in the cases of ultra-thin evaporated layers and surface analysis. The techniques are of great interest in various areas of surface technology and micro-electronics.

5. The short papers on method development can be organized under six main headings.

5.1 Progress and improvements in quantitative EPMA, which have already been discussed above (2.1).

5.2 Optimization and improvement of signals in EPMA include interpretation of line-of-flight back-scattering (BSE) and secondary electron emission (SEE), use of the Monte Carlo method for determining the intensity of characteristic X-ray emission and Bremsstrahlung in thin-layer and massive samples and the dependence of the efficiency of Si(Li) detectors on the incident angle of characterstic X-rays for small angles.

5.3 The development of new procedures for EPMA includes the possibility of scanning low-energy electron microscopy (SVLEEM) which works (from 0 to 20 eV) by means of a field-emission electron gun in ultra-high vacuum. The automatic analysis of weak X-ray spectra (oxygen *K*-spectra were used as examples) creates the possibility of explaining the nature of chemical bonding in oxides with molecular orbital theory.

5.4 Improvements in the limit of detection and reproducibility of quantitative analysis with EELS were claimed using a background-correction program. The back-scattering contrast in scanning AEM can be increased by subsequent signal processing.

5.5 Quantitative EPMA of biological samples, which has been discussed in 2.3.

5.6 Proposed developments in new apparatus included a combination of ion microbeam excitation and ion spectrometry as well as detection with a mini-cyclotron and acceleration mass spectrometry (MC AMS) in order to improve the current resolution and limits of detections of SIMS and AMS. A home-made EELS system was used with a TEM/STEM instrument.

6. Short papers on important applications in different fields can be divided into five groups.

6.1 EPMA, XPS and EELS have been applied for the investigation of thin layers of $(Y_2O_3)_x(ZrO_2)_{1-x}$, Ta_2O_5 and P, As and Sb on silicon or titanium substrates for electronics. Glass fibres for opto-electronics have been studied with a combination of EPMA and AEM.

6.2 In metallurgy the determination of carbon on the surface of steel plate is important in automobile production.

6.3 In mineralogy and solid-state chemistry the determination of the electronic configuration of the valence conduction bands of 3*d* metals in their oxides is possible with EPMA on the basis of X-ray emission and absorption spectra. The determination of molecular orbitals and of the influence of the crystal field on them is also possible.

6.4 In photographic technology, structural analysis of silver halides and epitaxial layers is very important.

6.5 Medical application of EPMA with EDS enables e. g. the Ca/P ratio in the human foetal fisses to be determined.

In conclusion, we are of the opinion that at the present time electron beam micro-methods and microscopy are exceptional techniques for obtaining both chemical and physical properties of microscopic and submicroscopic structures in solids. This is of great interest in different areas of scientific and technological research on inorganic and organic samples of various kinds: massive samples, thin layers and thin or ultra-thin sections. Elemental detection (from atomic number 3 upwards) ranges in power from depth profiling with high reproducibility (in most cases relative standard deviations of less than 1% in volumes of several μm^3 down to nm^3) to imaging ultramicrostructures at the atomic level. In addition to chemical composition, these techniques provide important information on elemental speciation, depth profiles, crystal structures and the nature of chemical bonding.

February 1992 *A. Boekestein* and *M. K. Pavićević*

GPSR Compliance

The European Union's (EU) General Product Safety Regulation (GPSR) is a set of rules that requires consumer products to be safe and our obligations to ensure this.

If you have any concerns about our products, you can contact us on ProductSafety@springernature.com

In case Publisher is established outside the EU, the EU authorized representative is:

Springer Nature Customer Service Center GmbH
Europaplatz 3
69115 Heidelberg, Germany

Batch number: 09625567

Printed by Printforce, the Netherlands